Traditions and Transformations of Habitation in Indonesia

Figure Inside Cover_View to the cityscape of West Jakarta, Indonesia 2016

Bagoes Wiryomartono

Traditions and Transformations of Habitation in Indonesia

Power, Architecture, and Urbanism

Springer

Bagoes Wiryomartono
Independent Scholar
Toronto, Canada

ISBN 978-981-15-3404-1 ISBN 978-981-15-3405-8 (eBook)
https://doi.org/10.1007/978-981-15-3405-8

This Springer imprint is published by the registered company Springer Nature Singapore Pte Ltd.
The registered company address is: 152 Beach Road, #21-01/04 Gateway East, Singapore 189721, Singapore

Desa mawa cara, kuta negara mawa tata

(Village has its own custom and tradition; urban settlement has its own law and order, from Javanese proverb)

Contents

About the Author

Dr. Bagoes Wiryomartono is an independent scholar in Toronto, Ontario, Canada. Bagoes earned his Doctorate in architecture and urbanism from the Aachen University of Technology in 1990, Germany. He was a postdoctoral fellow for architecture at the East West Center, Honolulu, Hawaii, and Fulbright Scholar at the Smithsonian Institution, Washington DC, USA. The area of specialization of his teaching and research experience is focused on history, theory, and design of urbanism of various cultures and traditions in Southeast Asia and North America. He was a senior lecturer at the Bandung Institute of Technology (1981–1983, 1993–2002), and visiting research associate at the Asian Institute, University of Toronto, Canada (2003–2005). He was an associate professor at the Department of Architecture, Faculty of Built Environment of *Universiti Teknologi Malaysia* (2010–2013). Most recent scholarly books by Bagoes Wiryomartono include: *Livability and Sustainability of Urbanism* (Singapore: Palgrave Macmillan, 2019), *Javanese Culture and the Meanings of Locality, Studies on the Arts, Urbanism, Polity, and Society* (Lanham: Rowman & Littlefield, 2016), *Perspectives on Traditional Settlements and Communities: Home, Form and Culture in Indonesia* (New York-Berlin-Singapore: Springer Business Media 2014). He published his research in various Scopus journals on history and theory of architecture, urbanism, and culture in Southeast Asia as well as philosophical studies within the phenomenological movement.

Chapter 1
Political Culture and Habitation: An Introduction

This book is about the pertinent practices of power in Indonesia with a focus on habitation. The book explores and unfolds the meanings of the relationship between people and place that constructs, develops, sustains, and conserves their culture and habitation in Southeast Asia; this relationship is regarded as the source and generator of cultural productions in terms of the built environment, artifacts, arts, works, and collective behaviors. One of the foremost sources of power in Indonesia is the patrimonial system of culture and tradition. This system categorically values and highly favors a strong leadership and dominant personality as father figure over a sociopolitically organized system and mechanism.

The study argues that the patrimonial system is the pivotal source and condition of the political tradition of Indonesia. The tradition maintains the patron–client relationship of sociocultural and economic production that characterizes, directs, manages, develops, maintains, and sustains the development of the built environment and populations. The system is considered effective because it resists direct opposition and disagreement. Challenges against the ruling government are considered threats and dangers of the seamless whole of the community as well as a nation-state. Accordingly, unity and integrity are necessary to be thought and perceived as a safe and healthy extended family.

Until today, the patrimonial system is commensurately functioning because the vertical mobility of populations is socioeconomically restricted with high cost education and an exclusive political career. Traditionally speaking, populations are characterized by the extreme dichotomy between the few dominating elites and massive submissive commoners. The patrimonial ordering system becomes a reasonably organizing framework of the most populous Muslims on the globe. Despite the democratic process of the public election, the system works efficiently in daily practical politics because of its top-down chain of command.

To a certain extent, the patrimonial system does not rely simply on legal formal mechanisms and apparatus but deeply rooted in the spiritual leadership and belief system of people so that the paternalistic leaders of the country are able to bring people together and mobilize the resources toward the unity in diversity. Moreover,

B. Wiryomartono, *Traditions and Transformations of Habitation in Indonesia*,
https://doi.org/10.1007/978-981-15-3405-8_1

the patrimonial system characterizes the country's lifeworld with the language of hierarchical gestures and social position. In other words, to understand the politics of Indonesia is to comprehend the culture and tradition of the political leadership of the country.

Substantially speaking, the book comprises several case studies concerning the relationship of power and habitation with specific geographic locations, certain socio-cultural groups, and historical contexts. The studies are considered interdisciplinary within the framework of environmental humanities—arts, architecture, urban studies—and social sciences—geography, cultural anthropology, sociology. In this system, the studies are expected to unfold, dismantle, and disclose relations, functions, constructions, and notions of history, tradition, and cultural production. However, these studies are not intended either to be critical or to be diagnostic. Rather, the studies are projected to be comprehensively divulging and unveiling the aspects of interactive connectivity between the patrimonial culture and socioculturally constructed system of habitation. The cases were carried out from the author's work and life experience and observation in Indonesia with a focus on the various aspects of environmental humanities that shape, develop, and sustain the essence and sense of habitation that includes history, arts, architecture, culture, and urbanism.

Considerably, patrimonial culture is understood as a sociopolitically ordering system that entails, upholds, and maintains the paternal patronage and leadership based on seniority, emotional stability, knowledgeable competence, and wisdom. In the Indonesian context, patrimonial tradition has been well known and practiced as *budaya paternalistik or bapak-ism.* The practice of this patrimonial culture is experienced as high respect and devotion to the persons who are charismatic and fatherly with a strong personality, knowledgeable by experience, seniority by social rank, and assertive and proactive by leadership.

By and large, the book comprises fourteen chapters and organized within the line of investigation and exploration for the relationship between patrimonial culture and the aspects of habitation. Chapter 2 is opened with a historical exposition on the relationship between patrimonial power and territorial claim in Indonesia. This exploration is expanded in Chap. 3 with the context and case of pre-colonial maritime kingdoms of Aceh and Demak; this chapter is an attempt to unfold and divulge the indigenous concept and practice of patrimonial power in relation to urban settlement. Chapter 4 is an exposition of how the patrimonial power of Western culture has been established and developed in Indonesia. This chapter is to depict and divulge about the imperialism and colonialism of the Dutch over East Indies with the establishment of the colonial urban settlement of Batavia; from this trading post, the Dutch colonial power was expanded to take control over other cities and towns of Java and other islands.

Chapter 5 presents the dynamics of the hegemony of urbanism under the Dutch colonial rule with Bandung as a case study. This chapter attempts to unveil the politically conflicting juxtaposition of spatial order between indigenous authority and the Dutch colonial rule. Indeed, the hegemonic power of the Dutch surpassed the symbolic representation of local authority. The colonial and imperial practice of power has not only been exercised by the Dutch on the native populations but also on other ethnic groups. Chapter 6 is about the Chinese settlement in Batavia. This chapter concerns the historical evolution of the Chinese quarters from the beginning of Dutch colonialism to the present.

The excess of patrimonial domination includes the practice of ghettoization and marginalization of populations based on racial identity. The exercise of patrimonial power is characterized by masculine identity with various contexts and actors. Chapter 7 explores the roots of patrimonial culture in the context of the underworld. Still, the relationship between masculinity and humanity is crucial for human work because it is about the vitality of life. Chapter 8 is presented to uncover masculine arts and architecture from the work of Silaban as a case study.

Nevertheless, the practice of patrimonial domination in urbanism cannot afford to ignore the marginalization of a certain socioeconomic group. Chapter 9 explores and discloses a group of people who are marginalized from the patrimonial status quo in urbanism; their settlement is well known in Indonesia as *kampung kota*. On the other side, patrimonial urbanism is architecturally characterized and signified by giving privilege to the ruling group of populations. The example for this is discussed and presented in Chap. 10 with the case of superblock development. Furthermore, the elaboration of patrimonial urbanism is presented in Chap. 11. The case is the development of the giant seawall of Jakarta Bay. This chapter attempts to unveil the political motivation of the reclamation development under the strategic anticipation of climate change.

Nevertheless, the consequences of patrimonial authority on architecture are obvious with the demolition of artifacts and buildings of those, which do not belong to the regime's preference. Chapter 12 presents the patrimonial drives of violence against architecture under the covers of social unrest, retaliation, abhorrence, and anger. Moreover, the studies on the relationship between patrimonial culture and habitation are not only about materiality and territoriality but also concern about spirituality in terms of adoration and reverence. Chapter 13 explores the relationship between the religious belief system within the patrimonial tradition and the historicity of place; this section divulges the practice of pilgrimage to venerated sites that recall the respect and adoration to ancestors and heroes as living beings.

The studies on patrimonial culture and traditions are never-ending stories. However, for this book, the studies are necessarily resumed with something worth for Indonesia communities in terms of patrimonial politeness and courtesy. Chapter 14 unfolds various traditions among peasant communities of Indonesia concerning local concepts and practices of propriety and respect. The exposition of the last chapter is to unveil the roots of the patrimonial practice of power in urban development under the veil of politeness and courtesy.

Culture has been an effective device for Western scholars to understand politics in Indonesia (Robison 1981, 1). However, few attempts have been made at a systemic study concerning either of the local concept of power or on its sustainable powerful impact on contemporary Indonesia (Anderson 2006, 17). Culture is not simply a context of political mindset and practices. Rather, culture provides people with a framework of ideological outlook and historical values and meanings for making decisions and actions. Since politics is not impartial from a set of leaders' personal traits, attitudes, and practices that shape and construct their political behavior.

Clifford Geertz (1960, 1965) is considered an influential exponent of theories on cultural politics in Indonesia; his categorical understanding of Javanese people and culture is incontestably valid until today because of the root of his theory on meaning and streams of consciousness regarding power. Accordingly, the Javanese people are characterized by their streams of consciousness into three groups: *priyayi, abangan,* and *santri*. The sociological issue is how does the relationship between the concepts of *priyayi*, *kawula*, *santri*, and *abangan* work in Indonesia?

Since the Javanese populations are demographically and socioeconomically dominant, Indonesian political culture is almost identical with the mindset of this majority. As a matter of fact, most of Indonesia's national leaders are Javanese such as Sukarno, Suharto, Abdurrahman Wahid, Megawati, Susilo Bambang Yudhoyono, and Joko Widodo. Only Burhanuddin Jusuf Habibie is a son of Bugis—Makasarese father—Alwi Abdul Jalil Habibie—and his mother—Raden Ajeng Tuti Marini Puspowardojo—is a Javanese woman.

From prehistoric to modern Indonesia, the country has been sociopolitically dominated by patrimonial cultural tradition. The familiar concept of patrimonialism has been well known as '*budaya paternalistik*' or '*azas ke-bapak-an*' (Subagya 1955, 82). The leader in this cultural context should be a strong personality with self-confidence, self-absorbed, and compassionless. The people in the patrimonial tradition do not prefer the leaders with friendly, accommodative, and approachable personalities but they lack charisma, firmness, and exude sexual energy; this does not mean that the people do not like the populist leaders. The patrimonial leaders should be more to be a protector than an administrator of public interest, to be more a director than a manager, and to be a conqueror than a caregiver. All these fatherly traits attract most of the people who value the masculine supremacy with discipline, strength, and a ceaseless appetite for power. Most nationalist leaders are characterized by these patrimonial traits of personality, such as Ki Hajar Dewantara, Sukarno, and Suharto.

The patrimonial status quo has been deeply foraging Indonesian tradition since ancient times that has established the noble social class known in Java as *priyayi*, or menak in Sunda, *si ulu* and *is ila* in South Nias, *orang kaya* in Banda Neira, *datuk* in Deli, *teuku* in Aceh, *daeng* or *puang* in Bugis, *marimba* in Sumba, and *dassi* or *dato* in West Timor. Even though the traditional social strata have been formally diminished by modernity, the sociocultural division of *bangsawan* or *priyayi* and *orang kebanyakan* or *rakyat* remains unchanged until today; they exist as modern elites in politics, religious clergy, and businesses.

Characteristically speaking, modern Indonesia's *priyayi* control the access to most resources in terms of trustee group or ruling group called *wali*. Meanwhile, the silent majority of populations is the *rakyat* or *wong cilik*. In Javanese tradition as well as in other customs of Indonesian ethnic groups, there is hard to find a similar concept with the notion of society because as a whole, populations are socioculturally divided into two functions: the ruling—*penguasa*, *pembesar*, or *pemerintah*—and the ruled—*rakyat*, *masyarakat*, or *kawula*. As a whole, both are conceived as the world. As a matter of fact, the world is not an egalitarian realm of individuals but a dualistic composition of cultural milieu groups. Ethnologically speaking, the symbolic dualism and antagonistic patterns characterize the functional differences of the Indonesian world (Kroef 1954).

In the pre-Islamic era, the common people were not identified with any religious affiliation but spiritually unified as the servants, followers, and supporters of the elite—*priyayi*. The arrival of Islam into the archipelago in the early fourteenth century had enriched, adjusted, syncretized, and assimilated the streams of consciousness of populations, especially in Java and Sumatra and other coastal areas of the archipelago, with Islamic element. Islamic influence introduced people to the egalitarian practice of socioeconomic interaction that engendered the spectrum of ideological aptness from extremely obedient called *santri* to tremendously noncompliant known as *abangan*. Thus, the *santri–abangan* spectrum works horizontally while the *priyayi–kawula* power relation works vertically.

The range of Islamic influence among ordinary people and *priyayi* creates two ideological identities: *santri* and *abangan*; both are actually not social groups but more about ideological characteristics. The *santri* are those who follow Islamic teachings and rituals while the *abangan* are those who are not strictly bounded with Islam or agnostics and free thinkers. Nevertheless, spiritually speaking, *priyayi*, *kawula*, *santri*, and *abangan* share something in common on the beliefs in unity and harmony as the existential condition for their world as a whole. Sukarno recognizes this sociological identities and characteristics of people in Indonesia that the economic classes do not work effectively for social and political transformations but the spiritual discontent and disintegration. He conceives the spiritual spectrums from Islam to agnostics should be united as a national spirit to find the common ground of ideological unity. Sukarno calls this ideologically unifying and syncretizing concept as *Pancasila*; Sukarno and other leaders of the national movement for independence formulated and drafted this concept with five principles, which were summarized, extracted, and synthesized from various Indonesian cultural values and beliefs.

The five principles of *Pancasila* include the beliefs in single Supreme Being, humanity based on justice and civility, the national unity of Indonesia, people's sovereignty with the leadership of wise deliberation and representation, and social justice for all. Since Indonesia's independence in 1945, *Pancasila* has become the ideological foundation of the country. From these five principles, there is no room for individual freedom without a commitment to the collective good: the unity in diversity. In terms of *Negara Kesatuan Republik Indonesia* or NKRI, whoever in power is necessarily committed to the principles and practices of democracy based on the national consensus of the state ideology Pancasila. Historically speaking, even though the body of the 1945 Constitution can be amended, the prelude to the Constitution, which lays down the form and ideological foundation of the state, must never be touched (Anwar 2010, 44). Accordingly, the civil liberty in the Indonesian context should be put in the sociocultural framework of unity in diversity—*keutuhan dalam keragaman*; solidarity—*kekeluargaan*; and tolerance—*tenggang rasa*.

As a matter of fact, the political world in Indonesia cannot afford to ignore the Javanese culture and tradition. This is not only because of the predominant Javanese populations but also the fact of the historical and sociocultural infiltration of this ethnic group in various levels and places of Indonesia. In short, to understand Indonesian politics, one should not overlook the influence of Javanese way of life and think about the world. The historical process began with Sukarno and other Javanese nationalist movements, Wahidin Sudirohusodo, Sutomo, Suwardi Suryaningrat, and Cipto Mangunkusumo, in the early 1920s. The language of political power in Indonesia is historically constituted by three languages: Dutch, Javanese, and revolutionary Malay language (Anderson 2006, 124); the Indonesian language is a syncretic acquisition of these languages. The predominant Javanese sociopolitical activists mentioned above played a crucial role in translating and assimilating the Dutch concepts through Javanese cultural perception in the Indonesian language.

Consciously and subconsciously, the Dutch and Javanese concepts concerning the interrelation, notion, context, function, operation, mechanism, procedure, means, and production of power have been synthesized by predominant Javanese nationalist leaders that constituted and established the political language of Indonesia. Most of the adoptions, assimilations, and acculturations of the Dutch, Javanese, and Malay languages had been linguistically constructed on printed media, and political discourses of Indonesian elites since Raden Mas Jokomono's or RM Tirto Adhie Suryo's Newspaper: *Sunda Berita* published on August 17, 1903, and followed by *Medan Priyayi* published in Surakarta in 1907 (Hill 2006, 25), *Oetoesan Hindia* by Cokroaminoto, *Halilintar,* and *Nyala* by Semaun, *Guntur Bergerak* and *Hindia Bergerak* by Ki Hajar Dewantara, and *Soeara Rakyat Indonesia*, *Sinar Indonesia*, and *Sinar Merdeka* by Sukarno.

There were also similar newspapers published with various local interests, such as *Fikiran Ra'jat* in Bandung, *Soeloeh Ra'jat Indonesia* in Surabaya, *Penghantar* in Ambon, *Sinar Borneo* in Banjarmasin, *Persatoean* in Samarinda, *Pewarta Deli* and *Matahari* in Medan, Sinar Sumatera in Padang. However, there had been several local printed newspapers with less political content since 1856 such as *Soerat Kabar Bahasa Melaijoe*, *Bintang Barat*, *Selompret Melayu*, *Dinihari*, *Bintang Djohar*, *Tjahaja Moelia*, and *Pemberitaan Bahroe*.

As a whole, the urban world of Indonesia cannot manage to ignore the influence of patrimonial Javanese culture; the patrimonial for political leadership fits the way the world is perceived and experienced by most Javanese populations (Mulder 1994). How does the patrimonial power work in the political world of Indonesia? From the ancient times until contemporary Indonesia, communities in various ethnic groups in the archipelago are sociopolitically organized by a dualistic system of antagonistically functional categories of the ruling elites—*pemerintah* and *tokoh-tokoh*—and the ruled commoners—*masyarakat* or *rakyat*.

Despite its diversity of customs and traditions, this dualistic and antagonistic system strongly exists and persistent to build the socially constructed system of Indonesian community (Kroef 1954). In the Indonesian context, there is less possibility to expect grassroots movements without involving the elites because most ordinary populations are spiritually bound with specific father figures—*pemimpin* or *bapak* or *tokoh* or *guru*—in terms of *wali* or *atasan* or *panutan;* otherwise, they feel socially dispossessed and lost.

Such dualistic function has been traditionally working for the Indonesian economy since pre-colonial times until today that divides the production and service activities into formal and informal transactions. As observed by Boeke (1942), such dichotomy can be resolved by modernization. As matter of fact, this dualistic economic system does exist and persistent that the formal is under the control of the government with legal documentation and taxation system while the informal works among the ordinary people based on unwritten agreement and mutual trust. Indeed, both mechanisms work with the same currency—*rupiah*. Until today, the informal practices involve almost 60% Indonesian employment system as observed by the governor of Bank Indonesia (Bandungkiwari 2018).

In Post-Suharto Indonesia, modernity and democracy do not imply the transformations of Indonesian communities that are sociofunctionally divided into two symbolic and functional categories: the elites and the commoners. Democratic election and free speech do not instantly change the world perception of the whole population about the necessity for Messiah for their sociopolitical and economic crises. Thus, the ultimate goal of general elections for traditional Indonesian populations is not only to choose the representatives they trust to be the best people in their job but also to consider and follow the advice and guidance of the local leaders of their spiritual stream.

Kiayi or *guru ngaji*—Islamic scholars—are the informal leaders for most Muslims with the strong stream of consciousness for practicing Islamic teaching while well-educated professionals or bureaucrats or retired army and police officers are the important figures for the ordinary people with the stream of consciousness for pragmatic development. The Post-Suharto democratic process of the general election is persistently overshadowed by the paternalistic tradition in Indonesian populations (Kompas.com 2018). Accordingly, political leaders are traditionally conceived by traditional populations that they are not only the secular select few but also spiritual bests with father figure quality, *tokoh-tokoh*.

The practices of patrimonial tradition are symbolically by the supporters and followers with paying a visit of respect, *sowan* or *kunjungan silaturahmi,* to senior leaders, *sesepuh*. Such visits are practiced to the tombs of senior figures in terms of *ziarah*. All these rituals symbolically show and express more than just paying homage. The practices are spiritual to count the blessing from the elders for their power of leadership and wisdom; all of these recalls the animistic traditions of the people in the archipelago. Paying homage to the ruling leaders in the patrimonial tradition is commonly practiced by the supporters and followers.

The obvious expression of paying homage is demonstrated with kissing the hand of juniors to their seniors, who are perceived as morally superior persons. This tradition begins at family home among Muslim populations. Children and women are due to kissing the hand of the men, who are considered senior persons and highly regarded father figures. In the Islamic boarding schools or *pesantren*, the ritual of kissing hand becomes an integrated part of patrimonial education. In the political practice, junior politicians commonly kiss the hand of their senior leaders. This is not only an expression of respect but also a demonstration of submission to their spiritually superior personages.

References

Anderson B (2006) Language and power: exploring political cultures in Indonesia. Equinox, London

Anwar DF (2010) Foreign policy, islam and democracy in Indonesia. J Indones Soc Sci Humanities 3:37–54.

Bandungkiwari (2018) Survei LSI: Elektabilitas Jokowi-Ma'ruf 52,2 Persen, Prabowo-Sandiaga 29,5 Persen. September 28

Boeke JH (1942/2012) The structure of the Netherlands Indian Economy. Institute of Pacific Relations, Honolulu, HI

Geertz C (1960) The religion of Java (Glencloe Il: Free Press)

Geertz C (1965) The social history of an Indonesian Town. The MIT Press, Cambridge, MA

Hill H (1996/2000) The Indonesian Economy. Cambridge University Press, Cambridge

Kompas.com (2018) https://nasional.kompas.com/read/2018/08/21/14123401/survei-lsi-elektabilitas-jokowi-maruf-522-persen-prabowo-sandiaga-295-persen

Kroef van der JM (1954) Dualism and symbolic anthithesis in Indonesian Society. Am Anthropol 56:847–862

Mulder N (1994) The Ideology of Javanese Leadership. In: Hans A, Sven C (eds) Leadership on Java: Gentle hints, authoritarian rule, by Hans Antlov and Sven Cederroth, Curzon Press, Surrey, pp 57–72

Robison R (1981) Culture, politics, and economy in the political history of the new order. Indonesia 31(1):1–29

Subagya R (1955) Pantja Sila dasar negara Indonesia. Basis, Yogyakarta

Chapter 2
Power, Territoriality, and Society in Indonesia: A Historical Overview

Abstract The aim of this study is to explore the concept of territoriality in relation to the modes, styles, manners, and forms of control over land and people throughout history. This paper is an attempt to dismantle and unfold the patrimonial practices of territoriality in Indonesia that have led to social injustice and dehumanization. The focus of the study is to examine the relationship between territoriality and hegemonic interest of each patrimonial regime. This study argues that political culture plays a significant role in the implementation of the territorial concept because territoriality is not simply a geographical boundary of political sovereignty. Rather, territoriality includes the cultural dominion and ideological realm of influence that construct the sense of safety and security of hegemony.

Keywords Territoriality · Patrimonial culture · Dutch V.O.C · Dutch colonial · Sukarno · Suharto · Post-Suharto · Indonesia

2.1 Territoriality and Patrimonial Culture

Territoriality is the concept for the claim and control over land and people based on the communicative form of boundary. This necessity constructs the relationship between society and their space-time concerning power and influence. There are three facets of the form: form of classification by area, form of communication, and form of enforcement of control (Sack 1986, 28). The genesis of form for territoriality is the necessity for spatial control that provides the sense of being in a safe and secure environment in the context of society. Territoriality is necessarily communicative so that its socially constructed system for social order is understandable, shareable, and normatively accepted. In the Indonesian context, territoriality is socioculturally perceived as the realm of influence—*wilayah*—that includes the land and its populations.

The manifestation of patrimonial territoriality varies as urban phenomena of new development, revitalization, rejuvenation, and redevelopment. The idea of patrimonial territoriality has been an integrated part of urbanism elsewhere that affects the establishment, resurgence, displacement, and transformation of settlement in favor of gentry. Obviously, patrimonial territoriality happens in towns and cities under

B. Wiryomartono, *Traditions and Transformations of Habitation in Indonesia*,
https://doi.org/10.1007/978-981-15-3405-8_2

patrimonial hegemony. Research on patrimonial territoriality has been familiar in urban studies with various cases, scopes, and interests (Smith 1979; Lees et al. Gentrification 2007; Brown-Saracino 2013).

Territoriality is theoretically a social field of domination that establishes control over the area as means of controlling access to things and relationships (Sack 1986, 20); this implies that territoriality is domination by virtue of a constellation of interest embedded in a spatially defined field (Oleinik 2011, 158). In the Global South, patrimonial territoriality becomes an area of interest of urban studies concerning its rapid population growth, economic disparity, social marginalization, and massive unemployment. Common issues, constraints, and opportunities of patrimonial territoriality in the developing world include but are not limited to displacement, dispossession, and social injustice (Lees et al. 2015).

In the Indonesian context, patrimonial territoriality has been a long-standing phenomenon that is less on the basis of the geographical location but more on the control of the populations; these phenomena have been traditionally nurtured by local patrimonial regimes as part of their consolidation of power. Patrimonial territoriality has been a political share of the regimes for establishing the exclusivity of their residential domain. The development of the socially gated community has been flourishing in various suburban areas in big cities of Indonesia in the 1980s until today. However, in this country, the roots of the phenomenon of gentrification as an architecturally concentrated settlement for elites have been an integrated part of a patrimonial culture that demonstrates the social class division between the elites—*tokoh-tokoh*—and the commoners—*rakyat*.

This chapter is a study on the relationship between power and territoriality in Indonesian populations from the pre-colonial to the post-colonial era. In Indonesia, power is understood as a spiritual energy that works in persons and beings; because of having such power, a person is able to lead others with a fatherly charisma—*wibawa*—strength—*tegas*—and exemplary excellence—*suri teladan*. Such paternal figures do exist in most ethnic groups; they are believed to be the trustees, *wali, sesepuh* or *tetua adat*, who unite and protect people from disintegration, *perpecahan*, and chaos, *kemelut*. In the today social or political organization, such persons usually sit in the board of advisors or *dewan pembina*. At the center of the patrimonial, regime is the executive leader, who represents the whole elites of the populations. The political relationship between executive leadership and the elites is crucial in the distribution of accesses to resources and wealth.

Patrimonial culture in this study is understood as a culturally controlling system that works in society for the domination of the access to and management of resources based on traditional privilege or power struggle. Patrimonial culture has been a long-standing tradition in Southeast Asia, especially in Indonesia; its practice controls and dominates the access to resources based on patron–client or lord–servant—*tuan–hamba*, *gusti–kawula*, *tokoh–pengikut*, and *bapak–anak buah*—relationship. Since patrimonial culture concentrates power in an exclusively organized system of polity, its practice is in favor of oligarch and patrimonial territoriality that potentially lead toward corruption, collusion, and nepotism. Patron–client relationship characterizes power relation in the political and sociocultural life of Southeast Asian populations since ancient times.

Patrimonial regimes exercise control and domination of resources based on patron–client relationships. In the Indonesian context, the patrimonial tradition is not necessarily associated with the gender of the person who is in power. Rather, the patrimonial tradition in this geographical area has been historically evolved to transcend the idea and form of masculine power. Accordingly, the person in the leadership is associated with a strong commander, cold-blooded executor, merciless lord, and malevolent protector. In order to achieve their goal, patrimonial leaders never hesitate in the use of violence of action.

The patrimonial system utilizes a pyramidal structure of the hierarchical organization from top to bottom for the decision-making process. In terms of organization, patrimonial culture values and upholds unity, stability, loyalty, conformity, and evenness. Practically speaking, patrimonial leaves no room for individual liberty, internal disagreement, and fair competition. All this is to set up the circumstantial conditions for benefitting and securing the position of the leader. In the patrimonial system, there is no egalitarian society but the dualistic division of the ruling regime and the socioeconomically and politically ruled group.

The patrimonial ruling group is not necessarily a fascist regime but a cluster of cronies in politics and businesses; they are the minority in number but they dominate most of the accesses to resources in terms of fund, energy, materials, information, and labors. Personal preferences and nepotism play an important role in the decision-making process of the patrimonial system. The majority of populations are the ruled called *rakyat* or *orang-orang kecil*. The dichotomy of the ruling and the ruled is not simply economic. Rather, this is a sociocultural division that establishes the superstructure and substructure of populations. The sense of this dualistic division is to uphold the historically established elite class of populations—*priyayi* or *bangsawan* or *para tokoh*—since the prehistoric times; the division was amplified by the Dutch colonial rules in their credo: *divide at imperia* in various aspects, such as economy, politics (Gouda 2008, 85), and culture/education (Willemyns 2013, 215). During colonial times, the Dutch practiced indirect rule by using the elites to exploit the ordinary populations. In the education system, only the children of the elites were eligible to attend the schools (Woodward 2010, 53).

The superstructure consists of the elites in various fields of politics and businesses; the struggle to power among the political elites is to prove themselves publicly as the best trustee—*wali*—of the floating mass of populations. Under the notion of *wali*, the elites try to gain their followers with the charm of fatherhood and spiritual aptness of their leaders. The concept of the *wali* is traditionally associated with the saint with fatherly traits. The language of the elites—*wali*—is not impartial from moral principles with the justification of religious texts or traditional values. They are respectful but unapproachable. Since ancient times, the bond between the politically ruling group and the religious clergy has been an integrated part of the conception and formation of the elites; they disengage and distance themselves from the mass populations with their customs, codes, languages, and comportments. Substantially speaking, the language of patrimonial leaders is pedantic and rhetoric; the leaders assume and assign themselves morally superior and better than the rest.

In the local Indonesian context, patrimonial culture is understood as *bapak*-ism or *kebapakan*; *bapak* means literally father and leader. Traditionally, patrimonial culture in Indonesia exists in various fields and agencies that creates and builds private and informal bonds in terms of patron–client relationship. The most common bonds involve the patron–client ties between the ruling politicians/bureaucrats and business people; they are called *pembesar* or *orang-orang besar*. Patrimonial culture establishes and develops a web of power relations from national, regional to the local level. To a certain extent, the practice of patrimonial culture is another word for the family business; while a father is in power, the son or relatives run the family business to get the privilege access for winning the tenders from the father's office. In many cases, politics and business have developed symbiotic mutuality.

Historically speaking, the patrimonial regime has been established by a traditional elite group of the populations who controls almost the whole wealth in the country. The patrimonial power of socioeconomic productions and services has been working for the cultural character and uniqueness of Indonesia. The practice of the power has been well known as *adat-istiadat*, which is incorporated deeply in various activities and forms through the collective streams of consciousness that matter for the lifestyle and cultural artifacts in Indonesia. In Javanese populations, the collective streams of consciousness are known as *aliran*. Clifford Geertz (1960) recognizes these streams in post-colonial Indonesia, which fall into three categories of elite sociopolitical groups: *priyayi*, *santri*, and *abangan*. The three main ideological pillars of Indonesian populations' identity characterize the country soci0political grouping because of the Javanese majority.

Despite modernity and democratization, the sociological grouping of the political elites is still valid until today with the adjustments of identity. The *priyayi* or *pembesar* or *penggede* group is customarily constituted by the well-educated persons from well-established families. The *priyayi* elites are mostly the leading nationalist figures of various organizations such as political parties, businesses, military, and police; they are ideologically stewarding and guarding the existence and establishment of Indonesia's nation-state. Since the majority of Indonesia's populations are Muslims, the elites of this group construct the soci0political cluster with a strong commitment to ideologically imbuing and incorporating Islamic teaching and values system into the political system of Indonesia's nation-state. The existence of the world is conceived as the integration of dualistic components: *priyayi* and *kawula*. Both are complementary, interdependent, and interconnected in terms of power relationships in the economic and political realm. How is the relation of the *priyayi–kawula* system to *santri* and *abangan*?

The spectrum of *santri* and *abangan* did not exist in Indonesia's pre-Islamic history. The influence of Islam has transformed the populations with the enrichment, assimilation, adjustment, and acculturation of their previous animistic, Buddhist, and Hindu culture and tradition. The Islamic influence has created and established a range and field of horizontal gradation of acceptance. The populations, who strictly follow and are strongly obedient to Islamic teachings and rituals, are characterized by the concept of *santri*. On the other hand, the populations, who are noncompliant or resist to strict Islamic principles and services, are identified with the concept of *abangan*. Thus, the concepts of *santri* and *abangan* are an ideological and horizontal spectrum in the populations. The non-Muslim populations are commonly similar to the *abangan* people. The spectrum characterizes the *priyayi* and *kawula* as well. Thus, there are *priyayi santri* and *priyayi abangan*.

Historically, the *santri* group is associated with the elites from *Nahdlatul Ulama* and *Muhammadiyah* organization; beyond these two dominant groups, there are some other small factions with their own Islamic ideologies. The *Nahdlatul Ulama* is ideologically syncretic that Islamic teaching and rituals and local culture and tradition can get along together to build a contextual civilization. The *Muhammadiyah* is ideologically selective that only a few local and indigenous culture and tradition are acceptable to be integrated into Islamic practices. In the daily world, the *santri* and *abangan* work and mingle together in various fields of economy, politics, and culture.

Beyond *Nahdlatul Ulama* and *Muhammadiyah*, there are fundamentalist Muslim groups, whose ideology strongly resist to adopt and assimilate the local and indigenous culture and tradition, some of them are affiliated with Islamic Wahhabism. Even though the followers of Islamic fundamentalism are not significant in number, their appearances and movements are often intolerant to the non-Muslim populations. On the other pole of the spectrum, there are the elites of abangan that ideologically free thinker, social democratic, nationalist, and liberal; they are independent individuals and factions with various backgrounds, occupations, and orientations. Indeed, there are the elites of *santri* as well, who are mostly obedient to practice Islamic rituals but moderately tolerant to other non-Muslim populations.

Among the Indonesian elites, the political contests for power are not a class struggle but a sociocultural fight for domination to be the best trustee of the floating mass of populations, *rakyat*; these populations the streams of *santri* and *abangan* mingle and socialize together with the other minorities that are the silent majority. The core of this struggle is to gain the trust of common people for the best regime with a strong leader that is able to unite all populations for a solid nation-state. Until recently, the national leaders of the country come from the elites of the *priyayi* group such as Sukarno, Suharto, Habibie, Megawati, and Susilo Bambang Yudhoyono. Only Abdurrahman Wahid is from the *santri* group and only Joko Widodo from the *abangan* component.

Historically speaking, the political stability of the nation-state is reliance upon the consolidation of the nationalist elites within the repressive and exclusive ruling regime. The New Order regime under Suharto is the best example of the practice of patrimonial power that represses and excludes the *santri* and *abangan* component; the military ruling group has not attempted to include the *santri* component but contained the opposition by depoliticization, by merging, co-opting, and manipulating the political institutions so that they lose their cultural distinctiveness (Emmerson 1978). Nevertheless, oppressing and repressing or embracing and reaching out the *santri* and *abangan* group has been historically always necessary to keep and defend the integrity and establishment of the nation-state. In the real politics of Post-Suharto

Indonesia, all these streams, the elites, *priyayi*, of *santri* and *abangan*, play an important role in the dynamics of Indonesia through the cooperation and contestation of political parties for all levels of executive and legislative government from the local, provincial, to the national scale.

In the Post-Suharto Indonesia, the struggles of power are inevitable to incorporate all ideological streams of *priyayi–kawula* and *santri–abangan* into an effective ruling regime for the economic growth and the integrity of the nation-state. Suharto's New Order has established a new socioeconomic force involving former high-ranking military officers and Chinese businesspeople. Under Suharto's regime, the patrimonial patron–client relationships between the ruling *priyayi* group and the business people had mostly dominated the strategic businesses in the country from agriculture, forestry, real estate, manufacture, mining, petrochemistry, pharmaceuticals, to foods and beverages. These ties are constituted within business groups taking the form of a complex series of joint venture enterprises in which, as a general pattern, the majority of the capital is provided by overseas partners, while distribution, subcontracting, and often management is provided by Indonesian-Chinese partners, and political protection by indigenous officials (Robison 1981, 27).

The impediment of the Post-Suharto Indonesia is the reality of patrimonial power and tradition that is still in existence among the elites of the country. They become a newly established group of *priyayi* in the political parties and bureaucracy. Post-Suharto's Indonesia is the era when everyone is able to run to be a politician from local, regional to the national platform as long as she/he has the support from their community based on the free public election. All this is impossible without financial support from people. Being elected as officials cost much money. The support from business people is not for nothing; they support a candidate to get paid off with privileged information and access to resources.

2.2 Territorial Control in Pre-colonial Java

Territoriality in this study is understood as an acclaimed domain in terms of geographic expression of influence and power that provides an essential relationship between society, space, and time (Sack 1986, 216). In its practical actuality, patrimonial power in pre-colonial Javanese states claimed territoriality on the basis of control over units of settlements. The establishment of territorial kingship was inaugurated as the founding and formation of the sovereign regime for being able to control agricultural production in adjacent areas. Such a regime was expanded within the social network of social elites, *priyayi* or *bangsawan*.

Territorial control in pre-colonial Java has been potentially possible through the establishment of the network of elites as landowners of the wet rice cultivation. Territorial control in this sense is not about to demarcate the land with strict geometrical boundary, but it is to control people within the sociopolitical organization of village in terms of *desa*. The expansion of state territorial control was made possible by the surplus of wet rice cultivation. However, the practice of territorial claim in early

Java was a politically integrated part of the patrimonial control of the state over populations; the land does not mean anything without populations.

Controlling people who work and cultivate it for agricultural production is more important than claiming the area of land. However, it does not mean that the early Javanese states did not have any concept of marking land for their territorial claim. Several copper-plate inscriptions from the eighth and tenth century Java show the practice of such territorial control under the concept of freehold *sima*; the notion is derived from Sanskrit *siman* meaning boundary or border (Day 2002, 179). In the area of freehold *sima*, commercial and religious activities were not allowed. The area was claimed as a state territory which was actually imaginary and not a geographical boundary. *Kuta* is a patrimonial realm of kingship with reference to the concept of the *mandala*, a Sanskrit term for the center of sociopolitically organized settlement Wolters (History, Culture, and Region in Southeast Asian Perspectives 1999, 27–40). Today in Indonesia, the concept of *kuta* or *kota* becomes the notion for urban settlement or city. In Javanese tradition, the idea of *kuta* is architecturally expressed with territoriality inside the enclosure made of brick or stone walls.

However, *kuta* is not any space but originally an archetype of the domain for habitation for the ruling group that represents the center of the universe on the earth. Religious rituals and ceremonies are necessarily conducted to transfer and connect the cosmic power into the microcosmic realm. Javanese territoriality is not about the geometrical plot of the domain on earth, but more about a socially marking place that is publicly claimed and announced with religious rituals and ceremonies. Only a set of Javanese ruling class and state officials was included in the *kuta*'s territoriality; merchants, commoners, and foreigners were excluded from this boundary. Outside the *kuta*'s, there are few domains constructed with an architectural boundary. Such domains in ancient Javanese kingdoms were known as *sima* or freehold lands. This territorial category was mostly for religious sites, sacred places, and other special uses. Endowing *sima* was rewarded by the state with the enticement to local chiefs and villagers who participated in the rituals and ceremonies of the *sima* endowment. In early Java, such land grants were commonly exercised as a political strategy of the state to bring populations and territory under control (Day 2002, 180). The more an early Javanese state gained *sima* land is understood as the establishment of cosmological center on earth. The idea of the city is architecturally articulated by the walled territory: *kuta*: the center without periphery. The rest of the land is the ocean of boundless area without a definite destination.

2.3 Territorial Control Under the Dutch V.O.C

In the early years of the sixteenth century, the presence of Europeans did not penetrate deeper into local affairs, but were mostly for practical matters of spice trade at coastal areas of the Southeast Asian archipelago. The penetration of maritime trade into local affairs came into being when the need for the monopoly of the spice trade was lifted up in the interest of larger profit-taking of the consignments for the European market. The establishment of the United Dutch East India Company or *Vereenigde Oost-Indische Compagnie* (V.O.C) in 1602 was in response to the growing demand of spices in Europe by consolidating the Dutch companies under one roof of operation and expedition. Seventeen Gentlemen or Herren of the Seventh served as the directors of the Company. Jan Pieterszoon Coen and Antonio van Diemen were well-known governor-generals of the V.O.C with a high contribution of profit (Parker 2010, 24).

Colonialism was the hardest practice of territorial control under the patrimonial system. The practice in Java and other islands was since the Dutch V.O.C regime-controlled and monopolized the economy of the archipelago from the seventeenth century to 1942. The establishment of Batavia by Jan Pieterszoon Coen of the Dutch V.O.C marked the beginning of exploitative domination of local resources by the European power. Even though the main mission of Dutch V.O.C was the monopoly of the spice trade, in practical reality, the Company controlled the whole economy of the archipelago. The consequence of this exploitation was not only the control of ports and the areas of spice plantations but also most local marketplaces and indigenous political centers. People of Banda Neira, Ambon, Bantam, Lampung, and West Java had to sell their commodities such as peppers, cinnamons, mice, and rice to the V.O.C with low price. As a trading company, the Dutch V.O.C exercised a patrimonial system of domination and extortions over local resources against indigenous and other Europeans.

One important figure of the V.O.C. patrimonial culture is Jan Pieterszoon Coen. As the founder of Batavia, Coen was well known in the history of colonialism as the butcher of Banda Island in 1621 (Benjamin 2007, 241) and Ambon in 1623 (Bown 2009, 49). Practically speaking, Coen's colonial patrimonial culture exercised ruthless control of economic resources from its cultivation and collection to its transportation and distribution; he was infamous for the use of violence and ruthlessness against others, English, Portuguese, and Spaniard. In order to achieve maximum profit, the V.O.C manages to dominate the spice trade in Southeast Asia with controlling the land for cultivation and collection of the crops. Under the V.O.C's control, people in the Banda Islands were compulsory to cultivate nutmeg, clove, and mace. Prior to the arrival of the Europeans, the people of Banda Islands cultivated such commodities for international traders from the Middle East, India, and China.

Jan Pieterszoon Coen inserted the patrimonial practice of estate and labor control for nutmeg and mace cultivation in Banda Islands; the practice was known under the concept of *perkerniersstelsel* (Winn 2010). Accordingly, the Dutch Company took the specific area under their control and employed migrant workers from other regions or countries, such as India and Africa. Banda Islands consist of five islands: Banda, Neira, Rozengain, Ay, and Run. John Crawfurd (1856, 36–7) describes the people of Banda Islands are Muslim traders with a patron–client social structure consisting of the elders—*orang kaya*–and the workers and planters—*orang biasa*. In many ways, they are more or less egalitarian; they do not practice slavery. Anthony Reid notifies the absence of long-term slave-based agriculture in Southeast Asia (Reid and Brewster 1983, 14–5). Indeed, in this region, agriculture has been run with family-based cultivation based on harvest sharing cooperation.

The Portuguese were the first European explorers who established their fort in the island of Banda in 1511; the Dutch expelled them and controlled the whole regions in 1603 (Encyclopedia Britannica 1842, 340). In order to establish their presence in the region, the Dutch V.O.C under Verhoeven built their fort of Nassau on the former Portuguese fortification. However, the Dutch wanted more; then they needed to control the trade of nutmeg and mace from upstream to downstream. The Dutch were not alone. The English Company had been in the area with a good relationship with the local patrimonial leaders—*orang kaya*—since 1609; the Dutch and the English Company had a trade agreement for nutmeg and mace from the Banda Islands.

Coen had brewed an idea of the colonization of Banda Islands since 1609. He wanted to control the cultivation of nutmegs and maces by depopulating the islands and replace with migrant workers (Bown 2009, 46). In order to achieve this goal, firstly, he defied the trade agreement with the English. Secondly, he managed to eliminate the resistance of indigenous people. Coen executed this idea with the help of Japanese mercenaries and his 2000 men in exterminating more than 20.000 indigenous people in 1619 (Lee 1999, 365).

Local sources of Banda Islands confirm that the massacre of Banda was ethnic cleansing in Indonesian history. The idea, approach, and action of Coen in the Banda Islands can be seen as a kind of patrimonial territoriality at its very raw and vulgar treatment. While J.P. Coen was back to the Netherlands, the Dutch V.O.C under Pieter Both established their presence in Banda Islands with the construction of fort of Belgica on the hill of Banda Neira. Later, the fort was developed and enlarged by V.O.C chief, J.P. Coen, in 1622. The practice of *perkeniersystem* in Banda Islands brought about significant profit for the V.O.C.

The milestone of domination with the patrimonial territoriality of European settlers was achieved during the establishment of the first Dutch town in 1619 at the Sunda Kelapa Port. From the territorial perspective, building this European settlement is imposing a geometrical abstract plan to the claims of the sovereign over the area as that appears to be the natural precondition for clearing a site for new community and authority and enriching more and more specific social organizations (Sack 1986, 127). The new settlement was given by Coen with reference to the plan of Leiden and his hometown in Horn. After the V.O.C had seized the indigenous ruler of Jayakarta and burnt down the town, J.P. Coen had to establish the V.O.C headquarters at the estuary of the Ciliwung River; territoriality is imposed on the former land of the Jayakarta town based on the map of the imagined V.O.C's headquarters settlement. Even though the directors of the V.O.C—Herrn the Seventeenth—did not have any intention for building a colonial town in East Indies but a place of trade *rendezvous*, J.P. Coen insisted on building the headquarters in the form of architecturally fortified town, Coen gave this town's name: Batavia.

The realization of this idea was not easy because there were not available human resources to do the construction job. After the capture of Jayakarta by the V.O.C, most of its populations run away and found another place of refuge. J.P. Coen asked his Chinese associates, So Bing Kong or So Ming Kang alias Ben Kon, for help (Parmer 2012, 28); he assigned Ben Kon as Chinese *kapitan*. All Chinese matters and affairs in the town would be managed and handled by Chinese *kapitan*. In doing so, Coen and the V.O.C did not interfere with the Chinese community. In other words, the Chinese populations should live and work in their own domain with their own customs and traditions. From the beginning, the Dutch V.O.C identified and established the domains of settlement according to its populations' cultural background. Under the patronage of the V.O.C, the territoriality of populations was associated with its economic and political resources. Chinese settlement in the Batavia area was mainly designated as the territory for sugar plantation and industry, while indigenous kampongs in urban area were the districts for general labors.

During the construction of the new town, the Chinese community mobilized their resources including skilled workers and laborers from the South China mainland. Coen instructed the V.O.C ship carpenters and other technicians to lead and work together with the Chinese master carpenters, bricklayers, and stonemasons for the castle and the fortified town with canals system. The form of the town was derived from the European Middle Age *bastide*. Deep canals were constructed to encircle and go through the town. The construction of the fortified town took several years from 1619. The completion of its architectural form was achieved in 1630. The construction of Batavia was not only to establish a trading post of the Dutch V.O.C in the Far Eastern region but also to insert the presence of politically dominating Dutch in the Southeast Asian archipelago. The end of Dutch V.O.C in 1799 was the conclusion of the patrimonial territoriality of Batavia *bastide*.

As a place of the patrimonial territoriality of European settlers, Batavia had accomplished its demographic and architectural perfection in circa 1630. The other gentrified area was for Chinese and other Asian populations; the area was situated adjacent to the Batavia *bastide*. Since then, the area has been known as the Chinese quarter or *pecinan*. Such territorial grouping in early Jakarta area was direct and indirectly related to racially segregating politics of employment by the Dutch V.O.C between 1612 and 1799. Each group has its own patrimonial governance. The Dutch and European group stood at the utmost ordering position, followed by the indigenous gentry and Chinese/another Asian group at the second layer, and the common native populations at the bottom of the system. The hegemony of the Dutch V.O.C was maintained and sustained by a patron–client relationship with the local elites and the Chinese group.

2.4 Territorial Control Under the Dutch Colonial Rule

The dissolution of the Dutch V.O.C in 1799 and the incorporation of the Netherlands under Napoleon's rule in 1808 changed the mission of the Dutch control over the Southeast Asian archipelago from maritime trade to land-based administration. The Dutch territorial colonialism began with the establishment of the Dutch East Indies in 1806 with Willem Daendles as the Napoleonic Governor-General. Territorially speaking, the Dutch supremacy of political economy over Java began officially under the establishment of the Dutch Indies colony in the Southeast Asian archipelago.

Unlike the Dutch V.O.C power, the Dutch Indies acclaims full control and supervision of land for the production and distribution of commodities.

The Dutch patronized the local rulers concerning the cultivation of export commodities they needed for the European market. Under the Dutch colonial rule from 1800, the patron–client relationship has been politically and economically practiced in terms of the indirectly ruling system. Local villagers did not have direct contact with the Dutch colonial apparatus; the contact was cushioned and muted by a layer of native civil servants (Stoller 2002, 25). The Dutch were the superior patron of the indigenous rulers, *priyayi*, the Chinese, *taukee*, and other European and Asian businessmen. The Dutch colonial rule never did and wanted any relationship with the farmers, collies, workers, and labors in agricultural production who were the clients of the local, Chinese, and foreign patrons.

During the mid of nineteenth century, the Dutch colonial rule opened plantation fields in Deli, North Sumatra. After the state monopoly of plantation enterprises for almost 50 years, the Dutch changed the policy and implemented the newly issued Agrarian Land Law in 1870. Accordingly, private enterprises were allowed to run their business in the archipelago. Deli region of North Sumatra was the appropriate land for such plantation businesses. This plantation region was considered one of the most lucrative sites for agricultural enterprise in Southeast Asia. Stoller brings into the light what happened in that profitable region during the nineteenth century with her studies (Stoller 2002, 26–40). The tobacco plantation in Deli is one of the agricultural private enterprises that have transformed the landscape of North Sumatra from its natural biodiversity to monocultural estate. The land for the plantation is located in Tanjung Sepassai nearby the town of Labuhan Batu. The Sultan Mahmud Perkasa Alam of Deli granted the concession of land for 20 years to Dutchman Jacobus Nieunhuys in 1865; he planted tobacco for the first time in Sumatra (Gould 2012, 28). The tobacco leaf of Deli reached the USA in 1873. Since then, Deli tobacco leaf has been well known as one of the finest tobaccos in the world, and the plantation area has been expanded to other regions of Deli. The tobacco plantation in Deli was made possible with the employment of Chinese migrant workers from the Straits Settlements: Penang, Malacca, and Singapore (De Silva 1982, 337).

Ill-treatment of labor often caused riots and assaults against the European supervisor. In the 1890s, the Dutch colonial rule intervened in the labor matter and supported the replacement of Chinese workers by the Javanese labors on a contract basis. The development of monocultural plantation in Sumatra is an introduction of territorial ity on the basis of the economic value of the land. The plantation in North and East Sumatra grew during the period between 1919 and 1925 with at least 342 locations comprising rubber, palm, tobacco, tea, and coffee plantations. The palm plantation dominated the area by 1919 with 9.000 ha and by 1925 with 31.600 ha (Corley and Tinker 2016, 6).

The Dutch colonial rule managed the territoriality of the land with a dual system of land tenure: for European and indigenous ownership. According to the Agrarian Law 1870, the lease of land for the agricultural estate was mostly for seventy-five years. The law regulated the agrarian concession for European enterprises in outer islands under the hand of indigenous rulers. However, the Dutch colonial rule took preventive measure for the apprehension of social order by declaring such lands as the state land and creating a mechanism that plantation estates obtained the leases (Kanō 2008, 283–4). Even though the plantation estates had the right to run their private agricultural business in their areas, the right to land was restricted only at the surface and not for mining activities.

Territorial control under the Dutch rule was intensified during the compulsory cultivation, especially in Java. The control did not only about the natural resources, but also the populations. The opening of plantations in outer islands needed workers. Javanese villagers drew from their homeland to be sent to Sumatra or other places like Dutch Suriname, South America. In order to control the land and its resources, the Dutch colonial rule administrated the land and its people into several administrative territories at three levels: governor-general, resident, and assistant resident. Each assistant resident was aided by several Controleurs, who did not have the executive authority but theoretically speaking, they acted as a colonial liaison for the native chiefs concerning local issues. Practically speaking, the Controleurs were the extension of colonial control that supervised and managed the native rulers at the village and subdistrict level of administration. At the district level in Java and Madura, the colonial official was represented by assistant resident, who indirectly supervised several district native chiefs, *wedana*. Traditionally speaking, *wedana* was the local ruler under the native head of the region, *bupati-*. At the regional level, several native rulers were supervised by a Dutch resident. At the federal level, the governor-general indirectly controlled several native Kings or Sultans.

The whole archipelago was divided into territorial governments, provinces, residencies, and regions/*Afdeeling*. The status of the territorial area was based on the number of populations. Java was divided into three territorial governments or provinces: West Java consisting of residencies of Bantam, Batavia, Buitenzorg, Preanger, and Cheribon, Central Java comprising residencies of Banjoemas, Pekalongan, Kedoe, Semarang, and Djepara-Rembang, and East Java including residencies of Madiun, Bojonegoro, Kediri, Soerabaja, Malang, Besuki, and Madura. However, there were so-called *Vorstanlanden* or the territorial government of native states in Java: Djogjakarta sultanate, Surakarta sultanate, and Klaten sunanate. Under the Dutch Indies, the outer islands were the islands beyond Java that were territorial governments of Sumatra, Borneo, Groot Ost or islands between Java and Borneo, Lesser Sunda Islands, Celebes, Maluku, and Western New Guinea. The provinces of Sumatra consisted of residencies of Atjeh, Tapanoeli, Sumatra East Coast, Sumatra West Coast, Riau and dependencies, Jambi, Bengkulu, Palembang, Bangka, and Lampung.

Prior to the culture cultivation in 1829, most lands, which leased by the Dutch East Indies to non-native Asians and Europeans were restricted for only less than twenty years. This condition was not favorable for many entrepreneurs in the agrarian sector. In order to attract more investment, the Dutch rule issued the Agrarian Act permitted the leasing lands to private companies and individuals for the periods not over and above 75 years. Accordingly, the land available for leasing should be the uncultivated area. The other areas that were not held by private individuals or enterprises were claimed by the Dutch authority as public domain. The notion of the public domain was introduced by the Dutch Indies colonial rule's decree in 1870 that was known as 'domain declaration.' As a matter of fact, there was no such land because most of the lands were under the realm of the native customary laws. As far as respect to the customary laws should be concerned, the use of land for public utilities should necessary to be consulted with the native rulers. However, this was never the case until the twentieth century.

The land policy was not without a challenge from various parts of the Dutch authority in the Netherland and in the East Indies colony. The most eloquent criticism of the Dutch land policy during the culture cultivation came from Professor van Vollenhoven of Leiden University, Professor Lagemann and Ter Haar from East Indies Law School; all of them are experts in Indonesian customary laws. According to these experts, the Dutch colonial land policy that provided the private sectors with concessions had violated the native property rights as described in the agrarian legislation of 1854, 1870, and 1874 (Vandenbosch 1944, 246). The concessions of land were necessary to be reviewed with a consultation of the native communities and paid them the fee for the reparation. According to van Vollenhoven, the territorial right to land under the customary laws was not only for the agriculturally cultivated land and settlement, but this included the wilderness in terms of the hunting area and graze.

In respect to the native customary laws, there was no no-man's land even though in the form of wilderness. In Article 50 of the Dutch East Indies Regulation 1854, forests with potential resources such as teak timber are managed and maintained by the state with no access to be sold to private individuals or companies. The origin of this land policy on forest conservation was actually from Governor-General Herman Willem Daendles, who was in power between 1808 and 1811 (Mizuno and Kusumaningtyas 2016, 34). Daendles introduced the management of teak forests by the state authority; he took away the native regent's authority of forests for timber demand. Even though all forests were the state's property, after the end of the culture system in 1870, the Dutch colonial authority contracted the lumber business to private companies. Due to uncontrollable logging and deforestation, the Dutch colonial government abolished the contacts with private enterprises for delivery and logging timber in Java by 1914. In the same year, the Dutch colonial government established the forestry agency under the ministry of agriculture. All this had been an integrated part of an administrative reorganization of the Dutch colonial government since 1903.

The reorganization of the Dutch colonial governance was impartial from the fact that was highly centralized bureaucratic with lack of legislative power. The burden of the governor-general was too much for one person in a vast territorial country. Despite the need for reorganization had been recognized since 1836 by the Council of East Indies; this was the body of legislative and executive authority in the Dutch Indies. The governor-general was designated as a member of this council. However, the legislative power of the Council was weakened by van den Bosch during the culture system; the Council was merely an advisory body. In 1854, the legislative power of the Council was restored and provided several recommendations. Regarding the native matters in Java and Madura, the Dutch colonial authority established a dual system of ruling power that the traditional rulers—*sultan* or *bupati*—should be assisted by the *patih*. Unlike the traditional system, the function of *patih* was directly under the traditional ruler. In the colonial system, the *sultanate* or *kabupaten* and the *kepatihan* were put at the same administrative level under the indirect supervision of the governor-general. As a matter of fact, a Dutch resident had mostly more interaction with *patih* in their region instead of with *sultan* of *bupati*.

Moreover, as the result of the Decentralization Act 1903, the territorial administration under the Dutch colonial rule was decentralized with legislative bodies for regional and residency councils, local and municipal councils. The establishment of municipalities began with Surabaya, Semarang, Cirebon, and Bandung in 1906, Medan in 1918, Bogor in 1921. Each municipality was provided with a major or *burgermeester* and city council. The establishment of municipalities signified the birth of modern urbanism in the Indonesian archipelago. Bearing the status of *gemeente* or municipality, cities were able to manage their own governance in terms of household, development, and maintenance. To a certain extent, the status of the municipality was identical with autonomous management for local economic and cultural matters. The executive power of major and the legislative authority of the city council played an important role in shaping and developing the quality of life of its populations.

The democratic movement in the political system of the Dutch East Indies was not impartial from what happened in the Netherland. The spread of liberalism had been working among the House of Representatives since 1840 for the constitutional amendment of parliamentary democracy. The amendment brought about the 1848 Constitution that established the chair of prime minister for Johan Rudolph Thorbecke. Even though the dominant liberal politicians in the House of Representatives did not significantly change the colonial policy in the Dutch East Indies, the need to the decentralization of power from the hand of the governor-general came into a movement. The outcome of this movement was the establishment of the representative body of people in the Dutch East Indies or Volksraad that came into the law in 1915.

However, the *Volksraad* was not fully a legislative house but an advisory body of the Dutch colonial government. The election of the representatives took place in 1917 and participated by several political parties and sociopolitical organizations for a total of 38 seats; the council consisted of 20 Dutch representatives of nine elected and eleven appointed members and 15 indigenous members; ten of them were elected and five appointed legislatures, as well as three Chinese appointed members (Schmutzer 1977, 57). The inclusion of native representative in the representative body was made as a realization of the Ethical Policy; the Governor-General J.P. Count van Limburg Stirum was sympathetic to the demand of indigenous nationalist movement (Pols 2018, 107).

The *Volksraad* was a Dutch experiment to deal with the diversity of East Indies populations. Ideologically speaking, the existence of the Council of People was not only to establish a self-regulating colonial administration but also to manage the check and balance of the governor-general's power. Despite the Council did not have legislative authority, the voices in this representative body had been mostly critical since its first session on May 18, 1918. The members represented a wide range of constituencies (Moon 2007, 78); In the Volksraad, there were capitalist Gerritzen and Birnie, social-democratic Cramer, Cipto Mangunkusumo from *Insulinde*, and

Cokroaminoto and Abdul Muis from *Sarekat Islam*. Ideologically speaking, the *Volksraad* was an attempt to represent the territorial and political landscape of the Dutch East Indies.

For the indigenous representatives, the council was an opportunity for their pollical practice of criticism against the colonial hegemony. The pressure of reformation was gradually increasing for the shifting of legislative power from the Netherlands to the East Indies. From inside the *Volksraad*, Mohammad Husni Thamrin, Sutarjo, and Wiwoho set three motions for using the term 'Indonesian' instead of '*inlander*' in official documents, for defining East Indies nationality, and for the reform of the body as a legislative house (Ricklefs 2008, 321). The motions were received with an unfavorable response from the Dutch authority. Meanwhile, the Japanese intelligent operatives had been penetrating the archipelago since early 1940. The Japanese troops invaded Borneo, Sulawesi, Sumatra, and Java in early 1942; the Dutch East Indies forces surrendered to the Japanese Army on March 9, in the same year. This marked the end of the Dutch colonial rule in Indonesia and the beginning of the Japanese occupation.

Nevertheless, the territorial administration under the Dutch colonial regime marked and established the historical formation of the township; in the Dutch language, the concept is understood under the notion of *gemeente*. Historically speaking, the idea of the township in Anglo-Saxon times was identical with the territoriality of the manor under the rule of a lord; however, every township was subject in demesne to a superior either King of ealdorman (Stubbs 2011, 89). This township was also the basis of parish territoriality, in which a priest was responsible for the church and land as a special demarcation. The establishment of township means the formation of territorial jurisdiction of a town of a city with its own elected local government. In term of township or municipality, the territory of the settlement will be divided into diverse land uses, such as housing, public facilities, road system, and utilities.

The formation of a municipality under the Dutch Indies administration was indivisible from the increasing European and Eurasian populations in the Indonesian archipelago. Between 1906 and 1930, the Dutch colonial authority had given a municipal status for most highly inhabited settlements with more than 2000 European and Eurasian populations in Java and other islands. Demographically speaking, the early 1900s only less than 6% of indigenous populations lived in urbanized areas of Java; European, Chinese, and other Asian populations dominate the urban life of the island with a monopoly of few trades (Vandenbosch 1944, 16 and 24).

As a matter of fact, the land uses of Indonesian cities under the Dutch colonial administration were racially segregated. The well-planned areas were designed for the European populations with villa-type residentials, public service buildings, schools, cinemas, stadiums, railway stations, hospitals, public parks, and amenities. All these settlements were provided and well supported with modern infrastructure systems and utilities, such as roads, electricity, drinking water, and telephone. The Chinese and other Asian populations were designated to occupy the interregional trading and commercial areas with high-density buildings. Meanwhile, the international and interisland port areas were mostly under the control of European traders and agencies. At the marginal parts of European and Asian populations were

Table 2.1 Urban populations in the seven largest cities of java and madura

Population group	1920	1920 in %	1930	1930 in %
Indigenous people	800,971	80.96	1,251,298	80.43
Europeans	72,235	7.30	104,788	6.74
Chinese	105,293	10.64	182,041	11.70
Other Asians	10,832	1.10	17,521	1.13
	989,331	100.00	1,555,648	100.00

Source Volkstelling 1930 adapted from Vandenbosch (1944, 17)

indigenous settlements. The native settlements—*kampong*—were mostly to occupy the watershed areas of the rivers or wastelands nearby the European settlements or commercial and port districts.

Despite the Dutch Indies did not transform the center of traditional towns and cities in Java, the symbolic representation of indigenous authority was architecturally marginalized and disrespected. The Dutch colonial rule established prison buildings and other military bastions and barracks nearby the indigenous city center, *alun-alun* or city square and *dhalem* or the seat of the rulers. The European urbanism in the tropical archipelago had achieved its vibrant peak in the period between 1920 and 1942. Most of the urban buildings were built with modern and European decorative design in big cities such as Batavia, Bandung, Surabaya, Semarang, Yogyakarta, Surakarta, and Medan (Table 2.1).

2.5 Territorial Control Under Sukarno

Under the Japanese occupation, the territorial administration in the Indonesian archipelago did not change that much but the adjustment to the Japanese concepts of the domain; the Japanese did not use the concept of the province in their territorial domain. However, they paid attention to control the territory indivisible from populations. At the neighborhood level, the Japanese introduced people to build a neighborhood association called *Tonari Gumi*; in Indonesian term, this neighborhood is known as *rukun tetangga*. Under this territorial concept, people in the neighborhood are informally organized as an extended family. They know each other and work together voluntarily to care for their common interest in safety, health, and security for their living environment. Moreover, the Japanese adopted the Dutch Indies organization domains with their version such as *Ku* for *desa* or village community, *Son* for *kecamatan* or subdistrict, Gun for *kawedanaan* or district, *Ken/Syi/Tokubetsu Syi* for *kabupaten/kota* or regent/municipality, and *Syu* for *karesidenan* or region.

The territorial governance under Sukarno was mainly derived from the Dutch East Indies' region based on native language and ethnic groups. The refinement and adjustment had been made to translate the Dutch terms into the Indonesian language. In order to support the daily activities of local governments, the newly established

Republic of Indonesia provided the governors and residents with *Komite Nasional Daerah* or National Committee of Regions; according to the Law No 1/1945, this body assisted the local governments with administrative matters. Due to political instability and financial shortage between 1945 and 1948, the local governments did not perform a full administrative function but they worked as the representatives of the Republic of Indonesia in the regions. As matter of fact, this body was nothing but the extension of central government so that each region was under a systematic control from Jakarta to provinces, residencies, regents or *kabupaten* and municipalities or *kota*, district or *kawedanaan*, subdistrict or *kecamatan,* and village or *desa*.

The consolidation of local governance was continued with the Law 22/1948 that provided the provinces with self-regulating authority for their executive and legislative body. Each provincial government was supported by the local administrative council and the local House of Representatives. The chief executive of the provincial government was appointed by the president of Indonesia while the members of the house were elected by people for a five year period in office. However, the implementation of the House of Representative in the provincial level did not come into reality until 1971; it was under the Suharto's New Order administration. Even though the first public election in Indonesia was held in 1955, there were not enough resources to establish the local House of Representatives at the provincial, regional, and municipal levels.

In terms of territorial organization, the parliamentary government, as the result of 1955 public election, managed to issue the Law No 1/1957 concerning the principles of local governance. According to this law, each local government was administratively autonomous. There was two kind autonomous status at provincial, regent/municipality, and district: regular or *swatantra* and special or *istimewa*. Under the notion of local government, there was an executive and legislative body. The head of executive body or *Dewan Pemerintah Daerah* was elected by the member of the local House of Representatives and legally confirmed by the president or by the ministry of domestic affairs. In reality, the implementation of this law was challenged by several issues, which mostly were caused by political instability and financial crises.

Sukarno understands the nation-state of Indonesia as the imagined realm of the historical revival of the Javanese emperor of Majapahit. Territorially speaking, the domain of this realm is not restricted by the former area of the Dutch Indies colonial territory, but it is expandable into the historical roots, linguistic similarities, and cultural identity of people in the Southeast Asian archipelago. However, the emergence of Sukarno's territoriality did not come up before his July 1959 Presidential Decree. For Sukarno, territorial control is not conceived as the expansionist exercise of his control, but the territorial claim is understood as part of the consolidation of his charisma and influence within the framework of Javanese concept of power, which is concrete, homogeneous, constant, and not about legitimacy (Anderson 2006, 17–77). Accordingly, the necessity for the territorial claim is not to control the area but the populations in terms of nationalism.

As a Javanese person, Sukarno believes in the expansion of the nation-state, *negara*, is not simply about the territorial claim and the domination of its economic resources. Rather, it concerns more about the accumulation of Sukarno's charisma as the ruler of the Indonesian nation-state (Anderson 2006, 78–93). In Javanese culture, the question concerning power is not about control over geographical territory but more about personal influence as ruler over other people as his/her followers in the context of nationalism.

The first move of Sukarno for this territorial claim was to include the West Papua into the nation-state of Indonesia. After the Dutch Police Action II in 1949, this region was demanded by Sukarno as an integrated part of the unitary Republic of Indonesia. Under the Hague Agreement on November 29, 1949, the Netherlands East Indies conceded the sovereignty of the Republic of Indonesia but still kept the West New Guinea under the Dutch control; according to the agreement, the status of West Papua would be decided in further talk to take place within a year (Saltford 2003, xvi). Sukarno mobilized military forces under the name: *Trikora*, the Three People's Commands, calling for the liberation of West Papua from the Dutch control. Indeed, the sovereignty of Indonesia over this region had been achieved on January 1, 1963, when the Indonesian flag was officially flown alongside the United Nations flag in West Papua.

Confrontation with Malaysia was another example of a territorial claim of Sukarno's personal ambition against any form of colonialism and imperialism. Even though the strong support for the crush of Malaysia was obvious from the Armed Forces and the Communist Party of Indonesia, most leaders of banned political parties under the Sukarno's Guided Democracy were not in the same page with the president. The political campaign against the British under the ideological label of neo-colonialism and neo-imperialism was not in the right direction in the interest of Indonesian nation-state, but that of Indonesia's Communist Party. By 1965, Sukarno's emissary and the high-ranking army officers had several exploratory peace negotiations with Malaysian (Weinstein 2009, 16). After the transition of power in 1967, the New Order government under Suharto put an end to the confrontation with Malaysia.

2.6 Territorial Control Under the New Order

During the period between 1965 and 1975, Suharto consolidated his power. The annexation of the former Portuguese colony in East Timor in 1975 was the outcome of this consolidation. The reasons for this annexation could vary, from the suspicion of communist danger to economic interest. However, one important cause was Suharto's personal need that was drawn from the perspective of the Javanese political culture. East Timor for Suharto is the test case for the solidity of his spiritually consolidated power, which has been remarkably signified by the economic success of the New Order in the 1970s; most of this attainment of economic development was made possible by the booming oil price.

The invasion into East Timor is considered by Suharto as his integrated part of a consolidation for his patrimonial power; this invasion is to restore his self-confidence in relation to the armed forces after the social and political riot of the Malari Affairs on the January 15, 1974. By invading the former Portuguese colony, Suharto believes that he is able to regain his powerful status and pride as the Javanese ruler as he has achieved in the period between 1965 and 1967. The territorial expansion of Indonesia's Suharto is a considerably spiritual necessity for his personal pride in terms of *wibawa*. Suharto does not have charismatic leaders, but as a Javanese ruler, he understands power as a personal captivation of authority. Suharto never talks directly on the political issues concerning the invasion of East Timor. The way he gains his powerful image from his people is constructed with his silence and socially distance from publicity.

Suharto's increasing remoteness from his people was balanced by an increased sensitivity to the matter that touched her personal life (Elson 2001, 209). To a certain extent, he builds the image of himself as a mysterious leader but if he wants to show his authority, he never hesitates to crush his opponents by using his power but with violence or subtlety. In dealing with his antigovernment protesters, without hesitancy Suharto ordered his security forces to put down ferociously hundreds of lives and to arrest several political and ex-military leaders of the Petition 50 group related to the Tanjung Priok incident in September 1984; most of these leaders were loyal supporters of Suharto in early New Order era (Masuhara 2015, 70). Any challenge against his government and policy is conceived as a threat to the integrity of Suharto's territorial power.

For Suharto, the territory is not merely about geographical domination, but more importantly, it is about an ideological realm that is necessarily under his control for its order and harmony. Suharto is like his predecessor—Sukarno—utilizes the state ideology of Pancasila and the fascistic 1945 Constitution as his ideological territoriality that is conceived as sacred and transcendence (Elson 2001, 240). Any political movement and opposition against the New Order's policies would have been identified as threat and danger to this ideological territoriality. Unlike his predecessor, Suharto utilizes the bureaucratic, military and police network and structural command system for securing his control over the territory from the village, regional, provincial, to national scope and level.

Under the New Order, the territorial administration of provinces, regents/municipalities, and districts became substantially strategic for its sociopolitical control and stable hegemony. The necessity for control had been well integrated with the urgency of national security for the economic and physical development of the country. In order to achieve this objective, the New Order consolidated the local authority at various levels, from village to national stage. The territorial concepts of district and subdistrict have been integrated into a single term: *kecamatan*; in other words, the concept of *kawedanaan* was dissolved. The dissolution was made as well for the territorial concept of residency; this term has been dissolved into regent or *kabupaten*. The consolidation was managed with the establishment of the board of local/regional leaders at the district, regent/municipal, and provincial levels. The local/regional leaders consists of the head of executive and legislative body, the military territorial chief officer, the local/regional police chief officer, At the district level, the board of local leaders or *Musyawarah Pimpinan Kecamatan* comprises the three essential components: head of the district, a military officer or *bintara pembina desa* or *babinsa*, and the head of the police sector or *kepala polisi sector* or *kapolsek*. Meanwhile, at the regional and provincial level, the board is filled with the three elements and other leaders: chief attorney, the head of the local House of Representative, and the head of the court.

Under the name of economic development, the Suharto regime perceives the nation-state territory of Indonesia as the site of natural resources. Aceh and Papua are perceived by the New Order as the Machiavellian target of economic development. The local populations in Aceh and Papua are not perceived by Suharto's ruling group; otherness, the regime overlooks and ignores historical affiliation of Aceh and Papua with Indonesian nation-state. During the struggle for Indonesia's independence 1945–1949, Aceh under the leadership of Daud Beureuh was voluntarily to join and support the newly established Republic under Sukarno–Hatta. The West Papua was integrated within the Republic of Indonesia after the 1963 New York Agreement between Indonesia and the Netherlands. Suharto is well-known as a Javanese ruler who is proud of his patrimonial tradition. The expansion and establishment of Javanese sociocultural influence over other ethnic groups of Indonesia are perceived by Suharto as his politically imagined nation-state of Indonesia.

2.7 Territorial Control in the Post-Suharto Era

The territorial management under Post-Suharto administrations is characterized by democratization and decentralization of power. All this is the implementation of the Law 22/1999 concerning the autonomy of local/regional governance. The law has been revised with the Law 22/2oo4. Accordingly, the heads of government and members of the House of Representatives, from the village, district, regent/municipality, and provinces, have been directly elected by the people since 2007; meanwhile, the direct election for president and vice president has been held since 2004. In 2019, 245,000 candidates for legislative offices competed simultaneously with the presidential election, which was participated by more than 80% of eligible voters of 160 million people.

In Post-Suharto's Indonesia, the ties between politicians and business circles are still part of the patrimonial political culture (Tomsa and Ufen 2012). The ongoing reformation has not gone further into the transformation of the state bureaucracy's mentality from ruling officer to public servant. Undoubtedly, it is understandable that patrimonial culture becomes one of worth noting approaches to the scholarship of Indonesian politics and culture (Bertrand 2013, 41). Despite the democratic system of public election, David Brown (1994) points out that patrimonial culture still works today in the governance of Indonesia and Southeast Asia; he characterizes the ongoing practice of such hegemony through three aspects: the local concept of power, religious tradition, and linkage to the past.

The traditional patron–client relationship has not yet been transformed by democratic reform has been deconstructing the structure and function of governance. In other words, in order to come to power, patrons need help and support their clients in traditional ways, especially during the preparatory works before the public elections.

After having been elected, the patrons owe to pay back their clients with accesses to resources. The patrimonial system is stabilized by a normative mandate that legitimizes bureaucratic assignments and draws the border of clientele (Vel 2008, 8). The necessity for being socially exclusive has been deeply established by local tradition. The ruling group sets aside from the common people.

In Southeast Asia, the genesis of gentrification is deeply rooted in the traditional and historical root of culture that has been working since ancient times as socio-cultural, political, and economic division of production: patron–client relationship. The outcome of this relationship is nepotism in the state bureaucracy and collusion ties have been subject to public criticism and scrutiny of news media. Even though several cases have been prosecuted by the Corruption Eradication Commission, the practice of nepotism and collusion is still strong and manifest in various forms. One important manifestation of the practice of patron–client relationship is the development of land for superblocks and gated communities; the practice of gentrification has been going on until today in various big cities of Indonesia in the interest of capitalist accumulation.

In this study, gentrification is understood as a phenomenon of concentration, accumulation and pull of gentry into a location for their urban settlement on the ground of socioeconomic privilege, primordial presumption, and hegemony; the British sociologist Ruth Glass coined the concept in 1964 in her unpublished report based on her observation on housing in the North Kensington in 1959 (Lees et al. Gentrification 2007, 4). Gentrification for Glass is associated with the idea of displacement of the poor in favor of the privileged rich. The idea of gentrification has been used to be understood as a concept of back to the city movement by capital (Smith 2010). As a matter of fact, most definitions of gentrification are focused on its consequences and outcomes rather than on its causes and character of the process (Brown-Saracino 2013, 13). Indeed, the genesis of gentrification needs unfolding and divulging. In doing so, we are able to go deeper to understand the concept in a broader perspective.

During the Post-Suharto era, the urban gentrified territorialities have been continued and perceived as an integrated part of urban development. As a matter of fact is that the main actors of urban developers are the same private developers from Suharto's New Order era. One important form of this gentrified territoriality is the superblock mixed-use development comprising residential and commercial components. Most superblock developments in big cities of Indonesia have been developed on the formerly abandoned lands. The development of superblock begins with the consolidation of lots by land acquisition. The acquisition of land is a time-consuming process. However, most superblock developers have a patient for this. Leaving the areas that look neglected and dirty area is a good strategy for transforming its image for new development. The development of the area for superblock needs a series of the regulatory process. Based on the recommendation from the local National Land Agency on the status of the land, the developer is able to acquire the principle permit for property development. For superblock mixed-use development, it is necessary to acquire at least 3 hectares area. Most superblock developments: Agung Podomoro, Lippo Kemang Village, Ciputra Land, and Gandaria City in Jakarta metropolitan area stands on 3–7.5 ha area.

From the perspective of territoriality, superblock development is a large scale of architecturally privatized urban space. This privatization abates the state and municipality to manage and control of the public sphere for street life urbanism. In other words, the superblock developments show the transfer of the right of urban citizens to public sphere from state territoriality to private estates; the private enterprises manage the safety, surveillance, and security inside the superblock territory. The consequence of this privatization of urban space is the affirmation of the patrimonial authority of the private estates over people who live, work, and play in their realm. In Indonesian urban context, superblock developments are stand-alone territorial realms. This is considered the decline of urbanism from its conceptual foundation because urban development fails to establish the flow of urban space within the network of streets and public transit systems. The common interest of the state and private sector for their partnership is obviously driven by a community need for socioeconomic justice on a daily basis. This need is practically accommodated with the available public domains, facilities, and amenities.

Since all activities in the realm of the superblock are under the patronage of the private enterprises, the sense of urbanism is put in question because the superblock development does not provide all citizens with free access and civil liberty for their creative works, labors, and activities. The business inside the territory of the superblock is a patrimonial practice that is based on mainly commercial interest; they take over the authority of the state and municipality to ensure that urban development is committed to all citizens for social justice and civil liberty, regardless their socioeconomic background and income situation. All this happens because the city is not able to establish its function as the steward of the right to urbanism for all citizens.

The private and public partnership does not work in the interest of the public good. This is not simply because the developments by private sectors are not well integrated within the network of street life and public transit. Rather, this is because the decision-making process within this partnership in the Indonesian context does not undergo the democratic process with the active participation of citizens and communities. The phenomenon of superblock development in Indonesia confirms Castell's thesis that the spatial transformation in the Age of Information does not reflect and express its society, but it is associated with the multidimensional changes of the global capitalist network; accordingly, good municipal governmentality is needed that is able to harness market forces and tarnish interest groups on behalf of the public good (Castell 2004, 83–92).

Urbanism in Post-Suharto era is characterized by the decentralization of power from central to local government. This includes the authorization of local development based on local needs and conditions. The laws no 22/1999, no 25/1999, no 32/2004, no 33/2004, no 12/2008, no 23/2014, and no 9/2015 provide the municipal government with the legislative power to plan, design, develop, manage, and sustain urban development in their area. The major and the members of the municipal council are publicly elected every five-year period. Historically speaking, the simultaneous public election for municipal mayors and regents took place in 2015. Despite the democratic process of the public election system, local governments need leaders with strong leadership. Power struggles for public offices have been demanding more and more financial support for political campaigns. Most politicians, who have gotten the seats of public offices, are mostly burdened with the campaign cost for their first and second year in their offices; they have to pay back indirectly their supporters with government projects. Patrimonial patron–client relationship with the former financial supporters of a political campaign is common practice with various cryptic tactics. Unsurprisingly to know that the public officials caught by the commission for the eradication of corruption were mostly municipal mayors, regents, members of a local house, and governors.

The other crucial issue concerning urbanism in many contemporary cities and towns of Indonesia is about the shortcoming of expertise. Most municipalities have not yet provided with sufficient personal resources for their planning and building department. In doing so, the local governments are not able to manage properly the development applications from private enterprises. Most of these real-estate developers come from big cities with massive financial power and professional supports. The development approach to the urbanism of the capitalist corporation is nothing but profit; the commercial developers prefer to build superblock malls and gated-community housing rather than pedestrian scale shops, row townhouses, and affordable housing. From this situation, the outcome of urban development is hardly in favor of socioeconomic justice for local traders, manufacturers, small businesses, and consumers. Legislatively speaking, local expertise is in shortcoming in drafting development regulations, zoning ordinances, and building codes based on local heritage and resources.

2.8 Concluding Remarks

Historically, the relationship between power and territoriality in Indonesia shows various forms of patrimonial practice of classification, language, and control. Accordingly, the ruling regime has the right to territorial claim in the framework of a patron–client relationship. The ancient Javanese ruling groups understand territory as the center of the cosmological realm on earth that is in the form of a territory for the gentry. The conspicuous manifestation of this center is the construction of architecturally walled territoriality that establishes the palace complex as the symbolic center of the city. Accordingly, territoriality is identical with the incorporation of the cosmic center as the core of the urban settlement. The ancient Javanese city is the center without definite periphery because it does not claim the territory on the land literally but allegorically. Accordingly, the center is flexibly contracted because it is the extension of the mind of the ruler that influences the populations.

During the colonial times, the European concept of territory is implemented by means of the map for marking and establishing domain in terms of possession and control over its resources. The Dutch V.O.C has implemented the concept of territory as the area under their control for the maritime trade of spices and other commodities. The territorial control under the Company is mostly to the waterfront vicinities and sea routes in archipelago until the end of the eighteenth century. The penetration of territoriality into the hinterland is expanded during the Dutch East Indies colonial rule since 1800. This period is the era when the concept of territoriality has been developed as the means of economic and political exploitation over land and people.

Moreover, territoriality under the Dutch colonialism is to establish the boundaries of settlements based on racial identity and social class on the geographical urban area. The dissolution of this territorial discrimination comes into being with the establishment of the Republic of Indonesia as an independent nation-state in 1945. Under Indonesia's government, the concept of geographical territoriality has been elaborated by Sukarno and Suharto with the idea of power in Javanese political culture. Accordingly, the expansion of territoriality to other parts of the archipelago has been understood by both as part of their personal influence that is not commonly accepted from international law and justice as well as from human rights' perspective.

However, the internal politics of territoriality under Sukarno and Suharto does not lead to the phenomenon of gentrification within the logic of capitalist accumulation. Gentrification comes into being in the Post-Suharto era as the superblock developments. The developments show economic progress and investment in most big cities in Indonesia. However, the essence of this kind of development does not care for the urbanism based on the democratic process and social justice. On the contrary, the superblock developments underscore the patrimonial relationship between patron and client at the elites of society that marginalize and exclude the role and function of urban citizens at large.

References

Anderson B (2006) Language and power: exploring political cultures in Indonesia. Equinox, London

Benjamin T (2007) Encyclopedia of western colonialism since 1450: A–E. Macmillan Reference, New York

Bertrand J (2013) Political change in southeast Asia. Cambridge University Press, Cambridge

Bown SR (2009) Merchant kings: when companies ruled the world, 1600–1900. D & M Publishing, Vancouver

Britannica E (1842) The encyclopaedia britannica, or, dictionary of arts, sciences, and general literature, vol 4. Adam & Charles Black, Edinburg

Brown D (1994) The state and etnic politics in Southeast Asia. Routledge, London and New York

Brown-Saracino J (2013). The gentrification debates: a reader. Routledge, London and New York

Castell M (2004) Space of flows, space of places: material for a theory of urbanism in the information age. In: Stephen G (ed) Cybercities reader. Routledge, London and New York, pp 82–93

Corley RHV, Tinker PB (2016) The palm oil. Willey Blackwell, Chichester

Crawfurd J (1856) A descriptive dictionary of the Indian Islands and adjacent countries. Bradbury and Evans, London

Day T (2002) Fluid iron: state formation in southeast Asia. University of Hawaii Press, Honolulu Hi

De Silva SBD (1982). The political economy of underdevelopment. Routledge, London and New York

Elson RE (2001) Suharto: a political biography. Cambridge University Press, Cambridge

Emmerson DK (1978) The bureaucracy in political context: weakness in strength. In: Jackson KD, Pye LW (eds) Political power and communication in Indonesia. University of California Press, Berkeley CA, pp 82–136

Geertz C (1960) The religion of Java (Glencloe Il: Free Press)

Gouda F (2008) Dutch culture overseas: colonial practice in the Netherlands Indies, 1900–1942. Equinox, Jakarta

Gould JW (2012) American in Sumatra. Springer, Berlin, New York

Kanō H (2008) Indonesian exports, peasant agriculture and the world economy, 1850–2000: economic structures in a southeast Asian state. NUS Press, Singapore

Lee KC (1999) A fragile nation: the Indonesian crisis. World Scientific, River Edge NJ and London

Lees L, Slatter T, Wyly E (2007). Gentrification. Routledge, London and New York

Lees L, Shin HB, Lopes'-Morales E (2015) Global gentrifications: uneven development and displacement. University of Bristol, Policy Press, Bristol

Masuhara A (2015) The end of personal rule in Indonesia: Golkar and the transformation of the Suharto Regime. In: Center for southeast Asian studies, vol 24. Kyoto University Press, Kyoto

Mizuno K, Kusumaningtyas R (2016) Land and forest policy in southeast Asia. In: Mizuno K, Fujita MS, Kawai S (eds) Catastastrophe and regeneration in Indonesia's Peatlands: ecology, economy, and society. Kyoto CEAS—Kyoto University Press, NUS Press, Kyoto and Singapore, pp 19–68

Moon S (2007) Technology and ethical idealism: a history of development in the Netherlands East Indies. CNWS University of Leiden, Leiden

Oleinik A (2011) Market as a weapon: the socio-economic machinery of dominance in Russia. Transaction, New Brunswick and London

Parker CH (2010) Global interactions in the early modern age, 1400–1800. Cambridge University Press, Cambridge

Parmer NJ (2012) Historical perspectives of Chinese populations in southeast Asia. In: Amstrong MJ, Amstrong RW, Mulliner K (eds) Chinese populations in contemporary southeast Asian societies: identities, interdependence and international influence. Routledge, London, pp 18–54

Pols H (2018) Nurturing Indonesia: medicine and decolonisation in the Dutch East Indies. Cambridge University Press, Cambridge

Reid A, Brewster J (1983) Slavery, bondage, and dependency in southeast Asia. St. Martin Press, London

Ricklefs, MC (2008). A history of modern Indonesia since c. 1200. Palgrave Macmillan, New York

Robison R (1981) Culture, politics, and economy in the political history of the new order. Indonesia 31(1):1–29

Sack RD (1986) Human territoriality: its theory and history. Cambridge University Press, Cambridge

Saltford J (2003). The united nations and the Indonesian takeover of West Papua, 1962–1969: the anatomy of betrayal. Routledge, London and New York

Schmutzer EJM (1977) Dutch colonial policy and the search for identity in Indonesia. Brill Archive, Leiden

Smith N (1979) Toward a theory of gentrification: a back to the city by capital not people. American Planning Association, pp 538–592

Smith N (2010) Toward a theory of gentrification: a back to the city movement by capital not people. In: Brown-Saracino J (ed) The gentrification debates: a reader. Routledge, London, New York, pp 71–86

Stoller AL (2002) Carnal knowledge and imperial power: race and the intimate in colonial rule. Berkeley, Los Angeles, London

Stubbs W (2011) The constitutional history of England, in its origin and development. Cambridge University Press, Cambridge

Tomsa D, Ufen A (2012) Party politics in southeast Asia. clientelism and electoral competition in Indonesia, Thailand and the Philippines. Routledge, London

Vandenbosch A (1944) The Dutch East Indies, its government, problems, and politics. University of California Press, Berkeley and Los Angeles

Vel J (2008) Uma politics: an ethnography of democratization in West Sumba, Indonesia, 1986–2006. Brill, Leiden

Weinstein FB (2009) Indonesia abandons confrontation: an inquiry into the functions of Indonesian foreign policy. Equinox Publishing, London, New York

Willemyns R (2013) Dutch: biography of a language. Oxford University Press, Oxford

Winn P (2010) Slavery and cultural creativity in the Banda Islands. J SE Asian Stud 41(3):365–389. https://doi.org/10.1017/S0022463410000238 (National University Of Singapore)

Wolters OW (1999) History, culture, and region in southeast Asian perspectives. SEAP Cornell Publication and Institute of Southeast Asian Studies, Ithaca NY and Singapore

Woodward M (2010) Java, Indonesia and Islam. Springer, Berlin and New York

Chapter 3
The Syncretism of Maritime Urbanism and Theater State in Indonesia 1500–1700: Banda Aceh and Demak

Abstract Maritime trade has characterized long-standing economic activities in Southeast Asian archipelago since the prehistoric times. However, the emergence of urbanism and state polity of *negara* in this region was not earlier than the fourth century under the influence of Sanskrit culture from India. During the period between the twelfth and fifteenth centuries, the state polity had transformed and adjusted by the maritime contacts and interactions with the Persian, Arab, Indian, and Chinese merchants. The outcome of this assimilation and enculturation between locality and foreign influences is potentially the hybrid concept of *negara* and *bandar*. This study is to investigate and examine how this syncretism works and is possible. Moreover, this study is to divulge and unfold local concepts and historical traces that underpin and conserve the practice of maritime urbanism in Indonesia since the fifteenth century, with reference to the pre-colonial, colonial, and post-colonial city of Demak, Java and the city of Banda Aceh, Sumatra.

Keywords Maritime · Urbanism · Samudra Pasai · Banda Aceh · Demak · *bandar* · *negara*

3.1 Maritime Urbanism

This study is about maritime urbanism and its historical roots and settings in the Indonesian–Malay archipelago. The focus of the study is to investigate and explore the historical formation and evolution of urbanism. Methodologically speaking, this study embarks its compilation of data from historical sources and fieldworks in the sites of the case study. The cases are focused on the geopolitical and historical context, urban form, landscape characteristics, and cultural geography of the cities of Banda Aceh and Demak. Both cities are considered historically important for understanding the relationship between Islam and urbanism in Southeast Asia, as well as the syncretism between foreign culture and locality. Moreover, the task of the study is to unveil and dismantle aspects, concepts, and elements of locality in dealing with foreign influences that build, develop, and sustain unique features and characteristics of urbanism in Indonesia.

B. Wiryomartono, *Traditions and Transformations of Habitation in Indonesia*,
https://doi.org/10.1007/978-981-15-3405-8_3

3.2 Theoretical Framework and Methodology

Based on the precursory surveys, rituals and ceremonies related to the maritime values system do exist in both cities: Demak and Banda Aceh. From this point, this study argues that urbanism in this geographical and cultural context is indispensable theatrical and festive for public esthetics. Geertz (1980) observes the presence of the polity of *negara* in the nineteenth-century Bali as a triadic system of themes: status, spectacle, and governance; pomp become an essential experience for the existence and presence of polity.

The maritime urbanism of *bandar–negara* must have been developed within the Geertzian theater state because of its historical roots in the Southeast Asian animism of the center, under the indigenous concepts of *pusat* and *semangat*; every being is contained with power. The necessity for the theatrical exhibition of splendid power is considered to invoke the magical sense and experience of domination in terms of *terkesima*. Under the notion of *terkesima*, one feels of being amazed and confused and subconsciously submissive to somebody who is for him/her very adorable, admirable, and respectable.

In terms of the center of power, the presence of polity in terms of *bandar–negara* as city and state is necessary spiritually appealing. At a practical level, this is sensually experienced by its populations with sacred rituals and festivity. From this spectacle, the role and function of the king/queen come into play as the conductor of the whole event and destiny of *bandar–negara*. Accordingly, urbanism in Southeast Asia is not simply a physical embodiment of the center of polity, but also the symbolic manifestation of power with splendid spectacle. The inherently urban King and court were regarded as the epitome of refinement; everyone else was ranked by their nearness to this ideal (O'Connor 1983, 33).

In the maritime region of Southeast Asia, waterfront urbanism is characterized by the establishment of the city as the center without the geographical boundary of the physical nation-state, *negara*; the stronger the state is the farther-reaching its urban center radiates its spiritual and political influence. This study argues that maritime urbanism in Southeast Asia is not about the concentration of populous settlement but the consolidation of the center for the performance of patrimonial power. This consolidation is represented and expressed in the rituals and ceremonies of the state that connect the relationship between the ruler and the ruled, land and sea as well as earth and sky. These practices have deeply rooted in society as the way of life that is based on the patron–client relationship as well as the lord–servant or *Gusti-kawula* dualism of society. Despite the political process of democratization and civil liberty has been transforming Indonesian society since the end of New Order in 1999, the resilience of patrimonial culture remains strong for urbanism.

3.3 Maritime Tradition and Culture

Under the notion of maritime tradition and culture, this study understands this as a system of the way of people's lives that are associated with the sea as their resources for livestock and trade. The Indonesian and archipelago, maritime and land-based way of life have been in coalition since prehistoric times. Rivers played an important role in maintaining the relationship between the hinterlands and coastal areas in Java, Sumatra, Kalimantan, and other islands. Rice, vegetables, fruits, and spices are the ancient export commodities of the hinterlands of Java and Sumatra, while the coastal areas are the markets for dried fishes, ceramics, beads, silks, and salt. By 200 BCE, the spread of bronze kettle drums in various parts of Southeast Asian archipelago showed how the maritime network of trade and exchange worked within the region prior to any significant trade with India and China (Bronson 1992). All this was made possible by the prehistoric native populations in Southeast Asian because of their expertise in sailing and astronomical navigation during the monsoons (Hall 2010, 4).

The obvious traces of maritime tradition and culture are apparent in the reliefs of Borobudur Buddhist temple from circa ninth century; the outrigger vessel crafted on the gallery is considered as one important means of the seafaring culture of Indonesian–Malay people since the prehistoric times. Archeologically speaking, the maritime-oriented tradition and culture of indigenous people of the Southeast Asian archipelago have been known since circa 5000 BCE; archeologist and historian Solheim calls these native people as *Nusantara* who speak a linguistically related Austronesian language (Solheim et al. 2006, 42). The maritime trade network between India and Southeast Asia has possibly begun in circa 700 BCE, based on the archeological evidence of beads and metal-forming tools while the contact between the *nusantao* and the people of Ainu in Okinawa, of Ryukyus, Kyushu and Chinese populations of Taiwan could have begun around 4000 years ago; the evidence for this is the spread of *javanica* rice—*beras bulu*—from Java (Solheim et al. 2006, 47) (Fig. 3.1).

Prior to the Sanskrit cultural influence from India, the maritime trade of the indigenous Southeast Asian inhabitants has been known from their settlement organization. The concepts such as *batin* for the village chief in the Malay populations have something to do with the boat (Hall 2010, 289). The Austronesian junk and the Chinese vessel inspired Southeast Asian boat makers to develop a hybrid form (Reid, Charting the Shape of Early Modern Southeast Asia 2000). Based on an archeological excavation, the Austronesian junk was the wooden construction of its hull with the width of 3.00 m and length of 15.00 m, while the Chinese ship of the Ming Emperor was a vessel with hull about 6–8 m in width and 25–30 m in length (Bacus 2004, 267). During the fifteenth century and sixteenth century, Semarang, Tuban, and Gresik had been port cities of Java with a high concentration of Chinese migrants from the mainland; in Semarang, they set up and established their ship-making workshop and overhaul (Muljana 2005, 197) (Fig. 3.2).

During the influence of Sanskrit civilization, the maritime-oriented culture was sociopolitically organized as the state, negara. Southeast Asia embraced the Sivaite

Fig. 3.1 Bronze kettle drum from circa 500 BCE, a collection of Museum Nasional Indonesia, Jakarta. Photo by author

Fig. 3.2 Javanese junk carved on the Borobudur's gallery from circa ninth century. Photo by author

conception of royalty based upon the Brahman-Kshatriya hybrid and articulated in the cult of the royal *lingga* (Coedes 1944/1968, 23). Historically speaking, the maritime landscape of Indonesian–Malay world under the Sanskrit influence can be seen as two regional and imperial powers: the Sumatra-based Srivijaya from seventh to thirteenth century and the Java-based Majapahit from thirteenth to the fifteenth century, and other local vassal kingdoms (Tan 2009, 136).

Maritime tradition in the Southeast Asian archipelago is apparent in various coastal areas with the rituals and customs of communal bathing at the estuary areas. In Sumatra's shoreline areas, from Aceh, Riau, Jambi, Palembang, and Lampung, this tradition has been being practiced since ancient times and known as *mandi shafar* of bathing in the Islamic month of *shafar*. In other coastal parts of Java, Sulawesi, Bali, Nusatenggara, and Molucca, people are used to the rituals and ceremonies of offering to the sea with foods and domesticated animal sacrifices; all this is to validate their relationship with the sea and their seafaring tradition.

3.4 Maritime Urbanism in Southeast Asian Archipelago

The Indonesian archipelago is geographically strategic between the maritime prowess of India and China. Prior to the fifteenth century, the trading connectivity between these two preeminent powers has been characterized by the sea silk route. The vibrant maritime trade in the archipelago has been beginning since the fifteenth century; the west part of Southeast Asia became the focal point of trade for gold, tin resin, spices, and precious wood (Maguin 2000, 409). The establishment of urban settlements in this region was not earlier than 358 AD (Vogel 1925). The new evidence of terracotta brick Buddhist temple in Batujaya West Java still has not determined the exact date of its construction, but it was for sure between the fifth and eighth century (Manguin and Indradjaja, The Batujaya Site: New Evidence of Early Indian Influence in West Java 2011).

Early state polity in the Indonesian–Malay world was the Hindu kingdom of Tarumanegara located in the hinterland of West Java or nowadays city of Bogor. The inscription of Ciaruteun says in Pallava script the name of Purnawarman with his footprint on the stone from circa the fifth century (Casparis 1986). Tarumanegara was not a maritime-oriented kingdom but land-based agricultural polity at the riverbank of Ciliwung River, in the hinterland of West Java. The kingdom was given its name from *tarum*, a name for a plant that its seed is used for coloring the fabrics or clothes made of cotton or silk with indigo color; this commodity was an important commodity for the maritime trade network from Southeast Asia to China and India during the fifth century (Devare 2009).

The emergence of maritime urbanism in the Malay-speaking region was signified and characterized with the concept of *bandar*; this is a loan Persian word for the marketplace or the place of trade and crowd (Clifford and Athelstane 1894, 141); the first and well-known maritime vicinity in the archipelago was *Bandar Bukit Siguntang*. Historically, the name of this place was associated with the capital city of the maritime kingdom of Sriwijaya in South Sumatra. The existence of this city is known from the local narrative, *Sejarah Melayu* (Reid, Understanding Melayu (Melayu) as a Source of Diverse Modern Identities 2004, 5). The advent and establishment of *bandar* as a port and urban settlement constitute a crucial event in the history of Indonesian urbanism. However, the establishment of *bandar*s was indispensable from the Chinese presence as main trading partner over most harbors of the archipelago between the fourteenth and fifteenth centuries; their presence was well-known with the existence of Chinese persons as the harbormaster of chiefs of port authority, *shahbandar*; this institution was an important body for urban waterfront. The origin of this institution is from Persian urbanism, the same root of the waterfront vicinity of *bandar*. The installment of Chinese *shahbandar* in most Southeast Asian ports must have had to do with Chinese interest in the spice trade and interaction with traders from Persia, Arabian Peninsula, and India.

Shahbandar was economically an independent institution with its autonomous authority for taxation and customs matters in relation to indigenous rulers. This relationship was a common practice as a result of the mutual interest between local indigenous rulers and the Chinese fleet under the Ming Emperor in Southeast Asia. The Admiral Zeng He was a high ranking officer of the Chinese Emperor who played an important role in the spread of Islamic teaching in most ports in Malay–Indonesian archipelago during the fifteenth century (Suryadinata 2005a, b); he and his fleet protected the Chinese interests for spice trade and silk trade.

The presence of the Chinese influence and their Islamic heritage in Malacca and Semarang were obvious (Suryadinata 2005a, b). In his 28 years of voyages and visits to more than 30 countries, Admiral Zheng He or Cheng Ho is considered one of the important explorers of the world. Although Zheng He was originally Muslim, he was a tolerant man with good knowledge on Buddhist faith; his Buddhist name was Sanbao, a great man who had done good things for establishments and restorations several Buddhist temples (Hum 2012, 255–256).

The relationship between the indigenous state and port authority was a patrimonial patron–client bond. Prior to the arrival of Europeans, ports and waterfronts vicinities in the archipelago were characterized with free and vibrant international markets for the Arabs, Persians, Indians, and Chinese merchants. The language they speak was mostly Malay, while the currency they used for their trades was Chinese copper cash or gold (Tarling 1999, 120). The integration of port, marketplace, and the state has sustained the relationship between agriculturally based states and waterfront cities. Intensive interaction with Middle Eastern traders and Zeng He's expeditions brought about Islamization of coastal areas of Sumatra, Java, Kalimantan, Sulawesi, and Molucca.

3.5 Bandar Samudra Pasai and Banda Aceh

In the thirteenth century, the Islamization had been intensified with the formation and establishment of the Islamic sultanate of Samudra Pasai in north part of Sumatra. The existence of this kingdom is known from the tombstones with Arabic inscriptions and local narrative *Hikayat Raja-raja Pasai* (Soekmono 2006, 43). From this local source, we learn that Samudra Pasai was not the first polity of the port centered royal court under Islamic influence, but the kingdom of Peurlak in East Aceh. The first ruler of Syiah Islamic state of Peurlak was Sultan Alaidin Syed Maulana Abdul Aziz Syah.

The polity of Peurlak was established with the help of the group of Persian merchant under the leadership of shipmaster—*nakhuda*—Khalifah who married with the daughter of the local ruler, Salman; his son, Meurah Peurlak Syahir Nuwi or Maharaja Pho He La became the first sultan of Peurelak (Hasymy 1990, 24). The center of this state is in Bandar Peurelak, later known also as Bandar Khalifah. The conversion of the state in Peurelak from indigenous beliefs system to Syiah Islamic state brought forth a rich cultural hybrid toward a strong sense of communal responsibility (Hall 2010, 288). The Persian concept of *bandar* or port city was firstly established at the center of polity in Peurlak region circa at the end of the nineteenth century.

Regarding its geographical location, Bandar Peurelak was possibly at the estuary of the river nearby today Peureulak (Feener et al. 2014, 29). The port town was described by Persians, Arabs, and Gujarati merchants as one important source of pepper. The seed of the spice was believed that was originally from Malagasy circa seventh and eighth century (Muljana 2005, 132); Aceh Peureulak region was the first site for the growing of pepper in the Southeast Asian archipelago. Later, Banda, Bantam, and Lampung became the next important locations for pepper planting. Persian traders played an important role in the development of Islamic polity and port city as well as the maritime power of indigenous kingdoms in Sumatra and the Malay Peninsula between the eighth and tenth century; many naval terms in widespread usage in the Indian Ocean were Persian, such as *nakhuda*, *bandar*, *bazar*, and *shahbandar* (Potter 2009, 15).

By 1292, Marco Polo observed that the Islamic state of Peureulak (Perlac) was a well-established community; he stumbled to witness a well-organized cemetery complex of Muslims nearby this port city (Andaya 2008, 113). From Persian, Arab, Gujarati, and Egyptian sources, the port city of Peureulak was described as a vibrant international hub between Chinese wholesalers and Middle Eastern merchants for silk trade (Muljana, Runtuhnya kerajaan Hindu-Jawa dan timbulnya negara-negara Islam di Nusantara 2005, 130); pepper of Aceh became the magnets of both sides for coming to Peureulak. The other important commodities of Aceh they need were rice, palm sugar, cloves, and coffee.

The Peurlak Sultanate enjoyed the participation of international trade with Middle Eastern Muslims from 840 to 1292. At the end of the thirteenth century, its neighboring Sunni Islamic state of Samudra Pasai was able to consolidate its political and economic dominance for the trade of pepper in the region of Aceh. This consolidation can be seen as the unification of upstream and downstream polity for pepper trade with the Turks, Arabs, and Gujaratis. In other words, the Persian influence on the trade of pepper became milder than before. The Venetian traveler, Marco Polo, reported in 1297 about the Islamic kingdom of Samudra Pasai and the King with Sunni Muslim title: Sultan Malik al Saleh (Tarling 1992, 513).

In the local narratives of *Hikayat Raja-raja Pasai*—Chronicle of the Kings of Pasai—and *Sulalatus Salatin* by Tun Sri Lanang, the establishment of Samudra Pasai was not impartial from the mission of Egyptian Admiral Nazimuddin Al-Kamil to secure the supply of pepper and other spices from Southeast Asia to the Middle East.

He conquered the port of Kambayat in Gujarat and then moved further east. Al-Kamil established a trading post in North Aceh between 1267 and 1297. Later, this post was entrusted to Marah Silu who became the first ruler of this city-state known later as Samudra Pasai with the title name: Sultan Malik Al-Shaleh. This polity is considered as the first Islamic polity in the Indonesian and Malay world.

At the center of this maritime polity is the king or the queen. The second layer of power is the Prime Minister—*perdana menteri*—and the state treasurer—*bendahara*. The third layer of the state is occupied by the port authority—*shahbandar*—admiral—*laksamana*—the army chief—*uelubalang*—and several ministers—*menteri*. The royal family members are recognized with the title of *tengku* for man and *pocut* for women before his/her name, while other state dignitaries are called with the title of *teuku* or *tuanku* for man and *cut* for women. Sociologically speaking, the main structure of pre-Islamic Aceh was constructed by two main groups: *Orang Kaya* or landowner or merchant—*saudagar*—and *Orang Biasa* or peasant farmers, fishers, and laborers. Both groups worked together in patrimonial ties for agricultural production and commercial trading. This patron–client relationship underlies the socioeconomic and cultural lifeworld of Aceh and most people in the Southeast Asian archipelago. Since this patron–client relationship works to build a social bond in terms of extended family, a patrimonial tradition in Southeast Asia does not go to establish a Master-slave system, but a big family business.

The location of *Bandar Pasai* of Samudra Pasai has not been archeologically revealed until today, but based on local sources, it was in Lhokseumawe area; this is the area where the gravestone of Sultan Malik Al-Shaleh is found. Malik Al Shaleh was the first king of Samudra Pasai who reigned from 1267 to 1297. His successor was his son, Sultan Al-Malik Az-Zahir, who was in power between 1297 and 1326. The Moroccan traveler Ibn Battuta (1304–1377) in his report *Rihlah ila l-Mashriq* confirmed the existence of the kingdom; he visited the city of Bandar Aceh in around 1345 on the way to China between 1345 and 1346 (Laffan 2011, 40).

Battuta was escorted by the deputy of shahbandar to the wooden-walled city which was a few kilometers away from the harbor; he was introduced to several Syiah scholars: Al Qadl al-Sharif Amir al-Shirazi and Taj al-Din al-Isfahani, as well as met with Mughal diplomat Amir al-Dawlasah (Azra 2006, 51). In his visit, Battuta met the From his visit, Battuta observed and participated in Friday prayers in the state mosque; he also found the similarities of the state rituals and ceremonies between Samudra Pasai and Ottoman Empire. The Host of Battuta was Sultan Al-Malik Az-Zahir II or Ahmad Laikudzahi; his wife spoke to Battuta in Turkish (Goeksoy 2011, 66). The city was described by Battuta as a wooden-walled territory with mosque and square at its center.

Several decades later after the fall of Samudra Pasai Sultanate, the bilateral relationship between Aceh and Turkish Ottoman continued for economic, military, and cultural exchanges between the sixteenth and the nineteenth century (Goeksoy 2011, 65–66). The other special relation to other Islamic states includes Saudi Arabia and South India. Samudra Pasai had been known by the Sharif of Mecca who sent a Royal regalia and a group of Islamic scholars to this state (Andaya 2008, 113). The contacts and interactions between South India Keling—especially Chola—and

Southeast Asian archipelago had been a long-standing trade and cultural connection since before the thirteenth century (Kulke 2009).

In matter of fact, the growing maritime power of Samudra Pasai in the western archipelago was not impartial from the decline of Srivijaya at the end of the thirteenth century; throughout the fourteenth century, the port kingdom of Samudra Pasai remained the most important center of Islamic learning in the Indonesian and Malay archipelago (Kratoska 2001, 95). Meanwhile, Majapahit was still to consolidate its power until the beginning of the fourteenth century and conquered Samudra Pasai between 1340 and 1345. The kingdom was in a vacuum from outside reports until 1406 when several Persian and Arab merchants were back in their trading with Samudra Pasai under the rule of a queen.

The pepper supply of Samudra Pasai to Ottoman Empire and other Middle Eastern countries underpinned the establishment of this port kingdom as a center of trade for Muslims in the archipelago. During the period between the fourteenth century and fifteenth century, the Chinese traders built their connection to the Javanese Majapahit for spice trade with their ports in Gresik, Tuban, and Surabaya (Reid, Aceh and the Turkish Connection 2010, 26). The Portuguese geographer and traveler Tomé Pires assumed the total inhabitants of Banda Aceh in the sixteenth century were about 20,000 populations; he visited port cities in Indonesian archipelago between 1512 and 1515 (Pires and Armando Cortesao 2005, 143).

Under the sultanate of Samudra Pasai of the fifth century, the role and function of *Bandar Aceh* was not the only trade center in the Straits of Malacca; the Queen Nahrasiyah reigned this Islamic kingdom between 1406 and 1428. Indeed, Samudra Pasai survived and recovered for its independence from the Majapahit's attack at the end of the thirteenth century, as told in the *Hikayat Raja-raja Pasai* between 1383 and 1390 (Andaya 2008, 113). The relationship between Malacca and Samudra Pasai was mentioned in *Sejarah Melayu* in a unique position; both royal courts were actually in the rivalry in spice trade with the Persians, Arabs, and Chinese merchants, but they shared something in common in Islamic learning. In many ways, Samudra Pasai was more advantageous than the pre-Islamic Malacca because of its direct stock of pepper, nutmeg, cardamom, and rice in its own hinterland while Malacca depended on its spices supply from Jambi, Palembang, Bantam, and Molucca. However, both kingdoms have something in common for their geopolitical security from the threat of the Siamese Ayutthaya expansion.

Since Malacca was in an economic pact with the Chinese state trade agreement under Yongle or Yong-Lo Emperor of the Ming Dynasty in 1405 (Andaya and Andaya 2001, 43–44), Samudra Pasai stayed away from any attempt to control Malacca. Unlike his father, Yongle or Yung-lo Emperor forbade private trade of Chinese merchants in foreign countries and fostered a special relationship with Malacca for the expansion of Chinese official trade overseas. It was a Persian powerful merchant, Haji Ma-Ha-Mo Chini, who persuaded the Ming Emperor for giving Malacca a diplomatic status under the protection of the Chinese power (Kratoska 2001, 97).

As a matter of fact, Samudra Pasai enjoyed its special relationship with the Ottoman Emperor. The advantage from this relationship for Samudra Pasai included the technological transfer of Ottoman nautical knowledge and skills for shipbuilding, armory system, and its navy arsenal. The Chinese official trade mission had visited Pasai with a good response and reception; the expedition of the Admiral Zeng He took place in 1407. However, the Chinese commercial activities in Southeast Asia was not much impressed and preferred to choose their trading representative in Malacca. The vibrant maritime trades in Samudra Pasai gradually faded away in the mid of the fifteenth century. To a certain extent, as an entrepot, Malacca was successful for several reasons, such as the successor of Srivijaya with its traditional support of *Orang Laut*, the special trade relationship with the Ming Dynasty, and the legal protection and conducive port regulations for international trade (Andaya and Andaya 2001, 44).

As a maritime vicinity with direct access to the Indian Ocean, Samudra Pasai had been gradually taken over by Malacca after the Chinese Emperor sent the trade mission and established the diplomatic status of Malacca as the extension of their permanent market in Southeast Asia. Such a diplomatic relationship was like what had been experienced and established during the eleventh century between Srivijaya and the Song court (Heng 2009, 98–101). In doing so, Middle Eastern and Indian wholesalers did not need to go to Shanghai and Guangzhou. From this point, Malacca became the meeting point between the Indian, Persian, Arab, and Gujarati products and the Chinese commodities and Southeast Asian spices. The absence of Chinese trading representative in Samudra Pasai was considered as one important reason for the decline of its economic performance at the end of the fifteenth century.

After the fall of Malacca into Portuguese control in 1511, Samudra Pasai became a growing center again as Muslim spice trade center from the Indian Ocean. This did not last long because of the increasing power of Aceh Sultanate with its polity in Banda Aceh circa 1520. Under the reign of Sultan Firman Syah, Aceh, and the Ottoman Emperor built a diplomatic relationship. The outcome of this diplomatic relationship was the support of the Ottoman for Islamic statecraft and military assistance to Aceh against the Portuguese sea power; Sultan Sulaiman Chan (1553–1566) of the Ottoman Empire sent a diplomatic mission to Aceh in 1545 (Madjid 2014, 101). In return, Aceh secured its supply of pepper and other spices to the Middle East market. The development of the capital city and the Islamic State of Bandar Aceh Darussalam had achieved its perfection under the rign of Iskandar Muda as the ruler of the sultanante of Aceh Darussalam.

The sources of this study on pre-colonial Aceh under the reign of Iskandar Muda mainly rely on the work of Denys Lombard: *Le Sultanat d'Aceh au Temps d'Iskandar Muda 1607–1636* (Lombard 1967) and Husen Djajadiningrat (1958/1987) and local narrative: *Sejarah Melayu* and *Bustan al-Sallatin* by Nuruddin ar-Ranieri. During the pre-colonial times, the state or Nangroe Aceh Darussalam sustained the trade of pepper to Muslim traders from the Middle East and India. The historical source of Acehnese Darussalam Sultanate under Iskandar Muda's rule was the travel reports from Turkish Ottoman scholar Saikh Davud Ismail Mustafa ar-Rumi or Baba Davud between 1650 and 1750 (Ozay 2011) and a French commodore, Augustin de Beaulieu in 1621 (Beaulieu 1996). The relationship between Aceh Darussalam and the Ottoman Empire continued with economic and cultural interactions. This included a technological transfer for shipbuilding; the Ottoman introduced

Acehnese shipbuilders, the galley type of ships equipped with heavy onboard artillery (Manguin, Lancaran, Ghurab, and Ghali 2012, 163–165).

Beaulieu observed and described in his report about the port city of Banda Aceh city, society, and its livability. The city was described by Beaulieu as a self-contained community inside the surrounding defensible wall made of stone and wooden construction; he was escorted to the city from his ship at the port by a small boat along the River Aceh, *Krueng Aceh*. During his visit, Beaulieu witnessed state ceremonies after the Friday prayer with the parade of Sultanate guards, the cavalry of 600 elephants and 200 horses passing the royal stage in front of the state mosque of Baiturrahman. Based on Beaulieu's observation, the Acehnese armed forces were supported by 900 elephants and thousands of cannons in various sizes. On the sea, the Aceh fleet was stationed in three naval bases: Banda Aceh, Pedir, and Daya. The total naval forces of Aceh in the seventeenth century included 100 galleys and 600 junks; the largest Acehnese galley was described to have the capacity to upload more than 800 people (Beaulieu 1996; Sufi et al. 1996, 48).

In the era of Sultan Iskandar Muda, the city was described by Beaulieu as a well-organized Islamic settlement with multiethnic populations from various parts of the archipelago; their existence was known from the names of their settlements such as *gampongJawa*, *gampong Keling*, and *gampong Kedah* (Madjid 2014, 47). Chinese quarters were not obviously mentioned in the travel reports of South Indians, Persians, Egyptians, Turks, and Gujarati merchants. Banda Aceh was also well known for her highly concentration of goldsmiths and silversmiths, in *gampong pande*; most of the skilled workers were believed to be trained by master goldsmiths from Ottoman Egypt. According to Beaulieu's account, the populations of Banda Aceh were guarded by fighting men more than 40,000 people (Beaulieu 1996, 72–74). The reason why this city was massively secured was that of its goldsmith industry. According to Beaulieu's travel notes, there were at least 300 goldsmith workers in Banda Aceh at the time of his visit (Lombard, Kerajaan Aceh Jaman Sultan Iskandar Muda 1607–1638 1966, 58).

Concerning the layout of the city, Banda Aceh under Iskandar Muda was developed with the indigenous concept of *kuta*. Accordingly, the center of the city was constituted by three main components: *dalam*, *medan*, and *peukeun*. *Dalam* is the core area for the royal palace. *Medan* is the public square in front of the mosque of Baiturrahman. The specific purpose of this square is for royal processions and state ceremonies. Meanwhile, the *peukeun* is located nearby the palace at the riverbank of *Krueng Aceh*. The sources of this layout are mainly from the maps of Manuel Godinho de Fridia (1563, 1617), Johannes Vingboons of Amsterdam, Bibliotheca Medicea Laurenziana, Dutch Military Expedition 1870 (Reid and Ito 1999), The improvement of this indigenous structure was developed with the assistance of Ottoman designers and builders; this included the landscape design of the garden of *gunongan* and the extravagant water pool—*taman sari*.

The landscape architecture with water element played an important role in Banda Aceh. Under Sultan Iskandar Muda, the River Deroy had been diverted from Taman Putreu Phang to the north-flowing into Krueng Aceh that defined the secure boundary of the court area at the western riverbank. This alignment was continued to establish a vast open space between the royal court area and the state mosque of Baiturrahman. This open space was an integrated part of the area for state processions and displays of warriors and hundreds of elephants and horses.

Bandar Aceh Darussalam or Banda Aceh was a compact settlement with a fortified enclosure made of stone and wooden wall. The main citadel of the fortified city was situated in the north area with direct access to the Krueng Aceh. Inside the city, units of the settlement were organized with a two-tiered roof mosque at the center of each district called *mukim*. In the inner part of the city, *dalam*, there were two-tiered roof mosques while the three-tiered roof mosque stood as the landmark of the city center with a public square—*medan khayalli*—as its extension for Ieds Adha and Ieds Fitr collective pray. The liveability of the city took place on regular every other five-day period at the marketplace—*peukeun*. From the maps mentioned above, the location of the marketplace of Banda Aceh was at the riverbank nearby the front square of the royal compound, *dalam keurajeuan*. Nearby this marketplace, the Chinese quarters were located.

During the reign of Iskandar Muda, the spatial and building structure of the city was intensively developed with sets of regulation called *qanun*. Such state regulations and rules have been basically founded with the help of the Acehnese statesman Tun Sri Lanang since the reign of Sultan Mughayat Syah (1496–1528). Iskandar Muda reinforced and elaborated the state rules and regulations in various aspects of statecraft including taxation, urban laws, codes, and public manners. All this was compiled in various titles, such as Mabaiyinas Salathin, Qanun Meukuta Alam, and Tazkirah Thabaqat Almajmu'us Sulthanis Salathin Al-'Alam. Dutch scholars, Drewes and Voorhoeve, organize and compile the state regulations and decrees under the title of Adat Atjeh (Drewes and Voorhoeve 1958). For urbanism, one important contribution of Iskandar Muda's governance was to develop a settlement district of an urban settlement with the concept of *mukim*. This territorially based unit of settlement was to integrate and consolidate the unity of people and their defensible territoriality. The purpose of this integration and consolidation was to build a strong and self-contained community with their own terms and local customary laws.

The distinctive livability of Banda Aceh as the center of Acehnese urbanism in the sixteenth and seventeenth century was generated by the state rituals and ceremonies associated with indigenous and Islamic festivals, such as the crowning throne of new majesty, Friday pray, Ied Fitr and Ied Adha, and other important days for the rites of passage for royal family members, the reception of foreign envoys, and other great days of Islam. Processions and parades of royal guards on elephants and horses belonged to the spectacle of the urban festivals that took place in the city square—*medan*—after congressional pray in the mosque (Reid 2005, 112–135). Splendid state rituals and ceremonies belonged to urban livability in Banda Aceh. The grand spectacle of state processions and festivals were sustained after the transition of Iskandar Muda. In 1641, the Dutch traveler and surgeon by training, Nicolaus de Graaff, witnessed the funeral ceremonies of Sultan Iskandar Thani. He experienced the grandeur of the procession of royal family members on 260 elephants with royal regalia and colorful banners. The funeral was followed by endless gunfire all night long after the King's body was buried. Graaff's report on the occasion was published in 1930 in the Netherlands (Warnsinck 1930; Reid 2005, 117).

During the sixteenth and seventeenth century, the indigenous days of celebration in Aceh Darussalam included *hari raya junjung duli*—the day of pledging of loyalty to the royal court—*hari ueleubalang*—the day of warrior leaders—and *hari mandi safar*—the day of ritual bathing for the King or Queen. However, the great state feasts and celebrations were not only for the Acehnese people but also for foreign travelers and envoys visiting Banda Aceh. Anthony Reid describes various occasions of the state feast and celebration including the reception of diplomats, Friday prayer, animal fight of elephants, buffaloes, and rams; his description is based on the reports of English and Dutch travelers visiting Banda Aceh in the seventeenth century (Reid, An Indonesian Frontier: Acehnese and Other Histories of Sumatra 2005, 112–135). From the observation of foreign travelers, state ceremonies and feasts in the urban center of Banda Aceh were to perform a theatric spectacle of power with the King as its chief conductor. For the reception of foreign envoys, the spectators and the actor of this state performance blended together in the course of processions, in the communal meal, and sometimes in the joyful frenetic of gun fires.

James Lancaster was one of the European envoys who was sent by Queen Elizabeth I to deliver the Majesty's letter to the Sultan Ala'ad-din Ri'ayat Syah in 1602; he describes that the reception of his diplomatic mission was escorted by six great lavishly decorated elephants and marching band to the royal court (Lancaster 1940, 91–92). In the royal court, his mission was ceremonially celebrated and entertained with abundant foods and drinks; the food for the envoys was served on the dishes, which were covered by gold and other shining metal. Cockfight was one of the attractive entertainments of the reception feast for foreign envoys in the Acehnese royal court under the reigns of Sultan Ala'ad-din Ri'ayat Syah, Iskandar Muda, and Iskandar Muda Thani (Reid, An Indonesian Frontier: Acehnese and Other Histories of Sumatra 2005, 127).

The other important event for Aceh Darussalam Sultanate was the Wednesday of the end of Islamic month Saffa. This was the Saffa bathing ritual and ceremony. The event was acclaimed as the state celebration. Sultan led the bathing ceremony at the River Deroy and followed by the royal family members and other state dignitaries. This bathing ritual and the ceremony had been a long-standing tradition among Acehnese people; the day of the bathing was known as *potemoreuhom* or *manoe rabu habeh*. Traditionally, they went to the downstream areas and believed in the ritual of purification toward well-being.

3.6 Demak and Islamic Negara 1475–1548

The main source of history of Demak Sultanate is the Chinese Chronicle from the Buddhist temple of Sam Po Kong in Semarang as compiled and presented by Slamet Muljana in his works: The Fall of Hindu–Javanese states and the Emergence of Islamic states in Java (1972, 2005, 2006). In the midst of fifteenth-century Java,

Demak was still a hamlet called Glagah Wangi, which is located deeply in the tropical forest of Bintara; the Javanese word *glagah* means reed and *wangi* means fragrant. The site clearing of Demak town began in early 1470 and led by the prince of Majapahit from a Chinese wife. The prince is Raden Patah who grew up in Palembang and back to Java at the age of 18. Being supported by the Nine Saints—*Wali Sanga*—The King of Majapahit, Brawijaya, granted his son, Raden Patah; he established the first Islamic state in Java in 1473. Patah consolidates his power with the help of the Nine Saints, *Wali Sanga*.

The core of urbanism in Demak was established with three components of Javanese *kuta negara* consisting of *dalem* or *keraton*, *masjid negara* or great mosque, and *peken* or marketplace. In pre-colonial Demak, these three structural components constitute the foundation of center or Javanese urban settlement. The spiritual center of the state—*negara*—was uniquely synergized by the royal court and the mosque; both these components were symbolically unified by the city square, *alun-alun*; this public space became the interface between the spiritual power of state palace and religious authority of the mosque. The manifestations of the synergy between spiritual and religious power were articulated and demonstrated through state processions, rituals, and ceremonies for annual state and religious festivals, such as *grebeg*.

The establishment of Demak as a polity was closer to Majapahit concept of the capital city rather than the concept of *bandar*. Patah and the Nine Saints developed Demak from the Majapahit's model. Even though Demak city is at the riverbank of Kali Dero, the main port of Demak was Semarang that acted as the entrepot of this first Islamic kingdom of Java. Semarang was like Tuban and Gresik for Majapahit's Trowulan.

Unlike the capital city of Majapahit, the geographical location of Demak is not well supported with the access to water; Trowulan was well served by the River Brantas with direct access to the Java Sea from the Eastern direction. Demak as a capital city is more about *mandala kuta* based on the Javanese concept of center, *mancapat* (Koentjaraningrat 1985, 449; Moedjanto 1987, 61; Ossenbruggen 1919/1977; Headley 2004, 147); the center of settlement is defined by its economically and socially confederated bond with other four centers in a cardinal compass direction system. In other words, the maritime power of Demak was still in line with Srivijaya and Majapahit. The naval base of the state and the capital city were in coexistence with significant distance.

In Javanese tradition, the establishment of the urban core is considered as the beginning of urban settlement that begins with the determination of the highest site for the seat of the ruler. This site is known as *siti hinggil*. From this area, the domain of *dalem* for the royal court area—*keraton*—and that of *jaba* for the public square—*alun-alun*—are determined. The royal court area should be on the south and the main public square on the north. On the west of the public square is for the site of veneration; the west is the direction of spirituality and life after death. On the north of the public square is the area for daily livability while the east for beginning reserve and green open area. The development of settlement units is left to grow along the accesses to the public square—*alun-alun*.

In the case of Demak, the pre-colonial structure of the city was mostly the same what is today. The only missing component of this structure was the *dalem* or the royal court area. The traces of this area are recognizable from its toponyms. Based on the relics of the surrounding bricklayer wall system of the royal court area and the foundations of large buildings, the area of the palace is located on the south of the today public square. The mosque of Demak and the royal cemetery are situated on the west of the public square. The marketplace—*peken*—of Demak was located on the north of the public square. Nearby this commercial center was Chinese quarters—*pecinan*—which played an important role in the trade between the upstream and downstream of central Java.

Even though the sultanate of Patah was a maritime kingdom, the capital city of Demak in the fifteenth century played an important role that was more as the spiritual center than as the center of Javanese maritime power; its naval concentration was in Semarang. From its onset, the polity of Demak had been a maritime kingdom. The infrastructural support for this potential was provided by Chinese shipbuilding industry and workshop in Semarang (Muljana, Runtuhnya kerajaan Hindu-Jawa dan timbulnya negara-negara Islam di Nusantara 2005, 193–194); the concentration of Chinese experts and skilled workers in this port had been developed by the Admiral Zeng He of the Ming Empire between 1405 and 1433 (Suryadinata, Zeng He, Semarang, and The Islamization of Java: Between History and Legend 2005b).

During the formative decades of Demak Sultanate, Semarang was still an important port for most Chinese traders who visited Java on the way from the mainland to Molucca; this port supported the Chinese ships with overhaul and repair facilities. The shipyard in Semarang employed Chinese technology of junk shipbuilding of Zeng He's expeditions between 1405 and 1445. During the expedition, Admiral Zeng He developed a diplomatic relationship with several indigenous rulers of Java to establish Ming's official trading posts in various coastal vicinities of Java and Sumatra, such as Bantam, Cirebon, Semarang, Lasem, Tuban, and Gresik. The Demak fleet in Semarang was led by Patih Yunus, the prince of Demak. The nautical forces were supported by several of the ships based on Chinese Zeng He's treasure vessel and traditional Javanese junks. Several times, Demak fleet attacked Portuguese Malacca and Johor without a victory.

Technologically speaking, the Demak battleship was derived from the Chinese Ming junk design with a length between 30 and 50 m. The hull of the ship is constructed with V-shape with a long keel and heavy counterweight. The heavy wooden keels were bound together with iron hoops that provide stability while sailing in various turbulences caused by waves; this stability is also improved with the floating anchors at both sides of the ship. However, like a battleship, Demak Chinese junk was less effective destroyer because it lacked an appropriate platform for more canons and battery. Compared to Portuguese *galleon* of the fifteenth century, the Demak Chinese junk was inferior in terms of maneuverability and arsenal system. By design, the Portuguese *galleon* and *carrack* were superior to the Ottoman galley because of its well-improved design for firepower for larger gun barrels and heavier projectile weights for naval artillery (Kraska 2011, 40). Most Acehnese ships in the fifteenth and sixteenth century were constructed with reference to the Ottoman galley. Technologically, Acehnese fleet and Demak naval power were not able to defeat the supremacy of Portuguese armada in Java Sea, Malacca Straights, and the Indian Ocean.

In the fifteenth century, the daily livability of the city of Demak took place mostly in the marketplace. However, the existence of Demak as a polity in terms of *negara* was signified by the state feasts and rituals that took place in its three components: *keraton*, *alun-alun*, and *masjid*. The most important occasions for the Sultanate Demak were the state rituals and feasts of *grebeg*. Most of these celebrations had to do with the Islamic Holidays, such as *Idul Fitri*, Idul *Adha*, the Islamic New Year, and *Maulud* Nabi Muhammad S.A.W. The Islamic and Javanese New Year happens at the same occasion. This special day has been celebrated as a sacred day for most Javanese people. The Demak's *grebeg* celebration was associated with the state ritual and festival of ancient Javanese tribute ceremony and feast of *sima*; it was the time when all vassal rulers, regents, and other leaders came to the capital city to pay homage and brought with them tributes to the King (Soemardjan 1982, 33). The *Wali Sanga*, who acted as the advisers and religious experts of the Demak Sultan Raden Patah, conceptualized and formulated the *grebeg* as the syncretic occasion of Javanese Animistic, Hindu, Islamic elements toward wholeness and harmony.

The great rituals and ceremonies for *grebeg besar* were instigated with the preparation in the royal court. Prior to the day of celebration, all royal regalia, heirlooms, and treasures should be ritually cleaned up with water full of flowers and burning incense. The royal priest and his men were in charge of all these processions and ceremonies of cleaning and conservation. In the night before the day of *grebeg besar*, the young princess and girls of the royal family performed the sacred dance of *bedhaya* in the royal court hall. This solemn performance was only attended by the inner circle members of the royal court and considered as the meditative congregation. Early in the morning before the event of *grebeg besar*, all foods for the communal feasts should be prepared and presented in the form of extra big *tumpeng* or *gunungan*; it is a mountain of rice with 2–3 m in height and 2–3 m in diameter, decorated with chicken meats, various side dishes, vegetables, and fruits.

On the day of *grebeg besar*, the group of 5–10 *gunungan*s of decorated rice was made ready in the front yard of the royal court. At 10.00 a.m., a royal priest, *penghulu*, led the ritual and ceremony for the procession from the royal court—*keraton*—to the mosque. Afterward, the royal guards carried the whole rice mountains encircling the front royal court hall and moved to the front yard of the state mosque. The ritual of the grand feast was commenced with the arrival of the sultan in the mosque. Then, they performed their congressional pray. The climax of the feast was to grab the *gunungan*s for a communal meal in the public square. In Surakarta and Yogyakarta, *grebeg* feast has been culturally enriched and more sophisticated. The *gunungan*s were prepared and built as a couple with symbolic male and female representation. During all rituals, ceremonies, and celebrations of *grebeg*, the sultan acted as the main conductor of this theatrical performance. The sultan enjoyed watching their people as the performers of his story and grandeur spectacle.

The ritual of *grebeg besar* has been surviving until today in the city of Demak. The disappearance of the sultanate of Demak in 1548 led to the discontinuity of the city rituals and ceremonies. However, the absence of a political center in the city of Demak did not shut down the spiritual and religious activities of populations. The mosque of Demak remains functional as the learning center of Islam for centuries after the dissolution of the sultanate. The revival of the ritual and ceremony of *grebeg besar* took place in Demak when this city became the regent administration—*kadipaten*—under the Dutch colonial control in the mid of the nineteenth century; the regent—*adipate*–was Aryo Purbaningrat. His name and title were inscribed on the wooden-carved wall with calligraphy—*maksurah*—of the mosque of Demak. The ritual and ceremony of *grebeg besar* were conducted and celebrated as it was in the past. The difference is that the procession of *grebeg* begins from the front hall, *pendapa,* of Kadipaten, regency seat, and moves to the mosque.

3.7 Concluding Remarks

The Persian *bandar* and the hybrid polity of the indigenous and Sanskrit *mandala* of *kutu negara* must have brought forth the new idea of town port with a unique form for Islamic maritime cities of Peureulak, Samudra Pasai, Banda Aceh, Malacca, and Demak. In other words, Southeast Asian *bandar* is a polity of the port authority and state capital in a well-established structure with two centers: *entrepot* or waterfront international market—*bazar*—and state ceremonial royal capital—*kuta nanggroe* or *kuta rajeu*. In the Indonesian archipelago and the Malay Peninsula, the concept of *bandar* is characterized by the coexistence of economic international market hub and spiritual and political center. Regarding the maps mentioned above, Banda Aceh was one of the port cities in the sixteenth century with a well-documented structure and form of *bandar* in Southeast Asia. The distinctive waterfront polity of *bandar* in Southeast Asia, which was under the Persian and Islamic influence, lies in the integration of upstream and downstream at the estuary area (Maguin 2000). Unlike the pre-Islamic capital cities, such as Bukit Siguntang of Srivijaya and Trowulan of Majapahit, the port cities of *bandar*, such as Malacca, Samudra Pasai, Peureulak, and Banda Aceh, brought forth the connection between upstream and downstream closer together that demonstrated the maritime orientation and openness of the state to international trades (Maguin 2000, 413–414).

The city of Demak was conceptually established to transfer the spiritual power of Majapahit. Most royal regalia and heirlooms—*pusaka*—of Majapahit were carried out from Trowulan to Demak; according to Javanese beliefs system, those precious things contain the divine power that enables the holder, in this case the Demak Sultanate, to be the center of the Javanese world. Unlike Samudra Pasai, Banda Aceh, and Malacca, the capital city of Demak was not derived from the concept of *bandar*, but this newly established polity was to take over the prestige and reputation of Majapahit. In doing so, Demak Sultanate was necessary to consolidate its power to control the maritime realm of Majapahit of the fourteenth century. The international commercial activities in Southeast Asian archipelago still under control of Muslim traders in the Indian Ocean and the Chinese Ming empire in the South China Sea. During the fifteenth century, the seafaring situation in the Indonesian archipelago has been transformed to be more vibrant with the arrival of European fleets for silk and spice trade.

As urban centers in the fifteenth century, Banda Aceh and Demak shared something in common for the meaning of urbanism. Both centers represented the importance of patrimonial rituals and ceremonies for the essence and existence of urbanism. Theatric festivals and grandeur spectacles belong to the necessity of urban settlement for the sense of community of urban inhabitants. Needless to say, the excess of such festivals was not free from hedonistic indulgence such as betting, gambling and other misdemeanors. To a certain extent, the urban festivals, rituals, and ceremonies are not only for joy and happiness for all but also for building the sense of belonging and identity in the context of the human condition for being *urbane*. Indeed, the excess of these festivals leads urban inhabitants toward hedonistic indulgence with betting, gambling, and drinking.

However, the excesses of the festivals do not discourage Demak and Banda Aceh to hold back the sense of abundance for what the city and state are about. *Grebeg* celebration in Demak and Ieds celebrations in Banda Aceh remind people of the necessity of the city and state of being inclusive. In other words, the city and state as a polity are not simply to accumulate resources for the sake of power and control. Rather, the city and state are responsible to reward its citizens or its inhabitants for collective joy and happiness as a communal experience in the public realm. Urbanism in this sense is experienced as the phenomenon of gathering that conjoins people and place in the ecstatic event of rituals and ceremonies.

The purpose of formalities is nothing but to demonstrate the cosmological order inconspicuous manner and fashion. During the ceremony, the seat and position of the person will be provided according to their title, status, rank, and privilege in relation to the center where the king/queen is. In other words, the structure, elements, and form of cosmological order will be unveiled that is to confirm and established the constellation of power in the state polity. The state rituals and ceremonies are performed by the king/queen to establish loyalty, hierarchy, and destiny of people for the cosmological ordering system. All this puts the necessity for unity, stability, and harmony as the state underlying principles. In doing so, there is no room for any action and motion against the authority of the state. The synergy between religious ideals and the state rituals is to justify and strengthen the domination and authority of the state—*negara*—as the source of orders and values in the lifeworld of people.

In most cases, the establishment of Javanese urban settlement is more about to define its center rather than a boundary. This implies that the urban center is symbolically represented by the emptiness of the public square. In other words, the Javanese city is a center without definite periphery. This principle is drawn from the beliefs that the center is the source of spiritual power that radiates to all directions without physical limitation. This concept of power has been explored and understood by the Javanese as something concrete, homogenous, limited, and beyond morally legitimation (Anderson 2006, 21–22). The radiating and vibrating capacity of power depends on the leadership of the ruler of the city or state, *negara*. Power in this sense is not about the inter-human relationship, but about spirituality that does not need any political legitimacy. The ruler is conceived by the Javanese people as the representation of the resourceful power of earth and sky. In doing so, the ruler is believed to be able to bring about people together in terms of spiritually unifying embodiment of nation-state, Negara. Unsurprisingly, the intensity of power is associated with a strong personality or charisma of the ruler. Based on this personal quality of leadership, he/she is able to mobilize the resources for his/her glorious spectacle. The purpose of negara is to exemplify this festive glory and phenomenal spectacle of all resourceful accumulation. In many ways, the Acehnese concept of power has something in common with the way the Javanese people believe in what power is; they share the idea that everything *is,* contains the spiritual power, *semangat*. The ruler is the representation of the source of resourceful earth and sky that is able to unify people together as a spiritual nation in terms of *negara*.

The contribution of historical maritime culture and tradition to urbanism is to uphold and sustain the livability of towns and cities in the coastal area. This livability is made possible by the synergy of the ecological and economic sustainability of the coastal region. Ecological sustainability needs to rely on the conservation of coastal biodiversity and the watershed area from upstream to downstream. Meanwhile, economic sustainability entails the growth of the maritime industry in alignment with the development program of integrated coastal management. The problem is how to manage the synchronization between environmental conservation and economic development in public policy in terms of spatial land-use planning, nature and park protection, and energy and water conservation.

References

Andaya L (2008) Leaves of the same tree: trade and ethnicity in the Straits of Melaka. University of Hawaii Press, Honolulu, HI

Andaya BW, Andaya LY (2001) A history of Malaysia. University of Hawaii Press, Honolulu, HI

Anderson BRO'G (2006) Language and power, exploring political culture in Indonesia. Equinox, London

Azra A (2006) Islam in the Indonesian world: an account of institutional formation. Mizan Pustaka, Bandung

Bacus EA (2004) The archaeology of the Philippine Archipelago. In: Glover I (ed) Southeast Asia: from prehistory to history. Routledge, London, New York, pp 257–281

Bronson B (1992) Patterns in the early Southeast Asian metals trade. In: Glover I, Suchitta P, Villiers J (eds) Early metallurgy, trade, and urban centers in Thailand and Southeast Asia. White Lotus, Bangkok, pp 63–114

Clifford HC, Athelstane F (1894) A dictionary of the Malay language: Malay-English, Parts 1–4. The British Malay Government Printing Office, Taiping Perak

Coedes G (1944/1968) The Indianized states of Southeast Asia (trans: Cowing SB). University of Hawaii Press, Honolulu, HI

de Beaulieu A (1996) Memoires deún voyage aux Indes Orientales 1619–1622 un marchand normand a Sumatra (Lombard D, ed). Maisonneuve & Larose for EFEO, Paris de Casparis JG (1986) Some notes on the oldest inscriptions of Indonesia. In: Helwig CMS, Robson SJ (eds) A man of Indonesiab letter: essay in honour of professor A. Teeuw. Koninklijk Instituut VKI 121, Dordrecht, pp 242–256

Devare H (2009) Cultural implication of the Chola maritime fabric trade with Southeast Asia. In: Kulke H, Kesavapany K, Sakhuja V (eds) Nagapattinam to Suvarnadwipa: reflections on the Chola naval expeditions to Southeast Asia. Institute of Southeast Asian Studies, Singapore, pp 178–192

Djajadiningrat PAH (1958/1987) Islam in Indonesia. In: Islam, the straight path: Islam interpreted by Muslims, by Kenneth W. Morgan (Morgan KW, ed). Motilal Banarsidass, Delhi, India, pp 375–402

Drewes GWJ, Voorhoeve P (1958) G. W. J. Drewes and P. Voorhoeve: Adat Atjèh, reproduced in facsimile from a manuscript in the India Office Library, with an introduction and notes. Verhandelingen van het Koninklijk Instituut voor Taal-, Land- en Volkenkunde, Deel XXIV, pp 2–176

Feener RM, Daly P, Reid A (2014) Mapping the Acehnese past. KITLV, Leiden

Geertz C (1980) Negara, The Theatre State in the nineteenth century Bali. Princeton University Press, Princeton

Goeksoy IH (2011) Ottoman Aceh relations as documented in Turkish sources. In: Michael Feener R, Daly P, Reid A (eds) Mapping the Acehnese past. KITLV Press, Leiden, pp 65–96

Hall K (2010) A history of early Southeast Asia: maritime trade and societal development, 100–1500. Rowman & Littlefield, Lanham

Hasymy A (1990) Sejarah Kebudayaan Islam di Indonesia. Bulan Bintang, Jakarta

Headley S (2004) Durga's Mosque: cosmology, conversion and community in Central Javanese Islam. Institute of Southeast Asian Studies, Singapore

Heng D (2009) Sino-Malay trade and diplomacy from the tenth through the fourteenth century. Ohio University Press, Athens

Hum SH (2012) Zheng He's art of collaboration. Institute of Southeast Asian Studies, Singapore

Koentjaraningrat (1985) Javanese culture. Oxford University Press, Singapore

Kraska J (2011) Maritime power and the law of the sea: expeditionary operations in world politics. Oxford University Press, Oxford

Kratoska PH (2001) South East Asia, colonial history: imperialism before 1800. Taylor and Francis, London

Kulke H (2009) The naval expeditions of the Cholas in the context of Asian history. In: Kulke H, Kesavapany K, Sakhuja V (eds) Nagapattinam to Suvarnadwipa: reflections on the Chola naval expeditions to Southeast Asia. Institute of Southeast Asian Studies, Singapore, pp 1–19

Laffan M (2011) Tracing the boundary of Jawi Eucomene: relections from distance. In: Hiruyuki Y, Milner A, Midori K, Kazuhiro A (eds) Bangsa and Umma: development of people-grouping concepts in Islamized Southeast Asia. Kyoto University Press & Trans Antlantic, Kyoto, Melbourne, pp 37–49

Lancaster J (1940) The Voyages of Sir James Lancaster to Brasil and East Indies 1612–14 (Foster W, ed). Haklkuyt Society, London

Lombard D (1966) Kerajaan Aceh Jaman Sultan Iskandar Muda 1607–1638 (trans: Arifin W). Balai Pustaka, Jakarta

Lombard D (1967) Le Sultanate dÁtjeh, au temp dÍskandar Muda 1607–1636. E.F.E.O (École française d'Extrême-Orient), Paris

Madjid D (2014) Catatan Pinggir Sejarah Aceh: Perdagangan, Diplomasi, dan Perjuangan Rakyat. Yasyasan Obor Indonesia, Jakarta

Maguin P-Y (2000) City-states and city-state cultures in pre 15th century Southeast Asia. In: A comparative study of city-state cultures, an investigation, by Mogens Hermann Hansen (Hansen MH, ed). The Royal Danish Academy of Sciences and Letters, Copenhagen, pp 409–430

Manguin P-Y (2012) Lancaran, Ghurab, and Ghali. In: Wade G, Li T (eds) Anthony Reid and Southeast Asian past. Institute of Southeast Asian Studies, Singapore, pp 146–182

Manguin P-Y, Indradjaja A (2011) The Batujaya site: new evidence of early Indian influence in West Java. In: Manguin P-Y, Mani A, Wade G (eds) Early interactions between South and Southeast Asia: reflections on cross-cultural exchange. Institute of Southeast Asian Studies, Singapore, pp 131–136

Moedjanto G (1987) The concept of power in Javanese culture. Gadjahmada University Press, Yogyakarta

Muljana S (1972) Islam before the Foundation of Islamic State "Demak". J South Sea 27(1 & 2):41–83

Muljana S (2005) Runtuhnya kerajaan Hindu-Jawa dan timbulnya negara-negara Islam di Nusantara. LKiS Pelangi Aksara, Yogyakarta

Muljana S (2006) Tafsir Sejarah Negarakertagama. Pelangi Aksara, Yogyakarta

O'Connor RA (1983) A theory of indigenous Southeast Asian urbanism. Institute of Southeast Asian Studies, Singapore

Ozay M (2011) Baba Davud: a Turkish scholar in Aceh. In: Kayadibi S (ed) Ottoman connections to the Malay world: Islam, law and society. The Other Press, Kuala Lumpur, pp 32–55

Pires T, Cortesao A (2005) Suma oriental of Tome Pires and the book of Francisco Rodrigues (Cortesao A, ed), vols 1 & 2, 2 vols. Asian Educational Service, New Delhi

Potter LG (2009) The Persian Gulf in history. Palgrave Macmillan, New York

Reid A (2000) Charting the shape of early modern Southeast Asia. Silkworm Books, Chiang Mai, Thailand

Reid A (2004) Understanding Melayu (Melayu) as a source of diverse modern identities. In: Barnard TP (ed) Contesting Malayness: Malay identity across boundaries. NUS Press, Singapore, pp 1–24

Reid A (2005) An Indonesian frontier: Acehnese and other histories of Sumatra. NUS Press, Singapore

Reid A (2010) Aceh and the Turkish connection. In: Graf A, Schroeter S, Wieringa E (eds) Aceh: history, politics and culture. Institute of Southeast Asian Studies, Singapore, pp 26–38

Reid A, Ito T (1999) A precious map of Aceh c.1645. Archipel, pp 191–208

Soekmono R (2006) Archaeology and Indonesian history. In: Soedjatmoko (ed) An introduction to Indonesian historiography. Equinox, London, pp 36–46

Soemardjan S (1982) Perubahan Sosial di Yogyakarta. Gadjahmada University Press, Yogyakarta

Solheim WG, Bulbeck D, Flavel A (2006) Archaeology and culture in Southeast Asia: Unraveling the Nusantao. University of the Philippines Press, Quezon City

Sufi R, Seno, Djuniat (1996) Sejarah Kotamadya Banda Aceh. Direktur Jendral Kebudayaan Republik Indonesia, Jakarta

Suryadinata L (2005a) Admiral Zheng He and Southeast Asia. Institute of Southeast Asian Studies, Singapore

Suryadinata L (2005b) Zeng He, Semarang, and the Islamization of Java: between history and legend. In: Suryadinata L (ed) Admiral Zheng He and Southeast Asia. Institute of Southeast Asian Studies, Singapore, pp 72–93

Tan TS (2009) Cheng Ho and Islam in Southeast Asia. Institute of Southeast Asian Studies, Singapore

Tarling N (1992) The Cambridge history of Southeast Asia: Volume 1, from early times to C.1800. Cambridge University Press, Cambridge, UK

Tarling N (1999) The Cambridge history of Southeast Asia. Cambridge University Press, Cambridge

van Ossenbruggen FDE (1919/1977) Java's Manca-pat: origins of a primitive classification system. In: de Josselin de Jong PE (ed) Structural anthropology in the Netherlands. Martinus Nijhoff, The Hague, pp 30–60

Vogel JP (1925) The earliest Sanskrit inscription of Java. Publicaties van de Oudheidkundiger Dienst in Niederlandisch-Indie I, vol I, pp 15–35

Warnsinck JCM (1930) Reisen van Nicolaus de Graaff, gedaan naar alle gewesten des werelds. s-Gravenhage

Chapter 4
Batavia, Dutch Indies 1602–1800: A Cultural History of Colonial Urbanism

Abstract This study is about the urbanism and society of otherness in confrontation with critical past and present. The focus of this study is to unveil and dismantle aspects of subjectivity and otherness concerning the history of social and cultural geography. The case study is the Dutch V.O.C urbanism in Indonesia between 1602 and 1800. This study argues the sense of subjectivity and otherness is not the reason for revulsion and resentment but an opportunity for acculturation within the framework of global fraternity, because the present urbanism, architecture, and society are the amalgamations of diverse sources, roots and circumstances of the past. The purpose of the study is to recognize and rethink imagery subjectivity and otherness in society, architecture and urbanism as a challenge of a historical necessity for openness toward multiculturalism.

Keywords Colonial heritage · Architecture · Urbanism: acculturation · Dutch V.O.C · Indonesia

4.1 Urbanism and Colonialism

Several scholars have done studies on Dutch colonial heritage in Indonesia. While most of them explore and document the artifacts, material culture, and historical records, they leave us the room of investigation concerning ideological traces and roots of Western modernity and culture in the politically colonialized archipelago. This study is an attempt to divulge and dismantle aspects and traces of Dutch architectural and urban heritage. The focus of this study is to unfold and unveil the relationship between the presence of the heritage and the memory of the past in texts, drawings, maps, museum collections, and census. In doing so, the study is able to explore the formative aspects and roots of multicultural urbanism in the former colonialized Dutch Indies. Until recently, studies on colonialism have been identified with the cases of exploitation, oppression and mistreatment by the European ruling regime against other people and their resources. However, less attention has been given by scholars to the phenomena of acculturation between the European, Asian and indigenous traditions in terms of material culture, culinary goods and happenings.

B. Wiryomartono, *Traditions and Transformations of Habitation in Indonesia*,
https://doi.org/10.1007/978-981-15-3405-8_4

Methodologically speaking, the study departs its investigation from identifying historical places and buildings that represent urban institutions and facilities of urbanity as well as modernity. The scope of the investigation is focused on publicly funded buildings or government institutions. The confrontation between the Dutch V.O.C colonial heritage and locality becomes a methodological foundation for critical analysis of this study. The exposition of the paper will be organized into three sections. The first section concerns the historical issues on Dutch architecture and urbanism in the tropical archipelago in the context of technicality and culture. The second part will examine the issues with cases in various circumstances and conditions. The third segment of the paper will deliberately wrap up the process of acculturation between Dutch architectural heritage and Indonesia's cultural development.

4.2 From Sunda Kelapa to Batavia 1602–1799

Studies on the history of Jakarta are indebted to several scholars (Abayesekere 1987; Heuken 1982, 2002; Blussé 2013; Burnett 2013; Surjomihardjo 1977) who have published historical events that have established the city since its formative years of Sunda Kelapa and Batavia . Batavia was an important site of multicultural settlement in Java since pre-colonial times. The seaport is situated at the estuary of the Ciliwung River. The town was originally developed from a seaport of the Pajajaran Kingdom. The name of the port was Sunda Kelapa. The seaport played an important role since the twelfth century, especially for spice trades between the indigenous people and Chinese merchants of the Song Dynasty (960–1279). The evidence of the contacts is well preserved in the National Museum in Jakarta (Dinas Museum DKI Jakarta Raya 1983). Sunda Kelapa was the *kadipaten* or vassal of the Pajajaran Kingdom with its capital town in Pakuan, now Bogor. As a *kadipaten*, the traditional town of Sunda Kelapa was provided with the seat of the ruler consisting of an Adipati's residence, a town square and a place of worship. Traditional settlements, *kampung*s, were configured along the routes between places. One important institution of the traditional port town was the fish market, *pasar ikan*; the presence of the market was shown on the painting of Andries Beeckman, 'the Castle of Batavia', 1661, oil on canvas, 108 × 151.5 cm. Rijksmuseum, Amsterdam (Rijksmuseum 1859). The location of the fish market was permanent that signified its relation to the port authority. Traditionally, the person in charge of the fish market was *mantri pasar* who reported to the port authority, *shahbandar* concerning tax collections, tolls, and gifts.

The direct connection between the port and the capital town of Pajajaran in the hinterland was made possible through the Ciliwung River. Geographically, the port of Sunda Kelapa is recognizable by its natural landscape. The view from the sea is directed by the river's flow leading to hinterland between two dominant mountains: Mount Salak and Mount Gede in the remote distance. The free spices trade in Southeast Asia was abruptly interrupted by 1511 when Malacca was sized by the Portuguese sea power. Portuguese control and interest in the monopoly of spices trade was expanded to have more trading posts. Sunda Kelapa was one of the Portuguese targets regarding its strategic position between Malacca and Halmahera. Moreover, the hinterland of Sunda Kelapa was well known for its commodities, such as pepper, indigo, cinnamon and rice. Unlike Malacca, Sunda Kelapa was not just an entrepot but also a hub port with strong hinterland for rice supply in Southeast Asia that was not under Islamic state control.

Tomé Pires, a Portuguese geographer, who went along the Portuguese sea power to Sunda Kelapa, wrote in his report on *Summa Oriental* (Leirissa 1995; Pires and Armando Cortesao 2005). He described his experience and observation about the port and town of Sunda Kelapa, its ruling persons, people and their land and culture. According to Pires, Sunda Kelapa was a port with a well-organized society under the vassal of the Pajajaran Kingdom. The King of Sunda Kelapa: Surawisesa was the Prince of Pajajaran with its capital town in Pakuan—now Bogor—and 120 km away from Sunda Kelapa. According to Pires's account, the settlement in Sunda Kelapa has been developed one kilometer away from the port and juxtaposed along the riverbank with a square and palace; the river was the frontage of Sunda Kelapa town. The way of managing multicultural society of Sunda Kelapa was commonly not regulated by law, but by contractual agreement between the local ruler and the chiefs of traders: Chinese, Gujarati, and Arab, Persian, and European group. Each group in town was not only granted a specific right of using the land for their trading and port activities but also to exercise their own customs and traditions. In doing so, each ethnic group had its own domain and customary jurisdiction for their social and economic affairs (Saidi 2010; Hanna 1988).

In the eighteenth century, most of Javanese port towns were populated by various ethnic groups such as Javanese, Malay, Arab, Gujarati, Chinese, and Bugis. Slaves were commonly found working in the port. The masters were mostly the local rulers and their families. Chinese businesspersons mostly controlled the agents of the slave trade. Stamford Raffles in his History of Java Volume 2 (1830: xcvi) gives a note in the appendices on slavery in the archipelago during the V.O.C controlled the spices trade from the Straits of Malacca to Banda Sea (Raffles 1817). According to Raffles' account, slaves were used as currency for the payment from native chiefs to the Dutch, along with gold and silver. Most slaves worked for building and ship construction and public utilities such as roads, bridges, sanitation, canals, and drainages.

The Portuguese sealed a treaty and commercial deal with Pajajaran King: Surawisesa for Portuguese trading post in Sunda Kelapa in 1522. The deal was marked with a stone called *Pandrao*. The stone is now the collection of the National Museum of Indonesia in Jakarta (Iguchi 2015, 154–155); the local Sundanese narrative of *Mundinglaya Dikusumah* describes the stone as *laying salaka domas.* The Portuguese sea power began to establish its fortress at the estuary of the Ciliwung River The fortress was constructed with warehouses, stores, workshops, and barracks for Mardijkers, as well as houses for the officers.

A few years after the deal mentioned earlier, the Portuguese stronghold was built with thick terracotta bricks and canon platforms at each corner of its cardinal positions. The Portuguese forces were mostly recruited from various places in Africa and Asia, known in Sunda Kelapa as Mardijkers; they were mercenaries who were trained by Portuguese officers and spoke Portuguese creole and had their own Catholic congregation and community. Such mercenaries were common to join with other European sea powers as well, such as the Dutch V.O.C, Spaniard, and the British during the Age of Trade in Southeast Asia between sixteenth and seventeenth centuries. By the 1522 treaty, the patrimonial relationship between the Portuguese and the indigenous Kingdom began to control the spices trade. Under this treaty, the traditional role and function of Chinese merchants and agents became less important; they gradually moved out to other ports such as Bantam and Cirebon.

One important quartier in Sunda Kelapa was the Chinese settlement called Pecinan. The presence of Chinese merchants in Sunda Kelapa dated back to the seafaring from Canton and Guangzhou since the Tang Dynasty (618–907) and the Song Dynasty (960–1279). During those periods, the pilgrimage to Java and Sumatra for learning Buddhist–Hindu was common among the Chinese seafaring people. In Sunda Kelapa, Chinese populations had been an integrated part of the seaport activities. Most Chinese inhabitants were traders who were lived and worked as a mediator between the native populations and the international merchants from Handramaut, Gujarat, Persia, Iraq, and South India. The workers for the Sunda Kelapa port activities were mostly people from various places in the archipelago. The Dutch scholars: Vollenhoven and van Leur have identified the forms of migration in the precolonial archipelago (Vollenhoven 1918, 100–101); one important of the movement was forced migration to urbanized areas. Then, they established colonization by a large group of fellow migrants from the same region who settled in another region; and establishment of authority in foreign regions (Hugo 2006, 57).

The Portuguese monopoly for spices international trade in Sunda Kelapa did not make the Sultanate Cirebon comfortable. For Sunda Kelapa of Pajajaran, the presence of Portuguese provided protection and security from the predominantly operating Muslim traders on the Java Sea. The protection from the sea power of the Chinese Ming Emperor had been withdrawn from the archipelago due to the internal crises in the mainland. The fall of Malacca to the Portuguese in 1511 was the pinnacle of the weak presence of the Chinese sea power. The Java Sea and the Straits of Malacca used to be under control of the Chinese navy led by Admiral Zeng He between 1371 and 1433 (Wade 2007; Tan 2009). In Macau, the Portuguese sealed the deal with the Ming Dynasty for their trading in 1516; later by 1557, the Chinese granted the Portuguese a permanent trading post there. The absence of Chinese protection and control in international trade led the Pajajaran Sunda Kelapa to get a new protector, the Portuguese sea power. Of course, Cirebon Sultanate did not let Sunda Kelapa under Portuguese control. The first move of Fatahillah of Cirebon Sultanate was to put the port of Bantam under control in 1526. From Bantam, Fatahillah asked the sultanate of Demak for helping by sending some troops to seize the Portuguese in Sunda Kelapa. Fatahillah's son and Demak forces attacked and seized the Portuguese stronghold in Sunda Kelapa on June 22, 1527. Fatahillah or Sharif Hidayatullah renamed the port as Jayakarta, meaning place of victory. Fatahillah built Jayakarta in a similar way with Surawisesa had done for Sunda Kelapa. As in many indigenous ports of the archipelago, traditional urbanism of Jayakarta took place mainly in the marketplace; this place is and ever the realm of women. Historical notes from foreign traders and

European colonial offices, such as Sir Thomas Stamford Raffles confirmed this that whatever the political situation may have been, the marketplace was dominated by women (Raffles 1817, 353; Blussé 2013, 355–356).

After the fall of Sunda Kelapa to Fatahillah's forces, the number of slaves increased. Then, Jayakarta was announced to be the vassal of Bantam Sultanate. The port was declared again as a free marketplace for spices of West Java. The Dutch merchants led by Cornelis de Houtman from Amsterdam visited the town in 1596 (Barrenveld 2009, 55–56); their fleet consisted of four ships: Mauritius, Amsterdam, Duyfken, and Hollandia. They departed from Amsterdam on April 2, 1595, with heavily loaded armed and 240 sailors and crews. The mission of the Dutch voyage was to find the sources of spices in the Far Eastern archipelago. The mission did not manage to reach Molucca, because of nautically inexperienced crews for such long-distance voyage. Cornelis de Houtman with his 87 men survived to go back to Amsterdam; they realized the situation of international trade of spices, which was mostly under the domination of Portuguese sea power (Barrenveld 2009, 31); they controlled the trading posts along the route from Indonesia to Europe. The Dutch trade expedition had learned from the Portuguese approach to establish their efficient trading posts and strongholds by recruiting local mercenaries.

The second voyage of Cornelis de Houtman to the archipelago was in 1599. Even though he did not make it because he was killed in a skirmish with Acehnese on the way to Bantam and Molucca. Admiral van Warwijk and the voyage continued the mission to Molucca; they visited Makassar, Ambon, and Ternate before they moved to Japan but unfortunately, they did not make it. After the success of the first and second Dutch expeditions to the Far East for spices, the companies in the Netherlands consolidated their effort and investment under the leadership of Johan van Oldenbarnevelt (Parthesius 2010, 31–50).

In the Netherlands, several companies gathered on March 20, 1602; they managed for the establishment of *Verenigde Ostindische Compagnie* (V.O.C), the United Dutch East Indies companies. By 1602, the V.O.C was able to consolidate and mobilize more than 65 ships ready to embark from Amsterdam to East Indies. The purpose of V.O.C's presence in the Southeast Asian archipelago was to establish the monopoly of the spice trade from Southeast Asia to Europe. The first move of V.O.C was to seize Portuguese control along the spices of trading posts in Goa, Malacca, Jayakarta, and Ambon. The Admiral Van der Heighen managed to take over the fort of Victoria in Ambon from the Portuguese in 1606. From Ambon, the V.O.C managed to seal the deal with the ruler of Jayakarta in 1611; the Dutch established and erected their stronghold or Kasteel in Jayakarta. The fort was constructed as a military garrison and trading post with warehouses and barracks for mercenaries.

In 1610, Jacques L'Hermite of the East Indies Company and the ruler of Jayakarta. Jayawikarta came to an agreement that allowed the Dutch to build their trading post on the east of the River Ciliwung adjacent to the ruler's compound. It was not only the Dutch V.O.C that was granted to establish their trading post by the Adipati of Jayakarta: Jayawikarta (Heuken, Historical Sites of Jakarta 1982, 27–28, 166) but also the English East Indies Trading company since 1602 when Sir James Lancaster of the English East Indies Company was granted a trading post in Bantam. As the vassal of Cirebon Sultanate, Bantam and Jayakarta were in good terms with the Dutch as well as with the English sea power. When tensions between the Dutch and the native escalated in 1618, the ruler of Jayakarta asked the sultanate of Banten: Abdulmufakir for help. Jan Pieterszoon Coen and his men were able to escape from Jayakarta to Molucca. Five months later, in May 1619, Coen and his V.O.C forces were able to attack Jayakarta and controlled its whole territory of Jayakarta. The Dutch V.O.C burnt down the royal compound of Jayawikarta at the East of the Ciliwung River. The V.O.C did not only manage to control Jayakarta from the sea but also all accesses from the hinterland.

By 1621, the V.O.C established Dutch supremacy of control for spices trade between Molucca and Jayakarta. After having flattened the town of Jayakarta, J. P. Coen began to build a Dutch Company town in East Indies in Java for the first time; in affirming his supremacy, he renamed the former town of Sunda Kelapa or Jayakarta to be Batavia. Renaming the town is nothing but claiming hegemony and control that its territory and populations are under the Dutch system (Taylor 2003, 146). Unlike the Portuguese that was granted to establish their trading post in Sunda Kelapa in 1522, the Dutch V.O.C captured and put the port and town of Jayakarta under the Dutch rule, politically, militarily, and economically. The capture of Jayakarta or Jacatra is a considerable milestone of the Dutch colonialization of Java. The Dutch V.O.C expanded its settlement outside the fort to the east river bank. The urban development of Batavia under J.P. Coen was not only to establish the V.O.C headquarters in the archipelago but also to build the Dutch sociocultural identity of its architecture and society. Coen worked together with the Dutch Reformed Church to underpin the presence of the Calvinist Christian congregation among European populations in Batavia. The V.O.C did not allow other Churches built inside the bastide, but the Hollandse Kerk. The cultural politics of the V.O.C for the Dutch-ness of Batavia did not make other Europeans feeling comfortable.

As a township, Batavia was not intended by the Dutch V.O.C as an independent municipality, but simply a company town and headquarters of the trading center. In the intention of the V.O.C directors—Herrn XVII in the Netherlands—Batavia was not more or less than a place of rendezvous of the Company (Taylor 2003, 11) that should pay its own expenses. In this notion, Batavia was not likely to be a site of urbanism, but simply a temporary place of meeting and arrangement as well as engagements for transactions and exchanges. It was understandable for the gentlemen who did not work and live in a distance away from their hometown. However, for those who were far and away from their home countries, such as for Jan Pieterszoon Coen and his V.O.C men, the establishment of Batavia was not a stationary post in a foreign land, but an extension of the homeland. Coen wanted to establish a home in East Indies; he initially wanted to call the new town on former Jayakarta as New Hoorn instead of Batavia; Hoorn is the hometown of Jan Pieterszoon Coen, he was born there on January 8, 1587 (Goor and van Goor 2004, 67).

Patrimonial culture of Chinese settlers in Java made it easier for the Dutch V.O.C to mobilize and utilize Chinese resources in terms of labor, land, and capital. The specific Chinese settlers were unique for urbanism because of their expertise and experience in building construction, trade, horticulture, ceramic industry, culinary, metallurgy, medicine, and acupuncture. Mixed-blood marriage between Chinese men and indigenous women from Bali, Makassar, Nias, Java, and other islands built, developed and sustained the acculturation, assimilation, and amalgamation of customs and traditions toward new cultures. Lack of sufficient skills labor for the construction of Batavia, Jan Pieterszoon Coen appointed his other Chinese associates of trade, Jan Con, who was a chief Chinese carpenter (Parmer 2012, 18–54); he brought the army of carpenters and stonemasons from Fujian of mainland China. His patron–client ties with Jan Pieterszoon Coen were confirmed by his role and function as the trade advisor of the Company.

On January 6, 1623, the English East India Company delivered a grievance to General Pieter de Carpentier of the V.O.C in the fort of Batavia concerning restraint trade for English inhabitants (Great Britain Public Record Office 1878 1878, 38). The English company filed the report to the court complaining about the unjust and violence of the V.O.C against English people in Batavia and their properties that happened on January 3, 1623. The English East India Company had been settled down in Sunda Kelapa since 1602; they established their lodge with clothing factories opposite the V.O.C's Castle of Batavia. From its very beginning, it was not easy for the English populations living and working in cohabitation with the Dutch. The ruling style of Jan Pieterszoon Coen (1618–1623) and (1627–1629) was infamously ruthless with his profit-oriented trade monopoly of spices (Burnett 2013, 108). Eventually, the English East India Company reduced their factory's production in Batavia in 1627 (Chaudhuri 1965, 67).

Jan Pieterszoon Coen had suggested the Heeren XVII—the director of the V.O.C—several times that the Company should manage the deals of trade more flexible in East Indies. Governor J.P. Coen would have suggested the V.O.C to restrict with the monopoly of spice trading and to throw the remaining business to other civilians such as Chinese traders who are not the servants of the company (Randeraad 1998, 23). His commendation was not without reason. The English East India Company did the patron–client ties for textile trades with the Parsis as intermediaries in Bombay.

The presence of V.O.C in Batavia was significantly marked by the construction of the fortress or citadel with a 35 cm thick brick wall system. The River Ciliwung divided the whole town into two equal areas: the West and the East part. Spatially, the whole structure of Batavia was configured with three zones of land use. At the Northeast, the position was an island for the chateau, while at the East and West sides of the River Ciliwung were designated for a self-contained urban settlement in support of international trading activities. From its very onset, Batavia was not intended simply as a Dutch trading post—such as Surabaya, Semarang, Makassar, and Ambon—but a Dutch capital and trading headquarters of the V.O.C in East Indies colony. The construction of canals and urban blocks with roads system was a projection of geometrical layout of settlement on the Javanese landscape for the first time.

The plan of Batavia was derived from a Medieval French *bastide* plan (Hilton 1995, 36–37; Morris 1972/2013, 119–120) that built a cluster of the well-established boundaries between Dutch settlement and its surrounding landscape and other settlements. The bastide was constructed with coral rock walls and twenty-two fortifications. The *bastide* occupied 270 ha area with a well-defined boundary made of wall, river, and ditch. Three characteristics of the *bastide* plan of 2250 in length × 1200 in width included in Batavia layout of 1623–1750 that consisted of a castle, gridiron blocks and 9.8 m wide roads system, and the privilege of stay in *bastide* system by granting access to economic activities outside the town such as farming land or other trades. In other words, Batavia was not simply a military garrison and trading post, but a Dutch settlement with the outreach of territorial control to the hinterlands of Java. On the Map of Batavia from 1669 (Princeton University Library 2013), the Eastern part of the *bastide* was developed for the V.O.C facilities and settlement, while on the Western part was mostly empty lots. The core components inside the *bastide* of Batavia were town hall, town square, Dutch Church. The settlement was arranged within blocks consisting of lots for two or three-story buildings. Shophouse was a common building type in each block.

On the Map of Batavia from 1750, the transformation of the town was obvious with the development of areas surrounding the *bastide*. Civic buildings mostly occupied the common lots at the Eastern part of the bastide: town hall, church, government offices, and housing for V.O.C officers. Most banks, insurances, and trading offices, as well as private houses, shophouses and warehouses, were designated for the lots in the Western part of the River Ciliwung. Military garrison and the barracks of mercenaries were located separately in the fortress of Batavia; its site was an isolated 0.8 ha area encircled by a wide ditch.

The south part of the bastide was the land for plantations and housing. The V.O.C claimed the land like the backyard of the bastide. Some areas were sold to Chinese and other Asian populations. The V.O.C allowed the Chinese populations occupying the southwestern land for their quartiers known later as Glodok with its center in Petak Sembilan. Chinese headman who reported to the V.O.C officer concerning taxes, gifts, and tributes controlled the Chinese settlement. Other Europeans lived and worked within the walled town; they were allowed by the V.O.C to fill the posts of town councilmen, public servants, and supervisors of charitable institutions; unlike other ethnic groups, the Europeans and Eurasians mainly of Portuguese origin were registered by the V.O.C as militias for town's defense and honor guards for urban ceremonies (Taylor 2003, 152).

For the construction of Batavia, Coen asked for help from his Chinese friend: So Bing Kong or So Ming Kang; his familiar name among the Dutch V.O.C officers was Ben Kon. Coen assigned Ben Kon as kapitan of Chief of Chinese communities in Batavia only a few days after the capture of Batavia in October 1619 (Parmer 2012, 29). Ben Kon was well *cukong*—entrepreneur, shipowner, and later V.O.C contractor—of Chinese Batavia. By the Dutch V.O.C, he was appointed as the Chief of Chinese affairs who was in charge of tax collections and tolls for trades, gambling, and coinage. Coen and Ben Kon were in a patrimonial patron–client relationship for years until the first governor-general of the V.O.C died in 1629. Coen had the help of Ben Kon as an intermediary person for the supply of workforces and trades with the Chinese and indigenous people, while the Chinese chief enjoyed the monopoly of trades and other businesses such as gambling, prostitution, and opium from the boss of V.O.C. Such patrimonial relationships had been familiar between the indigenous rulers and the Chinese merchants, prior to the arrival of the Portuguese and other European merchants in the archipelago. The reason was simply that the Chinese had established a system of international trade since the thirteenth century in Southeast Asia; they set up the coinage system and gold as currency as well as the network of points and hubs of exchanges in ports of Southeast Asian archipelago.

Batavia's *bastide* was an extension of Dutch Imperial authority in a foreign land. Four institutions characterized the presence of Dutch authority in Batavia: town hall, Dutch reformed church, town market square, and courthouse. In contrast to its surrounding indigenous villages that were built between trees, woods, and rice fields, the whole area for the bastide was flattened and cleared from trees and bushes. Dutch urban settlement prescribed open land, free for staking out the canal, lot, road, and drainage system. Batavia's town plan and landscape design were architecturally derived from the Dutch island towns of Low Countries such as Leiden, with parallel roads along the river and canals encircling the fortified area of the town (Price 2014, 283).

In order to establish its hegemony, the V.O.C established the presence of town hall and Dutch Reformed church in the center area of Batavia. The Dutch Church stood as the symbol of the social and moral authority of Europeans living and working in the town. Former Portuguese descendants, who were mostly Catholics, were morally obliged to convert into Dutch Calvinist Protestant Christianity if they were married to Dutchmen. Administratively, the Statutes of Batavia was issued in 1642 that mentioned the status of citizenship. One important point of the Statutes was to maintain the sustainable supply of labor for trades and port activities of Batavia. According to the statute, the V.O.C forbade Dutchmen to marry a native woman unless the indigenous was able to speak Dutch and read the Bible. As a matter of fact, the mixed-blood generations called Eurasians or Mestizos were more common to find in Batavia society of the pre-cultivation era, than full European blood generations. Having slaves as household helpers, workers for trade, agriculture and construction had been part of the European family lifestyle.

Beyond the *bastide*, there were private European houses, Chinese quartiers, indigenous *kampung*s, and slave slums. Since the thirteenth century, slaves were part of life and work in the port of Sunda Kelapa. Slaves were not always the prisoners of war or people without social affiliation. In pre-colonial Java, slaves were migrants from hinterlands who voluntarily move out from their home village because of warfare, famine, or natural disasters. Towns and ports were the common destinations of such migrants for instance from Bali.

During the construction process of the castle and the bastide, slaves played an important role. After 1619, several Javanese trade persons in Jayakarta fled to Bantam when the V.O.C seized the port and burnt down its town. The only inhabitants left in town were Chinese populations. The construction of the walls and canals needed most skilled workers for stonemasons, bricklayers, and carpenters. The V.O.C contracted the works to Chinese *kapitan*s and their associations—*kongsi*. Of course, some Dutch carpenters, who worked for the V.O.C ships from the Netherlands, directed and supervised the construction works. Unsurprisingly, most buildings in Batavia during the V.O.C era characterized traits and features of the seventeenth-century Dutch architecture, with a simple gable roof; the form was well preserved until today (See Fig. 4.1). Unlike the Dutch Indies colonial era, most V.O.C buildings were clear-cut and functional brick-masonry architecture but proportionally elegant with less embellishment and ornamentation. Building materials for the castle and walled town of Batavia were mostly carried out from the surrounding areas. Ancol and Angke are places for coral rocks while Tanah Abang is the area with potential limestone for brick building material.

Multicultural society lived and worked in Batavia since the sixteenth century with the demographic figures that slaves outnumbered the European and Chinese populations. Indeed, slavery had been a long history in Java. The local concept of the slave is *kuli* meaning labor or worker for dirty jobs. The concentration of slaves was mostly in ports of Northern coastal areas of Java, from Bantam, Sunda Kelapa, Cirebon, Tegal, Semarang, Jepara, Lasem, Tuban, Gresik, Surabaya, Probolinggo to Banyuwangi.

The demographic figure of Batavia in 1673 can be described as shown in Table 4.1.

Even though the architecture of V.O.C *bastide* of Batavia represented an exclusively isolated European society, the interaction between the Westerners and the Asians as well as indigenous populations was inevitably to happen on daily basis in the market. The most valid data for the lifestyle of Batavia's society in the seventeenth century are well depicted in the paintings in the collection of Rijksmuseum in Amsterdam. Kehoe (2015) analyses some relevant paintings by Andries Beeckman

Fig. 4.1 V.O.C warehouse building in Jakarta Kota area, 2014

Table 4.1 Demographic figure of the Dutch colonial town of Batavia 1673

Description	Populations	%
• European and Eurasian	2750	8.6
• Chinese	2747	8.6
• Mardijkers	5362	16.7
• Javanese	6339	19.8
• Balinese	981	3.1
• Malay	611	1.9
• Slaves/Kuli	13,278	41.4
Total	32,068	100

Sources Dagh Register (1674), Batavia (1902, 27–30), Castles (1967)

of 1661, Aelbert Cuyp of circa 1640–1660, and Jacob Coeman of 1665 concerning social milieu and hierarchy of Dutch, Eurasian, and indigenous people in the Dutch colonial town of Batavia (Kehoe 2015). Having overwhelmed by indigenous and other Asian populations, the Dutch Batavian inhabitants faced ambivalent appearance in public space for showing their socially superior status between being gentle European traders and being master of colonialized tropical land and other people. The need to display their higher status led the Dutch people to adopt local customs and traditions such as servant-held parasol, pearls necklace, gold and buckle button, silk clothing, outfitting of carriages, and going publicly anywhere accompanied by slaves in the background.

Various Dutch Governor-Generals introduced sumptuary codes during the company period (1619–1795) in efforts to manage this behavior (Kehoe 2015, 86). The regulations were not to restrain the lavish and snobbish lifestyle of the Europeans, but it was actually to manage their social status of the elites according to their rank and position in the Company. Accordingly, the outfits of carriage, jewelry, parasols, and a number of slaves should have adjusted in compliance with their eminence in the Company.

During the V.O.C period, Batavia remained the bastide of the Dutch East Indies with a strong mission for supremacy in spices trade. The mission of V.O.C was to control the access to resources: commodities and laborers, not for territorial acquisition. Practically speaking, dominating trade was impossible without territorial control. Interventions into local affairs by the Dutch V.O.C were unavoidable. Since the human resource of military forces for such interventions was limited, the V.O.C did not go deeper into indigenous affairs for territorial acquisition. Instead, the V.O.C implemented the principle of indirect rule and divide. Accordingly, the V.O.C managed to control the local rulers and native chiefs of ethnic groups. Each racial group had its authority and domain for practicing their own customs and tradition.

The V.O.C sustained the local tradition of Sunda Kelapa in managing the diverse societies in Batavia. In contrast to the pre-colonial multiculturalism of Jayakarta, the

Batavia under the VOC regime was populated by various racial groups with economically and socially discriminative affirmation with the European origin at the most privileged position. Regarding its town plan, the Dutch citizens and other Europeans occupied the area inside the fortified wall system, while the Asians: Chinese, Indian, and Persian, indigenous and others were outside and surrounding the fortification. While the lifestyle of European elites in Batavia was mostly focused on their comfort in their family homes and on alleyways or wharves of the walled town, common people used the front porch—*serambi*—of their houses for their social life. Ordinary Company's soldiers and sailors used to gather in their tavern inside the castle.

Practicing other traditions, customs, and beliefs but Dutch Calvinist Christianity was limited inside the town of Batavia until 1740. One of the important businesses of the Chinese populations in Batavia was sugar plantation; the other commercial activities included middleman traders at ports, shopkeeping for ceramics, porcelains, silk, and opium. Prior to the arrival of the Europeans, the presence of Chinese Hokkien had been part of port activities in most coastal areas of the Malay Peninsula, Java, Sumatra, and Sulawesi. Chinese traders and local rulers had a long-standing client of local rulers in terms of patrimonial relationship; they were mostly organized into units of society or associations (*kongsi*) led by *kapitan*s. During the period between 1602 and 1799, the patrimonial role and function of Chinese merchants as clients did not change, but the patron was not the local rulers anymore but the V.O.C. During those periods, the intensity of direct junk trade between Batavia and Xiamen China increased with an average eleven large ships annually (Kuhn 2009, 62).

The development of Chinese quartier, *pecinan*, in the southwest area of the walled town of Batavia, was generated by the influx of migrants from the mainland of China from 1619 to 1740. The V.O.C did not care too much for what happened in China quartiers concerning trades, prostitution, opium houses, and gamblings. The most important thing for the Dutch Company was the revenues from Chinese *kapitan*s to their agents secure and on time. To a certain extent, the patron–client ties between the V.O.C agents and Chinese traders were closely related to their personal fortune. Corruptions, smugglings, abuses, and frauds among the V.O.C agents were a common practice that was for the most part because the Company did not have a transparent auditing and controlling system for their financial administration.

The other important weakness of the Company was its human resources and recruitment system. The high ranking officers of the V.O.C. in Batavia were only for the Dutchmen; either experienced abroad or newcomers from the Netherlands. Despite their experience and expertise in local trading affairs and traditions, Eurosians, Mestizos, and Mardijkers did not have the chance for important keys in the Company (Bosma and Raben 2008, 184); unsurprisingly most of them often had engaged in illicit businesses with Chinese traders such as gamblings, prostitution, opiums, and alcohols. Frustrations and dissatisfactions among the low-rank V.O.C. officers cost the Company's integrity, business, and revenue in Batavia. Jan Pieterszoon Coen did his best to manage the Company's revenue from trade monopoly by consolidating the V.O.C. military power with ruthlessness and determination against other European powers and native rulers.

The Dutch V.O.C under Jan Pieterszoon Coen had defended Batavia from Sultan Agung's attacks in 1628 and 1629. After the war with Javanese Mataram, Coen died in Batavia of cholera on September 21, 1929. Even though the memory of his presence in Dutch and Indonesian history is still alive, nobody is able to find his grave in Batavia until recent days. The most striking historical reminder of Jan Pieterszoon Coen with Indonesia was the association of his name with the first Dutch colonialist leader who had established the Dutch urbanism in Batavia with political, sociocultural, and economic hegemony over the Indonesian archipelago. Indonesian history never forgets Coen's role and involvement in the massacre of Indigenous populations of Ambon in 1921 (Schefold 1991, 145); the massacre was executed mostly by the Japanese macenaries under J.P. Coen's order. This event put an end of Anglo-Dutch cooperation for the spice trade in Southeast Asia.

For Jan Pieterszoon Coen, it was clear that the supremacy of spice trade was necessarily military hegemony on the land and sea. Afterward, the V.O.C seized Malacca in 1641 and expanded its domination of international trade in Southeast Asia. Machiavellian Jan Pieterszoon Coen bequeathed his patrimonial leadership to the Dutch V.O.C until the end of the eighteenth century. After his death, the company maintained and sustained its political and economic hegemony by practicing the politics of 'divide and rule' on indigenous rulers (Drakley 2005, 33).

Inside the *bastide*, the urban development of Batavia was marked by the establishment of the town hall and its town square in 1710 (Ward 2009, 93); the building was designed by W.J. van der Velde that was to replace the old town hall on the same site. The Old Dutch Church stood side by side with the old hall on the same site. The construction of the hall affirmed the founding of Dutch architecture in the Far Eastern tropical colony. Batavia Stadhuis is probably a revelation of the Dutch Enlightenment that reflects the liberal values of the Dutch Republic (Kusno 2012, 259). However, the presence of the town hall is likely a determination of the Dutchness and its colonial domination in Southeast Asian archipelago, instead of unveiling the spirit of Dutch liberalism. The 1770 engraving of Johannes Rach of the scene of the town hall and its square shows the V.O.C Batavian society in public space with its racial discrimination and snobbish European inhabitants; their appearance in public place was featured with local attributes for their high social eminence, such as parasol, army of slaves, and fancy carriages. What is striking from the Rach's engravings on town square and Tigers Gracht is the absence of native children in social life in Batavia public space; as a matter of fact, the indigenous populations were inside the *bastide* as workers, slaves, and street vendors.

Architecturally speaking, the town hall represented civilian authority in contrast to the military outlook of the Castle of Batavia. The form of Batavia's town hall reminds people of Amsterdam's town hall designed by Jacob van Campen and built between 1648 and 1665 (Harris 2005, 330), with traces of Tuscany vertical windows with wooden split shutters and classical wall with the sobriety of Greece Antiquity. The barock watchtower and attic windows embellish and enhance the iconic components of the hall. The scheme of the two stories hall is based on a rectangular East-West plan of 20 × 65 m. The entrance of the hall is crafted with a modified Palladian porch facing to the North. As a whole, the town hall reveals as a well-crafted Dutch architecture with an appropriate adjustment in tropical climate and setting. During the V.O.C. period (1602–1799), the town hall included the court of Aldermen and a prison at its wing and basement.

The architecture of Batavia had been completely achieved in the 1730s. Abeyesekere estimates the Golden Age of Batavia in around 1730 where all urban institutions the military garrison, port, warehouses, shops, offices, town hall, town square, hospitals, schools, orphanage houses, and residentials were in place (Abayesekere 1987, 38). Collective memories from travelers and inhabitants of the town between 1730 and 1740 called Batavia was one of the magnificent towns in Asia or Queen of the East (Blussé 2013, 356). As mentioned above, the livability of Batavia was well depicted by the Danish artist: Johannes Rach showing scene at its parallel promenade streets along the *Tigersgracht*—now Jalan Lada and Jalan Cengkeh—inside the bastide (See Fig. 4.2). The function of the canal was not only for drainage but also for economic and recreational activities; the canal created waterfront promenades at its banks where Batavians enjoyed their leisure with children and pets and served by native workers, slaves, and servants.

The successors of Jan Pieterszoon Coen did not make any significant improvement for the town of Batavia in terms of demographic proportion between the Europeans and others. Since 1619, the growth of Chinese populations in Batavia outnumbered significantly the Europeans and Eurasians. By 1740, there were 2500 shophouses owned Chinese people with populations not less than 15,000 in Batavia (Ricklefs 2001, 112), while the total employee of the V.O.C was only 5628 (Lommerse 2008, 315). The V.O.C. imposed to regulate the influx of Chinese immigrants from the mainland by a quota system, but the Chinese *kapitans* were always able to evade it with the assistance of the Company's officials. It was not surprising to read from the report of Van Rechteren who visited Batavia in 1623 that five or six thousands

Fig. 4.2 Batavia Town Hall, now Museum Fatahillah of Jakarta 2014

of Chinese populations lived inside the bastide; the Batavian Chinese people were described as chronic and reckless gamblers, frequently staking their properties, businesses, wives, and children (Lach and Van Kley 1998, 1338). Most of them were the victims of the 1740 massacre. Beyond the town walls, the Chinese quarters expanded southward. Sugar plantations and mills attracted more and more Chinese and indigenous people to work in the sugar industry. Even though Chinese, Eurasian, and European populations lived together in the *bastide*, they hang out with their own flocks and associations. Racial origin and cultural identity determined their costumes and attires. Mixed marriages between European men and Eurasian women were common practice in Batavia.

In the case of marriage between Europeans and mestizo, the women were required by the V.O.C policy to convert from Catholicism to Dutch Calvinist Christianity. Unlike other Asian populations, the Japanese people were treated by the V.O.C's regulation of citizenship at the same eminence with European people. Chinese men usually married with Balinese or Makasarese female slaves. Most Dutch and European men in Batavia were bachelors; only a few of them coming with wives and children. Some of them married with Eurasian women, some got their wives in the motherland, and then, they came back to East Indies. Education for their children in the half seventeenth century Batavia was mostly organized by the Company (Borchardt 2014, 42–45).

Instead of the racial identity of the father, baptism by the Dutch Reformed Church was the key requirement for the eligibility of a child for their seats in the classroom of the Batavian school. Children of mixed-blood parents were able to attend the school as long as their parents were members of the Dutch Church. According to the V.O.C policy in Batavia, only children of Dutch burghers and other European high ranking V.O.C officers had the access to the Company's school, while Euroasian children mostly went to the school of the Orphanage house. Elite families in Batavia commonly sent their children for their higher education in the Netherlands.

Even though the Chinese merchants brought good commodities for European markets, the V.O.C revenues from Batavia to the motherland significantly decreased. Meanwhile, in the areas surrounding Batavia, the Chinese businessmen gradually developed strong domination for sugar plantations and its productions in the market; this raised concern of the V.O.C that would challenge the Dutch trading monopoly. The Company did not treat the Chinese merchants as their business partners, but simply as their subordinates in charge of taxes and tolls collection. As a group of immigrants, Chinese populations in Batavia were an organized community of sociocultural and economic systems; their well-being was maintained by a long-standing tradition of economic and professional societies and guilds. As a community, Chinese populations in Batavia had their own hospitals, orphanage houses, clubhouses, and temples. The leaders of Chinese settlers—*kapitans*, lieutenants, and secretaries—were able to mobilize their resources efficiently that had been in favor of the V.O.C for the urbanism in Batavia, such as the construction of the castle, canals, and middlemen of trades with indigenous people.

Socially and culturally, the Chinese populations did not integrate into the framework of the Dutch sociocultural system; under the V.O.C racialist politics, there was impossible in Batavia (1619–1740) for the Europeans and the Chinese to grow together for the process of acculturation (Randeraad 1998, 22–23). Even though such acculturation did not work overnight, Dutch and Chinese populations in Malacca were able to manage their cohabitation toward cultural assimilation. Such assimilation was well known as the phenomenon of *baba-nyonya* culture (Ooi 2004, 200).

As a result of high unemployment among Chinese inhabitants, organized crimes and gangs became part of daily life in Chinese quarters. On one side, the V.O.C. treated the Chinese matters with brutality and ruthlessness. On the other side, the V.O.C personnel—mostly Euroasians—engaged in illegitimate trading and large scale of fraud that gradually alienated them from the mainstream of the Dutch elite society of Batavia. The unhealthy and distrustful cohabitation between the Europeans and the Chinese group escalated in 1740 when Chinese vagrants outside the *bastide* of Batavia killed Europeans. The V.O.C bloodily crushed the Chinese uprising movement; the massacre of Chinese populations was considerably an outcome of Dutch ignorance of the necessity for acculturation in a multicultural urban settlement such as Batavia. It was not the case for the English in Bombay in the seventeenth century (Randeraad 1998, 48).

The English Company in Bombay integrated the Parsis into their sociocultural system so that they were able to work together not simply for the business but for the integration of Parsis into the European values system. In doing so, British colonialism was able to manage the cultural tension between the Parsis merchants and the English Company at a low level. In Malacca, the Dutch V.O.C did not have any issue with the Chinese populations probably because Dutch private traders and Chinese merchants were able to do business together. In Batavia, such possibility had been denied by the directors of the Company in the Netherlands; Jan Pieterszoon Coen pleaded for more room for the private burghers of Batavia 1623. The absolute control of trade monopoly by the V.O.C did not give benefit for the Batavian private citizens, but for the Chinese traders.

The Chinese massacre in Batavia was not likely impartial from the internal struggles of power in the V.O.C—between Gustaaf Willem Baron van Imhoff and the Governor-General Valckenier. During the period between 1740 and 1795, the integrity of its military organization had been weakened by the penetrations of the V.O.C into various indigenous affairs and struggles for power in Java and outer islands of the archipelago. The acceleration of the V.O.C's capital and the military decline was made worse by the Anglo-Dutch war (1780–1784). Moreover, mismanagements and corruption had led the V.O.C to its bankruptcy in 1795; the witty spell of the V.O.C was known as *Vergaan Onder Corruptie* (Ward 2009, 88). The end of the Dutch Republic at the end of the eighteenth century marked the dissolution of the V.O.C from the international stage of trade. The fall of the V.O.C was the beginning of urban restructuring in Batavia for new development. In 1807, Napoleon Bonaparte of French appointed Willem Daendels as the Governor-General of the Dutch East Indies; his assignment was to defend the territorial control of the archipelago from the British imperial forces (Ricklefs 2001, 145).

In order to establish defensible position against the British attack, Daendels withdrew the Dutch-French capital city from waterfront area to suburban region; his first move was to demolish the Castle of Batavia and reused the materials for a new town of Weltevreden and a new citadel called Castle of Cornelis which was located on higher ground of today Jatinegara. The second move was to connect all Javanese ports and towns with a post road of 1200 km length. Gradually, the urban institutions and European settlement were moved from Old Batavia at the waterfront to a new settlement in Gambir–Menteng area. Even though some buildings in the old town were left empty or demolished for new constructions, the function of the port of Sunda Kelapa remained the same as it was in the seventeenth century. Sunda Kelapa became an important seaport of Java for exporting commodities from the Dutch compulsory cultivation system between 1830 and 1870 (Fasseur 1992).

4.3 The Old in the Present: Jakarta Lama

The Dutch colonial urbanism did not break up its patrimonial relationship with the traditional indigenous authority. This relationship had been working well since the V.O.C times. The Dutch did not care for the process but the outcome of revenue and tolls from the local indigenous rulers. In return, the native chiefs enjoyed their status quo under the Dutch's protection. In other words, the patron–client ties between the Dutch and the native rulers were a kind of symbiotic mutualism.

Most of Dutch colonial urbanism took place in the existing site of the indigenous urbanized areas. Characteristically, the native towns are provided with a public place called *alun-alun* (town square), *dalem/kraton* (palace), and masjid (mosque). Environmentally speaking, the development of indigenous towns and villages accommodates and adjusts the existing terrain and its flora into their built environment system; site clearing for building sites and roads is done with limited effort for cutting terrain and trees. On the contrary, the intention of Dutch V.O.C colonial urbanism was an attempt to set in the power of reason for human dominance over nature; the geometrical form of *bastide* and artificiality of built environment with canal and fortification system actualize this architecturally humanized domination. The plan of their colonial town of Batavia reflects their seventeenth century European Enlightenment. The Dutch V.O.C urbanism cleared the existing terrain and its flora for their building sites, roads, and open spaces. In other words, colonial urbanism did not negotiate and assimilate with the existing locality and traditional system, but burnt them all down and set in their own mindset on the landscape.

A German traveler and cartographer, Selberg (1840) reported and documented the development of Batavia and its surrounding with a map in 1840 (Selberg 1840). Since the Chinese massacre, the town did not make significant development in terms of demography, architecture, and urbanism. Demographically, the European populations of Batavia declined slightly due to the outbreak of malaria, typhoid, and dysentery (Blussé 1986, 26–28). Today the artifacts, relics, and traces of Batavia are still there; few are still functional and useful, some are traceable and recognizable but most of them have gone without a trace. However, the site of Batavia remains the same but it has transformed from every aspect of urbanism demographically, administratively, socioculturally, economically, and environmentally.

One important aspect of these transformations is acculturation that works in various fields. The core area of historic Old Batavia or Jakarta Kota is now still in good shape. The town square and the town hall are now a public space and a museum with several collections of the colonial past. In today's Indonesian democratic society, the public place of Fatahillah square becomes part of the livable place of Old Jakarta. For decades, the Old Jakarta square and its historic building have been part of the site of urban rituals for the celebration of the 1527 founding of Jakarta by Fatahillah. Certain international events and festivals take place in the Fatahillah square and the museum. The municipal section for cultural affairs and tourism of Jakarta works together with local and international communities for the annual agenda and events of Old Jakarta livability. Concerning locality, Jakarta local arts and culture show diverse traces and roots of Portuguese, Dutch-European, Chinese, Arab, and Indian origin. During the performance of dances and orchestral music, the audience will experience acculturation of multiculturalism alive, while culinary festivals offer visitors various tastes and smell from diverse places on the globe, from Asia, Middle East, and Europe. The best examples for such acculturation are *Keroncong*, *tajidor*, and *gambang kromong Betawi* music (Fig. 4.3).

During the Dutch V.O.C times, the square was a weekly market of Batavia that was like what happened in every Dutch town in the Low Countries. The square of Batavia might have been a vibrant public place. In such a public place, most Dutch and European citizens, Chinese and other Asian settlers would have been able to gather for exchanges. Such gatherings would have led to the acculturation of various cultures and traditions toward new ones; it did not likely happen because of the V.O.C's racially segregated policy of urbanism based on the premise of European cultural preeminence.

At the Fatahillah square, some V.O.C Batavian buildings still stand in their authentic form and construction, such as Toko Merah and other shophouses at the Grand River area (now Kali Besar). The Nieuw Hollandse Kerk of 1736 at the town square had been destroyed by the earthquake in 1808; the relics of the Church were preserved in Museum Wayang at the Fatahillah square (Dinas Museum DKI Jakarta Raya 1983). The other V.O.C heritage in Old Jakarta is the complex of Shahbandar building and the Company's warehouses at the Sunda Kelapa port area. The complex is known today as Museum Bahari or Maritime Museum consisting of two buildings: the Westzijdsche Pakhuizen and the Oostzijdsche Pakhuizen. The watchtower of port authority called *de Uitkijk* stands tall not far away from the warehouse building mentioned earlier. The historic site of the V.O.C warehouse complex includes the remaining town wall of the Batavian *bastide*. What has been from this whole of historic V.O.C artifacts is the story of the spice trade and Dutch V.O.C in Old Batavia or Old Jakarta.

The striking question of the present for the past V.O.C heritage buildings is why Indonesia as a nation-state transforms those artifacts into museums. The museum is conceptually a house of memory. For what kind of memory are the V.O.C buildings?

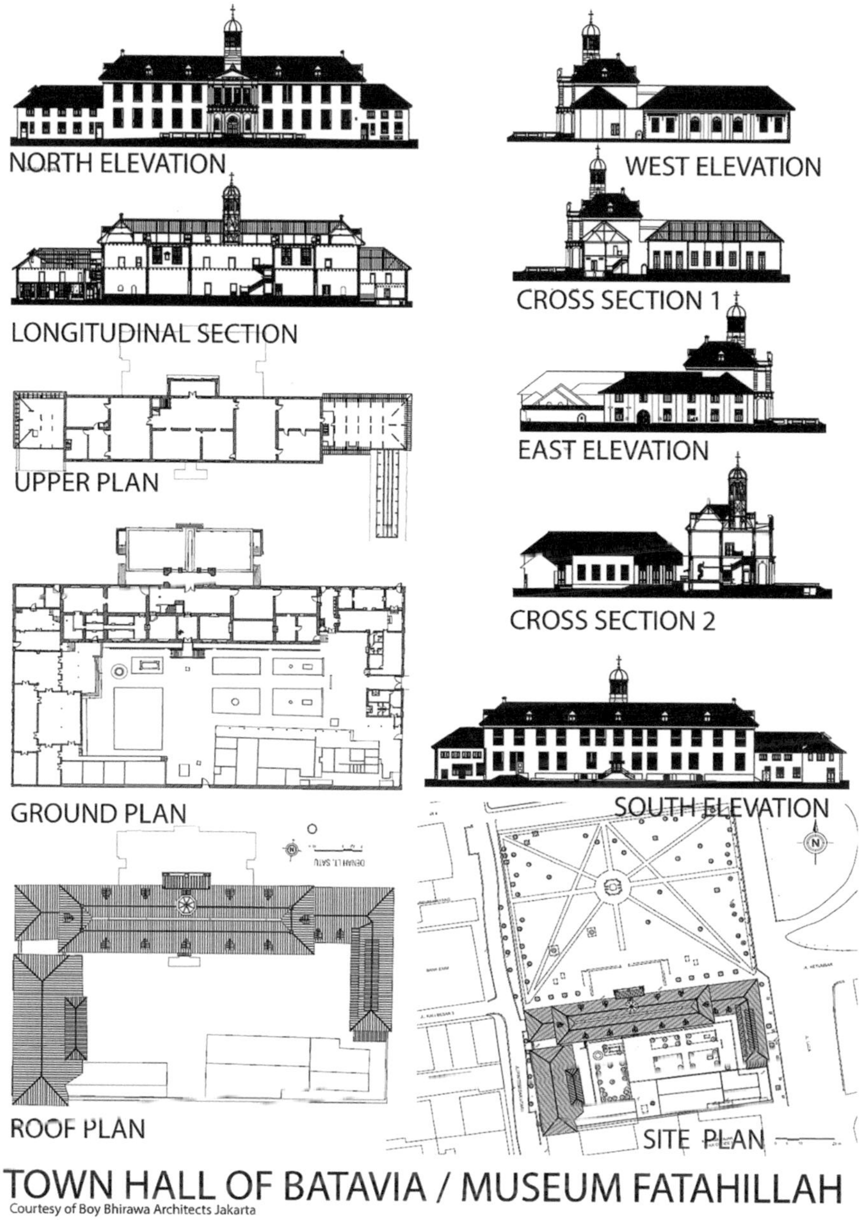

Fig. 4.3 Scheme of Town Hall of Batavia

In the case of the Maritime Museum, the transformation is not simply for the practical reasons that the V.O.C artifacts have been installed and stored in the buildings and its open spaces. Rather, the transformation is ideologically a cultural process of historical understanding. During the Sukarno (1945–1967) and Suharto (1967–1998) era, the intention of programmatic transformation from the warehouse to the museum was an integrated part of the nation-building process of Indonesia as a nation-state. The V.O.C legacy in terms of architecture and its artifacts inside has been utilized as the background of local maritime pride from the glorious ancient kingdoms of Srivijaya and Majapahit. Contradictory relationship between form and content undermines the ideology of the Maritime Museum as a learning institution from its very foundation. The presence of V.O.C in the past was erasable from Indonesian history, based on whatever values and perspectives people see and judge. Subconsciously, the integration of maritime artifacts of Indonesia, Portuguese, Chinese, Japanese, and Dutch into the former V.O.C warehouses is the act of acculturation that matters for the sustainability of memory from the past to present. Under the notion of acculturation, the mission and function of the museum become loud and clear as the educational resource of learning from the multicultural heritage.

Despite its vulnerability to flooding and environmental pollution, the Old Town district of Jakarta has been one of the most attractive tourist destinations in the country. Restoration and revitalization of its built environment and infrastructure are necessary as its integral part of historic preservation and conservation; the necessity for such programs is not simply for tourism. Rather, the program of historic preservation and conservation is the generator of sustainability of local arts and culture as well as its supporting communities.

4.4 Concluding Remarks

Colonialization is the concept of domination over otherness in various aspects and levels of existence. The notion of colonial architecture seems to be impartial from the state of affairs and events of domination over other's aesthetics and lifestyle. Domination in this sense is not just only about the control of resources—land, people, and capital—but also concerning aesthetics and culture. Colonial architecture does not speak about quantitative domination, but it is more about the qualitative supremacy of Western civilization over indigenous and other traditions. From a historical perspective, the Dutch colonial architecture was not simply the product of colonial domination. Rather, it was the means and representation of modern urbanism over the traditional settlement system, comprising politics, economy, arts, and culture. In other words, colonial architecture in the Dutch East Indies was a cultural necessity of modernity.

Despite its collateral violence and exploitation, Indonesia inherited from the Dutch colonialism the territorial boundary and political definition of what the country is about (Oostindie in Oostindie ed. 2008, 68). Indeed, the Dutch colonial architecture is not simply historical artifacts. Rather, the former Dutch buildings provide an infrastructural systems of modern governance, public service institutions, utilities, and educational facilities. The end of Dutch colonialization in East Indies archipelago did not mean the end of modern urbanism. Even under the Pacific War between 1942 and 1945, the Japanese occupation authority utilized and made the most institutions from the Dutch colonial rule in the same function and service.

The Dutch colonial rule introduced to the local states in Indonesia the concept and practice of bureaucracy that separate the household from the workplace. Bureaucracy replaces the power relation in the organization based on personal loyalty with impersonal rule and written rules and records. The birth of architectural institutions originates from the expansion of a patrimonial environment that establishes the dualistic system in urbanism: public and domestic realm. Modern bureaucracy sets patrimonial business and state of affairs as an independent system from family household matters. In other words, bureaucracy does not only change the cultural scripts but also a transformation of system in the how and where of the organization (Collins in Adams and Charrad, eds. 2011, 19).

The Dutch architectural heritage has granted most cities and towns in the Indonesian archipelago with modern institutions and spatial structures. Typologically in Java, the spatial division and land use of the Post-V.O.C colonial urbanism superseded traditional and indigenous form and structure with a rational system of communication/transportation and sanitary infrastructure. This case includes Jakarta, Surabaya, Bandung, Semarang, Yogyakarta, and Surakarta. In many ways, the modern approach of Dutch colonial urbanism secluded European settlements from the native kampung habitations. The interface between the European and indigenous settlements was the port, commercial and business areas. Most other Asian inhabitants, such as Indian, Chinese, and Arab populations used to live in commercial districts. Morphologically speaking, the logic of geometrical network and order had led to the structure and pattern of European quarters. Inside the lot system, the buildings were juxtaposed with freestanding structure while the commercial areas followed the logic of the supply-distribution function with a linear building block system. At the margin of the architecturally planned area was the domain of indigenous settlements with the irregular configuration of form and space.

References

Abayesekere S (1987) Jakarta: a history. Oxford University Press, Singapore

Barrenveld D (2009) Henry Hudson and the rise of fall of New Amsterdam. Lulu.com.

Blussé L (1986) Strange company: Chinese settlers, Mestizo women and the Dutch in V.O.C. Batavia. Foris, Dordrecht

Blussé L (2013) Port cities of Southeast Asia 1400–1800. In: Clark P (ed) The Oxford handbook of the cities in the world history. Oxford University Press, Oxford, pp 356–362

Borchardt N (2014) Growing up in the VOC Batavia; transcultural childhood in the world of the Dutch India Company. In: Jabowski C, Max T (eds) Childhood and emotion: across cultures 1450–1800. Routledge, London, pp 42–56

Bosma U, Raben R (2008) Being "Dutch" in the Indies: a history of creolisation and empire, 1500–1920. National University of Singapore Press, Singapore
Burnett I (2013) East Indies. Rosenberg, Kenhurst
Castles L (1967) An ethnic profile of Jakarta, Indonesia III, pp 153–204
Chaudhuri KN (1965) The English East India company: the study of an early joint-stock company 1600–1640. Taylor & Francis, London
Collins R (2011) Patrimonial alliances and failures of state penetration, vol 636. In: Adams J, Charrad MM (eds) Patrimonial power in the modern world. The Annals of American Academy of Political and Social Science, Washington DC, pp 16–32
Dinas Museum DKI Jakarta Raya (1983) Sejarah Singkat Gedung-gedung Tua Pemda DKI Jakarta. Pemda DKI Jakarta, Jakarta
Drakley S (2005) The history of Indonesia. ABC CLO, Westport
Facebook (n.d.) Kota Tua Jakarta: calendar of events. https://www.facebook.com/notes/lwg-dmo-kotatua-jakarta/calender-events-kotatua-jakarta-2014-update-412014/264821590341999. Accessed 2 Mar 2015
Fasseur C (1992) The politics of colonial exploitation: Java, The Dutch, and the cultivation system (Studies on Southeast Asia) (Elson RE, Kraaf A, eds; trans: Elson RE, Kraaf A). SEAP Publication Cornell University, Ithaca, NY
Goor J, van Goor F (2004) Prelude to colonialism: The Dutch in Asia. Uitgeverij Verloren, Hilversum
Great Britain Public Record Office 1878 (1878) Calendar of state papers of colonial series, East Indies, China and Japan 1513. Longman, New York
Hanna WA (1988) Hikayat Jakarta. Yayasan Obor Indonesia, Jakarta
Harris AS (2005) Seventeenth-century art and architecture. Laurence King, London
Heuken A (1982) Historical sites of Jakarta. Times International, Singapore
Heuken A (2002) The earliest Portuguese sources for the history of Jakarta: including all other historical documents from the 5th to the 16th centuries. Yayasan Cipta Loka Caraka, Jakarta
Hilton RH (1995) English and French towns in feudal society: a comparative study. Cambridge University Press, Cambridge
Hugo G (2006) International labour migration and the family: some observations from Indonesia. Asian Pac Migr, 273–301
Iguchi M (2015) Java essay: the history and culture of a southern country. Matador, Leicestershire
Kehoe M (2015) Dutch Batavia: exposing the hierarchy of the Dutch Colonial City. J Hist Neth Arts 7:1
Kuhn PA (2009) Chinese among others: emigration in modern times. Rowman & Littlefield, Lanham
Kusno A (2012) Jakarta's City Hall: a political history. In: Chattopadhyay S, White J (eds) City halls and civic materialism, towards a global history of public space. Routledge, London, pp 255–275
Lach DF, Van Kley EJ (1998) Asia in the making of Europe, Vol. II. A century of advance: Southeast Asia. Book 3. University of Chicago Press, Chicago
Leirissa RZ (1995) Sunda Kelapa sebagai bandar jalur sutra: kumpulan makalah diskusi. Departemen Pendidikan dan Kebudayaan, Direktorat Jenderal Kebudayaan, Direktorat Sejarah dan Nilai Tradisional, Proyek Inventarisasi dan Dokumentasi Sejarah Nasional, Jakarta
Lommerse H (2008) Populations figure. In: Oostindie G (ed) Dutch colonialism, migration and cultural heritage: past and present. Brill, Leiden, pp 315–316
Morris AEJ (1972/2013) History of urban form before the industrial revolution. Routledge, London, New York
Ooi KG (2004) Southeast Asia: a historical Encyclopedia, from Angkor Wat to East Timor. ABC CLIO, St Barbara CA
Oostindie G (2008) Historical memory and national canons. In: Oosindie (ed) Dutch Colonialism, migration and cultural heritage: past and present. Brill, Leiden, pp 63–96
Parmer NJ (2012) Historical perspectives of Chinese populations in Southeast Asia. In: Joselyn Amstrong M, Warwick Amstrong R, Mulliner K (eds) Chinese populations in contemporary Southeast Asian societies: identities, interdependence and international influence. Routledge, London, pp 18–54

Parthesius R (2010) Dutch ships in tropical waters: the development of the Dutch East. Amsterdam University Press, Amsterdam
Pires T, Cortesao A (2005) Suma oriental of Tome Pires and the book of Francisco Rodrigues (Cortesao A, ed), vols 1 & 2, 2 vols. Asian Educational Service, New Delhi
Price JL (2014) Dutch society 1588–1713. Routledge, London
Princeton Library (n.d.) Map of Batavia 1669. http://libweb5.princeton.edu/visual_materials/maps/websites/pacific/spice-islands/map-batavia-1669.jpg. Accessed 18 Mar 2015
Princeton University Library (2013) Map of Batavia 1669. Map of Batavia 1669. http://libweb5.princeton.edu/visual_materials/maps/websites/pacific/spice-islands/map-batavia-1669.jpg. Accessed 18 Mar 18 2015
Raffles STS (1817) The history of Java Volume I & II. John Murray, London
Randeraad N (1998) Mediators between state and society. Uitgeverij Verloren, Hilversum
Ricklefs MC (2001) A history of modern Indonesia Since c. 1200. Stanford University Press, Stanford
Rijksmuseum (1859) The Castle of Batavia, Andries Beeckman, c. 1661. https://www.rijksmuseum.nl/en/collection/SK-A-19. Accessed 4 Jan 2020
Saidi R (2010) Sejarah Jakarta dan peradaban Melayu-Betawi. Perkumpulan Renaisance Indonesia, Jakarta
Schefold R, Dekker V, de Jonge N (1991) Indonesia in focus: ancient traditions, modern times. Kegan Paul International, London
Selberg E (1840) Über die Vergangene und Gegenwärtige Lage der Insel Java: nach hollandischen und englischen Quellen und Eigener Anscahuung berabeitet. Literarisch-Artistiches Verlags-Institut, Ritteln, Leipzig
Surjomihardjo A (1977) Perkembangan Kota Jakarta. Dinas Museum Pemda DKI Jakarta Raya, Jakarta
Tan TS (2009) Cheng Ho and Islam in Southeast Asia. Institute of Southeast Asian Studies, Singapore
Taylor JG (2003) Indonesia: peoples and histories. Yale University Press, New Haven Connecticut
van Vollenhoven C (1918) Het Adatrecht van Nederlandsch Indie, vol 3. Kolff, Leiden
Wade G (2007) Southeast Asia-China interactions: reprint of articles from the Journal of the Malaysian Branch, Royal Asiatic Society. Royal Asiatic Society of Great Britain and Ireland, Malaysian Branch, Kuala Lumpur
Ward K (2009) Networks of empire: forced migration in the Dutch East India Company. Cambridge University Press, Cambridge

Chapter 5
The Patrimonial Culture and Urbanism in Bandung Since 1800

Abstract This chapter concerns the relationship between the patrimonial politics and its spatial transformations that have led to urbanism in Java since the nineteenth century, with special reference to the city of Bandung, West Java Indonesia. The paper examines and unfolds the aspects of the patronage of morally superior over others in the context of society, culture, and urbanism. In this study, patrimonial culture is considered as historical and cultural root of power relation in Indonesia. The focus of the study is to explore and examine the evolution of this patrimonial tradition in the socioculturally constructed reality from the indigenous setting, colonial experience, and post-colonial practice.

Keywords Bandung · Public space · Political culture · West Java · Dutch colonialism

5.1 Place Memory Identity

The construction of settlement has been associated with the establishment of the center of the world or axis mundi since ancient times. Mircea Eliade observes and highlights the significance of axis mundi as the symbolic connectivity between the profane and the sacred, as well as between the mortals and the divines (Eliade 1959). Building the place of dwelling is nothing but setting up the center where the sense and experience of departure and arrival are defined and provided with reference to a structure or spatiality. Heidegger speaks of this sense and experience in terms of the gathering of the fourfold (Heidegger 1971, 147–9); this is the happening of the communion of mortals, divines, earth, and sky, the fourfold. The gathering finds a place that enables them to gather and fold into the site of the world for a sustainable future.

The urban settlement is the location where the act and form of construction find an essential connection to humanity because of its habitable spatiality in a permanent location. The phenomenon of center in term of axis mundi is to mark out and signify the establishment of connectivity between earth and sky as a place for the world of habitation where mortals and immortals dwell. In ancient times, the beginning of building construction is considered a sacred happening; it needs a ritual with offering

B. Wiryomartono, *Traditions and Transformations of Habitation in Indonesia*, https://doi.org/10.1007/978-981-15-3405-8_5

because it is to denote the birthday of a communion of people and location when joy and happiness are experienced. The collective memory of the event is established with the ritual of the commemoration of the even on a periodical basis.

Maurice Halbwachs (1877–1945) speaks of collective memory as a socially constructed notion that involves a specific group in drawing on the local context to remember or recreate the past as a collectively shared experience (Halbwachs 1992). The continuity of collective memory is indispensable from people who remember so that the communion of people, place, and event comes into being as the presence of the absence in the form of commemoration. The ritual and ceremony of commemoration are to endure and preserve the collective memory of a specific event. Such an event brings about the sense of belonging and communion among people who participate in the ritual of commemoration. In this commemoration, one participates with other persons in establishing with them the horizontal community that lies perpendicular to the vertical commonality that participants establish the *commemorandum* proper (Casey 2011, 184). The purpose of this commemoration is nothing but the construction of the communion between the commemorators and their common *commemorandum*—the commemorated thing or person.

The ritual and ceremony of commemoration are not simply in honoring the *commemorandum*. Rather, this is considered as the process of the historically constructed notion of Anderson's imagined community (Anderson 2006). The relationship between people and place is not simply constructed, developed, and sustained with daily habitation. Rather, this needs to thrive the sense of communion for the larger context of people concerning the care for what the *commemorandum* is for them. Urban society is not a village community; villagers know each other by the first name and by each other's business. Urban citizens do not know each other. Even though so, they feel to have something in common and share the same interest to live anonymously together in the proximate location we call it town or city. Anderson coins this group of populations as an imagined community. The relationship between urban dwellers and their locality is circumscribed by historically overlapping memories generation to generation in open-ended stories.

The city square is considered one of the public places that play an important role and function in building collective memory among urban dwellers. This is the place where people are anonymously able to hang out and visually interact with others without having to know each other's business. Theoretically speaking, such a public place is potentially the site of power relation (Bourdieu 1987, 1984, 1986; Bourdieu and Wacquant 1992). This study attempts to frame the public realm with Bourdieu's theory of habitus and social field. This is the site where social behaviors are constrained by social conditions of relation, position, disposition, production, and reproduction in terms of Bourdieu's habitus. In indonesia context, this behavioral sustem is known and understood as *adat*. The habitus takes place in the locally contextual platform, which is understood by Bourdieu as a social field; it is considered as the microcosmos of social practices that culture reproduces social structures. In this field, unequal power relations are unrecognized as such and hence accepted as legitimate. This is also the field where such relations are embedded in the system of classification of daily habitation.

5.2 Bandung in the Colonial Era

Data and related information concerning pre-colonial Bandung have been compiled and studied by a number of scholars. Since ancient times, the Sundanese populations in Bandung have been socially divided into two groups: the aristocrats, *menak* or *juragan* or *priyayi* and commoners or *kawula*. The interdependence between the ruling and ruled is circumscribed within the framework of the patron–client relationship. The ruling group is mostly associated with those who control access to resources, while the ruled is the group of people who do the business and work on resources. In traditional Sundanese tradition, the concept of society is problematic because the dualistic division between *priyayi* or *juragan/menak* and *kawula* group is historically established within the patrimonial system. This condition has been experienced as a common issue in most Indonesian communities, especially the Javanese group; until recently, the Javanese language has no word meaning society (Anderson 2006, 6). Vertical mobility in society becomes less motivated as a collective movement. This does not mean that there is no vertical mobility of individuals from the *kawula* group or peasants. The modern educational system and bureaucratic system enable this mobility. However, such individuals' mobility is considered for their own sake and interest and barely an integrated of a social movement. They might come to power at various companies and state institutions, but their role and function prescribe them to behave like the new *juragan/menak*. Sociologically speaking, the highly-stratified system in populations remains within the traditional and patrimonial system (Ardjo 2012, 398).

Geographically speaking, the terrain of Bandung is characterized by its highland climate. Prior to the establishment as a center of polity in the mid of the fifteenth century, Bandung was known as Tatar Ukur region, a territory of the Kadipaten of Timbanganten, a vassal Sundanese Kingdom of Pajajaran. One most well-known vassal King of Timbanganten was Dipati Ukur; under his reign, the kingdom's territory controlled most area of Priangan Highlands of West Java. The spread of Islamic influence of the Banten Kingdom and the internal crisis in the Pajajaran Kingdom brought about the Hindu vassal kingdoms in West Java in danger. Some of them adopted Islamic teaching and were under the influence of Banten and Cirebon Kingdom. Politically, Bandung was integrated as the vassal of Cirebon Sultanate. The consolidation of Bandung as a regency, *kabupaten*, took place within the period between 1579 and 1580.

5.3 Alun-Alun Bandung

City square or *alun-alun* area in Java has been always the place where the traditional destination of the urban center is spatially constructed and established. In Javanese tradition, the *alun-alun* is conceived as the architectural incorporation of *axis mundi* of polity and urban settlement, *kota*. Regarding this meaning, the establishment, reposition, rearrangement, and readjustment have been never impartial from who is at the apex of patrimonial hegemony of this town or city. Under the Dutch colonial rule, the idea of the city has been dispersed and deconstructed from spiritually symbolic center to daily functional place.

City square in Javanese urban tradition is the site where the ritual and ceremony of commemoration take place. This place is where the notion of center and *axis mundi* of urban settlement is perceived. This site is left empty in daily life. In Bandung, the city square or alun-alun has been historically associated with the center of polity, and by its territoriality the sense of departure from and arrival to the city is socially and geographically defined. The indigenous construction of the city square is derived from the ancient Javanese idea of the center. Accordingly, the square is conceived as the spatially mediating structure between political and religious center; *alun-alun* is the spatially meeting place between the regency, *kabupaten,* and mosque, *masjid.* The establishment of city square is symbolically to mediate, negotiate, and synergize the political power and socioreligious authority. This meeting place represents the Bourdieu's power field between the traditional ruler, *bupati* and the Islamic cleric, *ustadz* or *kiyayi*. The indigenous courtesy refines and polishes the struggle and contestation between both authorities their dominant influence on populations. Traditionally speaking, this is not simply about economic, cultural, and social capital. Rather, it concerns the symbolic capital for pride, *marwah*; this is the pride of being somebody who is highly regarded because of his/her vanity—*kasohor*—for having done something worthy for his/her community.

For the elite Sundanese populations, *marwah* is regarded as the highest achievement in life. For this, they mobilize their economic, cultural, and social resources to get the acknowledgment from whole populations concerning this pride. State rituals and ceremonies are nothing without the struggle to achieve vanity that is expected to build a legacy that makes his dynastic generations and followers to be proud of such an achievement. The actualization of this achievement includes the construction of the monument, *candi* or other architectural structure, such as regent palace, *kabupaten* or *masjid agung* or *alun-alun.* The construction of the alun-alun began under the regent Wiranatakusumah II between 1794 and 1829 (Pemerintah Kabupaten Bandung 2013). The Dutch Governor-General Herman Willem Daendels (1808–1811) put the city square of Bandung on the 1000 km Post Way between Anyer in West coast to Panarukan in East coast of Java Island.

The history of this square has been inseparable from the establishment of its polity under the Dutch colonial rule since 1811, shortly before the Dutch colonial rule moved some of their railway company to this city in 1894 (Yong 1973, 4). The connectivity of the city to Batavia had played an important role in the growth of Bandung's European populations. In 1917, the Dutch colonial administration moved their Department of Defense and followed with the Department of Communication in 1923. Prior to the Pacific World War, Bandung was one of the most important cities in the Dutch East Indies colony. The construction of *alun-alun* signifies the commencement of the urban settlement in the location. The area of this city square is determined after the highest area for the regent's residence has been located.

The square is located at the north direction of the location for the regent palace; on this site, the establishment of the regent of Bandung was symbolically represented by the construction of its palace and mosque. The structure of the urban core is derived from the Javanese concept of *mancapat* for cardinal cosmological orientation (van Ossenbrurggen 1977). In urbanism, the Hindu–Islamic–Javanese idea of the cosmological center is distinguished from the Balinese concept of *nawa sanga mandala* by their position of the market place and the landmark of the center. For the Javanese, market place, *pasar* or *peken*, is not necessarily located as an integrated part of the center, while its center is signified with the city square.

The Hindu–Balinese idea of the urban center is actualized with the intersection of *kaja-kelod* and *kauh-kangin* streets. Despite the Javanese and Balinese spatial and cosmological orientation has been conceived as the traditionally imperative model of spatial layout in terms of the *mandala*, the role and function of agencies in the implementation and exploration have been less unfolded in various cases (Hauser-Schäublin 2004, 285; Schulte-Nordholt 1999). This study attempts to divulge the relationship between the agencies and the designs of the public square in the context of Bandung urban center.

Wiranatakusumah II initiated and led the mobilization of the regency from Krapyak in Citeureup to today location (Ekajati et al. 1985); he built the hall on the highest area of the site as the first building in the location. The center area of Bandung is located at the site with an open view to Mounts of Tangkuban Perahu and Burangrang on the North. The eastern area of the center is left unbuilt for reserved green open space; this area ends up with the River Cikapundung. The North area, *Bancey*, of alun-alun, was dedicated to the place for the stalls of horses. The regent palace compound consists of several buildings: the front hall, *pendapa*, the main building, *dalem*, and the supporting building.

After the construction of the regent palace, they continued to build mosque and market place, *pasar*; both edifices were constructed on the western area of the center area. In this constellation, the mosque, *masjid agung*, was located in the north area of the city square, while the market was on the north side of the reception hall, *pendapa*. The architecture of the buildings was characterized with steep roof form covered by palm fibers material. The mosque stood out with three-tiered roofs structure that reminds people to the Chinese Buddhist pagoda or the Balinese Hindu *meru* shrine. The buildings of the ancient *pasar* of Bandung were constructed with steep gable roof form made of palm fibers covering material. The structure and form of the urban center had been completed by Wiranatakusumah II shortly before the construction of the Daendels' Post Way came to Bandung area in circa 1812. The development of the urban center was followed with the growth of *kampung* settlements around the regent palace and mosque.

The city square of Bandung has been always the social field of West Javanese patron–client system since pre-colonial times. The patronage of indigenous elites is displayed on the city square with the rituals and ceremony of *seren taun*; literally, *seren* is the word devotion or consecration for thankfulness, while *taun* means year. Practically speaking, this is the annual rice harvest celebration and communal meal of Sundanese peasant communities. The origin of this feast is from Pre-Islamic

West Java since the Tarumanegara Kingdom. The symbolic form of these rituals and ceremony is to deliver the harvest from the peasants to their patron through the hand of the priest. In return, the patron will give the peasants the seeds of rice for the next planting season. Most villages in West Java perform this celebration as the moment of unity between the patron—*menak*—and the commoners—*kawula*. The patron becomes the sponsor of the feast, while the commoners participate in the event. The feast of *seren taun* had been adopted as state rituals and ceremonies of kingdoms of Pakuan and Pajajaran.

Traditionally, the patron of state and village sponsored the rituals and ceremonies with a communal meal and ritual accessories. The high priest used to lead the rituals and ceremony in open public place, while the king or village chief used to act as the inspector or supervisor of the whole performance overseeing the spectators from high ground. The ceremony of *seren taun* used to take place in the city square in front of the reception hall, *pendapa* or *paseban* of the ruler's residence, while the celebration of the communal meal took place in the city square with various entertaining attractions. The other rituals used to take place in the city square were the ritual of *ngarot* or *ngaruat jagat*. This ritual is performed mostly as communal care—*salametan*—for the well-being of the village and the world. Long dry seasons, recovery from natural disasters, and dealing with epidemic diseases are mostly the reason to perform the ritual of *ngarot* in the public square. Offering domesticated animals for the rituals and ceremony of *ngarot* used to be part of this event.

Throughout history, the alun-alun of Bandung has evolved since the nineteenth century with various designs and performances. The development of the square had begun with the transformation of Bandung from a traditional regent town to a regional center of colonial plantation industry. The beginning of Dutch colonial intervention in restructuring the city was developed by the Dutch colonial rule since 1810. The construction of the Post Way or *Postweg* was continued with the development of railways from Batavia to Bandung circa in 1880.

The imposition of European sports and other social activities into the indigenous city square was made possible with the patronage of the Dutch colonial rule. It seems for a practical reason because the city square is daily empty space. The Dutch do not see the symbolic significance of the city square because of its emptiness until the care—*pangeling*—for cosmological unity calls for the spiritual gathering—*ngariung*—of whole people for the safe, healthy, and prosperous world. The sense of this gathering is signified with rituals and ceremony that underscores the importance of wishful thinking—*eling*—and gratitude to the source of life.

The evolution of designs and constructions of the Bandung's city square goes on until today. The design and construction have been always associated with the sign and signature of Bandung's chief officer. During the Dutch colonial rule, the city center area had been kept as a symbolic representation of indigenous authority. The Dutch colonialism applied indirect ruling politics to manage local affairs through the power of indigenous rulers (Furnivall 1956/2014). In the West Javanese context, this Dutch indirect rule was applied at the highest level of administration only, while at the intermediate and lower levels were employed (Schrader 1997, 263).

In the nineteenth century, Bandung was a regency of the Cirebon Sultanate. During the Daendels' era, the Dutch colonial rule, Bandung was regarded as a remote West Java place with scenic landscapes and mountains with a pleasant atmosphere. Economically speaking, Bandung and its Priangan highlands had been explored by the Dutch V.O.C as the region for coffee plantation. During the British interlude, the attention to Bandung and its region was less than to Bogor area; Raffles developed a tropical botanical garden in this town, on which her wife—Olivia Mariamne Devenish—was buried on November 26, 1814, with a memorial monument.

Since the beginning of its urbanism, the city square of Bandung has been the object of politics that concerns the presence of the ruling person of the city. The transformations include the design of the mosque and its surrounding. Despite historical changes, the former indigenous regent's residence—*kabupaten*—remains in the same form and position. Only parts of this building complex have been renovated and remodeled but still in compliance with its original design. The building complex comprises three traditional domains: *paseban*, *tengah*, and *dalem*. The front hall—*paseban*—is dedicated as formal reception place, while the middle and rear part are for the office and a living area of the regent—*bupati*.

5.4 The Politics of Indigeneity and Colonialism

In Java, the British Interlude introduced the concept of land as a private property to indigenous people. Raffles introduced the land-rent system in Java in 1813; it was a system of taxation adapted to a monetary economy with predominant agriculture (Klaveren 2013, 89). Merchants and bankers were to deliver cash crops and raw materials to the ports for export to the European market. Raffles was fascinated with the way the indigenous people in their market, especially the role of female merchants. The British treated the indigenous and other non-European traders equally, while under the Dutch, regardless their skills in commerce, rural Javanese entrepreneurs who tried to expand their business had to experience unequal partnership with the Chinese (Schrader 1997, 262). However, the British Interlude had only less successful in engaging indigenous traders with the market; it was because of the stagnation in society, the absence of the dynamic capitalist sector, and the deficit of administration (Schrader 1997, 262).

Unlike the monopoly of trade practiced by the Dutch V.O.C, the British colonialism applied territorially based control of the indirect administrative system to collect revenue from people based on land registration; the local rulers acted as the collector for their own people from city to village level. The Dutch colonial rule implemented the British administrative system with slight modification in the way they collected revenue. The system has been still working until today known in West Java as *girik*. Based on this system, the Dutch colonial administration worked together with the indigenous rulers to lend the land for private investors for growing and developing plantations and other private estates. After the end of the British Interlude 1816, there were vacant lands in Java that were conquered by the Dutch since the seventeenth century, had been leased, sold, and granted to private persons. These estates were privately run by their owners or their representatives so that the Dutch colonial administration had neither influence in nor control over these domains (Schulte-Nordholt 1999, 57).

5.5 Bandung and Compulsory Cultivation System

Prior to the formal implementation of the cultivation system, the indigenous ruler of Bandung and Priangan region—Wiranatakusumah IV—renovated and developed the great mosque of Bandung with three-tiered roof system in circa 1826. The construction was to provide the mosque with a front hall and two towers with the wall surrounding the area made of the terracotta structure. The architecture of the mosque had been well documented in the collection of the Amsterdam Tropical Museum Collection dated from 1929. As a whole, the architecture of the mosque was constructed with a similar idea of form and spatial structure with the great mosque of Demak, Cirebon, and Banten in the seventeenth century. The Dutch colonial regime in the mid of nineteenth century did not intervene on the architecture of the mosque. However, the public square was not left without use as a public amenity of the Dutch populations.

Priangan highlands of West Java are suitable areas for tea, coffee, quinine, and rubber cultivation. Under the Governor-General Johannes van den Bosch, the Dutch Indies was financially in mess at the end of Java war in 1830. In order to cover the financial shortcoming, van den Bosch introduced the cultivation system in Java. As a former officer of the V.O.C and as the owner of a private estate, he was familiar with the way of making a profit from the agricultural resources, and then, he came to the idea of trade monopoly and compulsory cultivation.

Van den Bosch saw the economic advantage of agrarian productivity of Javanese slaves in Suriname West Indies 1827. Profit-taking from free labor crossed his mind for implementing compulsory cultivation in Java. Peasant populations were due to plant certain part of their land for export commodity plants. He treated the Javanese village heads as foremen of the compulsory cultivation system. Fasseur (1992/1994) described in details how this compulsory cultivation in Java worked from 1840 to 1860. This system was designed by van den Bosch with indirect rule and in the partnership with Dutch private enterprises. This is the reason why the cultivation as such did not exist in any regulatory document applying in Dutch East Indies. The indigenous rulers were instructed to manage the implementation of compulsory cultivation on their own people; the Dutch administration did not interfere in the distribution of burdens concerning the imposition of the cultivation system among the village inhabitants (Fasseur 1992/1994, 29).

Moreover, the cultivation system in Java invaded mostly the land for wet rice cultivation, *sawah*; this implies the Dutch colonial rule did not only exploited the free labor and the productive land of Javanese agriculture but also violated the land right of indigenous populations. The issue concerning this right was brought up by van Vollenhoven with his extensive work on customary laws in Indonesia; the work—*Adatrechtbundels*—consists of 40 volumes and 3600 pages (Otto and Pompe 1989, 242). According to van Vollenhoven, Javanese village is not only the political and spiritual center of Javanese community but also the center of the agrarian area, which is protected with customary law—*adat* or *tata cara*.

During the 1830s, the Priangan areas around Bandung attracted European investors for plantation business. The Daendels' Post Way became an important infrastructure that was not only as of the line of communication with Batavia and other places in Java but also as a symbolic dividing line between the European and the indigenous settlement. Since 1820s, most European settlers built their settlement at the northern part of the Post Way. The plantation estates and their business drove urban development of Bandung as European settlement with the unique township, in terms of political administration and urban design. Politically speaking, Bandung was symbolically under dualistic administration with two city centers. The indigenous center is represented by the traditional Javanese city square—*alun-alun*—with *kabupaten* and *masjid* with kampung settlement on the southern area of the Javanese Post Way, while European central settlement was concentrated at the north area. The image of urban zone remains until today that the north of Bandung is associated with a formally planned community, while the south of the Asia–Africa street or former De Groote Postweg is related to *kampung* settlement.

After the compulsory cultivation was ended in 1870, Bandung areas had been the destination of European private investment for plantations and urban services. The Dutch colonial administration established the regency of Bandung comprising areas within 60 km from the city at the beginning of the twentieth century. The concentration of European settlers in Bandung was growing in number and diversity. The European settlement in Bandung had reached its maturity as an urban center in the 1930s. By 1930, the urban inhabitants of Bandung were 166,815 (Ingleson 2014, 12; Volkstelling 1936, 122–3); 11.8% of them were Europeans and 77.1% indigenous populations.

By 1870, the Dutch colonial rule introduced the Agrarian Law that implemented the liberal policy for European investors in Java and Sumatra for plantation estates. This law inaugurated the claim of wastelands under Dutch control. Furthermore, the law granted the plantation estates by the private sector based on lease from indigenous authorities. Under the Dutch colonial regime, the European investors and planters insistently advocated a change in the Agrarian Law to permit them to be the owners of the property that had been granted to them or which they had rented (Robequain 2001, 28). In order to support the labor of cultivation, the Dutch colonial rule issued the Coolie Ordinance—*Koolie Ordonantie*—in 1880; the purpose of this legislation was to set up condition of workforce and its mobilization in favor to the European employers over the indigenous labor force (Houben and Thomas Lindblad 1999, 13–5; Kat Angelino 1931/2012, 508–9).

The European settlers and the business owners of tea, coffee, quinine plantations in Lembang, Patengan-Rancabali, and Malabar-Pangalengan areas contributed to the urbanism of Bandung; they established urban societies and enriched the architecture of the city with European modern buildings within Arts Deco movement. By the end of the nineteenth century, urban development began to grow with the establishment of the railways and telephone network between the city and Batavia. During the Daendels' governorship, several indigenous regencies—*kabupaten*—in Java were administratively grouped into Dutch residencies—*karesidenan*. West Java highlands region was known as *Preanger Regentschappen* or *Karesidenan Parahyangan* consisting of regencies of Cianjur, Bandung, Sumedang, Ciamis, Garut, and Tasikmalaya. The seat of Dutch regent was originally in Cianjur, then moved to Bandung in 1864 after the volcanic eruption of Mount Salak in the same year. Each Dutch regency was supervised by a Dutch resident, while each indigenous regency was supervised by a Dutch assistant regent. The territorial control under the indigenous *kabupaten* was district—*kawedanan*—that consisted of several villages—*desa*; the Dutch supervisor at the district level was known as *controleur*.

At the end of 1903, the Dutch Parliament enacted the *Locale Radordonanntie* for the Dutch Indies colony. This law regulated the decentralization of authority of the Governor-General to regions and municipalities to take care of their domestic affairs and programs for the implementation of Ethical Policy (Schmutzer 1977, 20). In the following years after 1903, several big cities in East Indies were administratively consolidated with regional and municipal governance status. Each region and municipality was provided with a regional or municipal council, consisting of representatives from various associations and districts; some were assigned by the Dutch administration, some were elected from communities. The role and function of this council were an advisory body to the local Dutch administrator for domestic affairs, public health, and education.

By 1906, Bandung was granted by the Dutch colonial administration as a municipality, *gemeente*. However, the seat of Bandung regency remained in the same city. The formative municipality entailed the concentration of new urban institutions, such as the new town hall, public park, cultural institutions, and amenities. All these were constructed between 1885 and 1930. Based on this development, Bandung's administrative center was territorially divided. The indigenous authority was represented by the traditional regency center at the *alun-alun* area. Meanwhile, the center of the modern municipality was developed around the today Taman Merdeka area; this site was chosen by the Dutch as a new urban center with European concept in the tropical colony in terms of *Tuinstad* within the influence of Ebenezer Howard's Garden City Movement in England. This area was called Pieter Shijthoff Park. The name was derived from the assistant resident Pieter Shijthoff who was in office between 1885 and 1898; he mobilized plantation owners and business people in Bandung and Priangan areas to build the city as a Dutch tropical urban center.

The mobilization of the plantation was followed with the establishment of Bandung societies, one of which was known as *Vereeniging tot Nut van Bandoeng en Omstreken*. The other society that was initiated by Shjithoff was *Toneelvereiniging* Braga; a club of moviegoers and appreciators of cinematic culture. The prominent members of this society included business people, scientists, medical doctors, and lawyers. In dealing with urbanism, Bandung's philanthropy mentioned earlier was to develop the city with tropical gardens. The integration of such an idea had been implemented by the Dutch architect Pieter Berlage for the southern extension of

Amsterdam's plan in 1906. This plan had inspired the members of Bandung municipality in 1910 to develop this city without being strictly associated with the old concentric city center at the *alun-alun* area.

The approach to the new development of European settlement in Bandung was to establish several subdivisions with parks as their spatial orientation and identity. There are several parks that characterize the city of Bandung with variety of tropical trees and flowers, such as Pieter Sijthoffpark or today Taman Merdeka, Ijzermanpark or today Taman Ganeca, Molukkenpark or today Taman Maluku, Insulindepark or today Taman Nusantara/Taman Lalu Lintas, and Jubileumpark or today Taman Sari or Kebon Binatang. Most of the public buildings from 1910 to 1942 were designed to integrate its architectural structure with scenic outdoor for a monumental view. For the Dutch architects, such as Charles Prospher Wolff Schoemaker, van Gallen, Albers, Gerber, and Maclaine Pont, Bandung was the site of their architectural experimentation of modernity in tropical setting; their legacy becomes landmarks of the city until today, such as Villa Isola by Schoemaker, Gedung Merdeka or Societeit de Concordia by van Gallen, Homann hotel by Albers, Gedung Sate by Gerber, and Aula of Institut Teknologi Bandung by Maclaine Pont (Fig. 5.1).

Bandung in the period between 1900 and 1942 was the best place to stay in the Dutch East Indies; Bandung is geographically in the interior of West Java. The location was considered as a strategically beneficial for defense against overseas invasion. Unofficially speaking, the Dutch colonial administration had a plan to move their headquarters from Batavia to Bandung in 1920s (Colombijn 2013, 25). Indeed,

Fig. 5.1 West Aula of Institut Teknologi Bandung, designed by Mclaint Pont 1920. Photograph by author

several departments had been relocated to Bandung, such as the Army headquarters, Public Works, and Telecommunication. The relocation brought about Bandung as a vibrant city with European populations. During weekends, the Braga street became the promenade of Bandung's urbanism with restaurants, cafes, and cinemas. The elites of the European populations gathered in a clubhouse called Sociëteit de Concordia. This place was provided with ballroom, bars, and restaurant. At home, European families enjoyed gramophone music player and radio; by 1910, most of the European homes on North Bandung and plantation areas were served with electricity and line phones. The radio set was sold in music stores and piano shops and considered as a kind of a new knickknack—a novelty of furniture—without which no modern house was thought to be complete (Mrazek 2002, 166).

By 1920s, Bandung was also a city where the young indigenous generations were educated within the Dutch system for medical and technical fields. The establishment of Bandung technical college in 1920 was the milestone of Bandung for its reputation as the source of local engineers and scientists in the Dutch East Indies colony. Many of European populations, who lived and worked in those periods, felt at home in Bandung; this city was regarded as the Dutch tropical Parijs van Java. Among prominent Europeans who were well known lived and worked in Bandung were Boscha and Junghuhn. K. A. R. Boscha was one of many who inherited the city and region with the astronomical observatory building in Lembang. Boscha was a tea plantation magnate who owned the business in Malabar area; Boscha's telescope observatory, even before it began to work, was hailed as a monumental of Dutch culture in this land (Mrazek 1999, 43–44).

Regarding its pleasant climate, Bandung did not only attract European Business people but also scientists and Engineers. Franz Wilhelm Junghuhn (1809–1864) was a German botanist and surgeon. He was enlisted as a Dutch colonial army as a medical officer. He arrived in Batavia in 1835 and explored the landscape of the Bandung area and backed to the Netherlands in 1849 for medical treatment. Junghuhn returned to Java in 1855. He introduced to Java the plant of chinchona from Peru. Despite the controversy around its process, Junghuhn managed to grow chinchona in Lembang. He and his friend, J. E. de Vrij, were able to establish a quinine laboratory in Bandung. The success story of quinine from Bandung was achieved by what had been started by Junghuhn since 1856; by 1930, 97% of quinine of the globe came from Java (Bosma 2016, 192).

In 1901, Queen Wilhemina delivered the speech from the throne concerning the Ethical Policy for the colony, especially the Dutch East Indies. This speech was not something from thin air but the follow-up and response to the political debates in public media and in the Dutch Parliament on the exploitative of the cultivation system and the impact of indirect rule of the Dutch colonialism on indigenous people. The publication of *Max Havelaar* and *Een Eereschuld*, —a novel and a journal article of *de Gids* by van Deventer in 1860 and 1899—, brought up the issue of social injustice in the Dutch East Indies; he argued that the Netherlands owed to indigenous people in East Indies for all wealth, which had been drained from their land (Ricklefs 1991, 183). The patrimonial Ethical Policy was implemented with the massive development of infrastructure for transportation, irrigation, public health, and education.

For Bandung areas, the Ethical Policy brought about the improvement of educational facilities for indigenous civil servants (*Opleidingscholen voor inlandische ambtenaaren*, OSVIA) and engineering school (*Technische Hoogeschool*). Needless to say, the Dutch schools were only eligible for the children of indigenous elites—*priyayi* or *menak*. During the 1920s, Bandung was the city where the well-educated indigenous people were in a clash with the Dutch colonial administration and their traditional indigenous authority. Most young indigenous students inspired by the nationalist and liberal movement as well as socialist democratic ideology. Some of them worked together in various social and political groups with strong affection against colonialism and patrimonial culture. In anticipation and response to the movement, the Dutch colonial rule amended the 1854 Criminal Code with the rights to banish any political activities against authority and government in 1919 (Ingleson 2014, 24–5). Members of sociopolitical organizations such as *Sarekat Islam, Indische Social-Democratische Vereeniging,* and *Partai Komunis Indonesia* were under the eye of the Dutch secret police service, PID; *Sarekat Islam* held the first national congress in Bandung 1918 (Yong 1973, 29).

The tension between the plantation business owners and indigenous labors came into public attention because most of the workers were suspiciously registered as members or followers of *Sarekat Islam*. Internally within the *Sarekat*, they did not have a final word or resolution of their conflicting goals between the improvement of welfare and struggle for a political agenda against the establishment. In Bandung areas by 1922, most leaders of *Sarekat Islam*—Gunawan, Mohammad Sanusi, Subakat, and Sugono—were indicted by the local Dutch newspaper *Preanger Bode* as extremists and communists (Yong 1973, 30). During the implementation of the Ethical Policy, the State Law 1854 and the Criminal Code were amended to make an exception for the public meetings for the election of the members of municipal councils and *Volksraad*. Even though some indigenous populations were elected as members of the council, they did not have any political power than as advisory voice. On the contrary, the interest groups of prominent European personalities played an important patrimonial role in shaping and developing urban forms and activities in the city of Bandung.

Architecturally speaking, the urban fabrics and street network of Bandung had been enlarged and elaborated with modern infrastructure for electricity, telephone, and drainage system. The traditional city square was left without public activities. Most European inhabitants came to the city square only for a special occasion when the baseball match took place. The street sides of the Braga area were the main destinations of casual and social interactions for the Europeans. How did the Dutch colonial regime think on the land they encroached for the development of European settlement in the Bandung area? Did they the same way what they did to the conquered lands around Batavia under the V.O.C regime?

In the eighteenth century Batavia, the Company seized the indigenous kingdom and took over the territorially conquered lands between Cirebon and Bantam Sultanate, controlled, rented, and utilized those areas for their need or sold the lands to private persons (Taylor 2004, 228). The relationship between Dutch colonialization and the legal status of land had not been put in question until Cornelis van

Vollenhoven (1874–1933)—a Leiden University professor of law—brought up the necessity for respect to the locality in terms of customary laws in East Indies instead of the European/Western attitude of cultural superiority over indigenous people and tradition (Otto and Pompe 1989, 243). Indeed, the Dutch colonial regime in the early twentieth century East Indies was not the Dutch V.O.C. However, the massive University of Leiden's study on *Adatsrechts* under van Vollenhoven was politically considered as a detrimental element of economic interest of the Dutch.

In response to the Leiden's school of thought in favor of the rights of indigenous people on land, the Dutch government asked for alternative and counter-attack opinion from the Utrecht University's faculty J. C. Nolst Trenite' with his *Domeinnota* in 1916, who asserted the superior right of the Dutch colonial administration to administer and allow the lands of Indonesia for development by outsiders (Burns 2007, 75). Respect to local customs and traditions in terms of *beschickingsrecht* had been advocated by members of the Volksraad in 1925 with further progress by 1930 within the Agrarian Commission that the courts of the Dutch East Indies would recognize indigenous rights. Despite their greater chance for the indigenous rights of land on the underdeveloped and uncultivated areas, the Adatrecht project came to a dead end. All this was, for the most parts, because the idea of Van Vollenhoven was principally to treat the rights of indigenous populations that led to his position to endorse the indigenous independence as a nation with their own. This is nothing but a move against the Dutch colonialism in Indonesia.

Van Vollenhoven speaks of *beschickingsrecht* or the rights of allocation as the translation of the Indonesian concept of *hak ullayat* (Burns 2007, 84). In modern Indonesia, the right has been well established as the supplement to the Agrarian Law 1960, paragraph 5 article 3. The right has been established with the decree of the Ministry of Land Development and the Head of National Land Agency no. 5/1999. The dormant issue on the right of apportionment is the living community that upholds the right still dependent upon the resources from the land. This condition makes this right in perils for its validity because the concept of dependency is, from legal perspective, interpretative uncertainty (Wallace 2008, 205).

5.6 Bandung Until the End of the Pacific War

During the colonial era, the livability of urbanism in Bandung had been sociopolitically shifted from the indigenous activities in *alun-alun* and *pasar*—marketplace—to the streets of Braga areas with various parks in North Bandung. Parks and boulevards became important public realms of Bandung in the period between 1920 and 1942. The presence of indigenous public realms had been treated by the Dutch colonial rule as the other world but under the control of their patrimonial supervision. The policy of racially segregated domains was applied with access restrictions and dress codes. The native, Chinese, Indian, and other Asian populations were due to dress with their traditional and native outfits. The dress code was enacted for the access to Dutch domains; within the framework of Ethical Policy, the children of native aristocrats—*priyayi*—to attend the Dutch schools were allowed to attend the Dutch schools with their local costumes. Gradually, the policy on dress code was not strictly regulated especially for college students and Eurasians in 1930s. However, modern urban amenities and entertainments such as clubhouses, cafes, and bar, restaurants, theaters, and cinemas, as well as Churches, parks, and swimming pools were mostly restricted domains for indigenous and other Asian populations.

Despite some restrictions and under observations of public meetings, the nationalist movement in Bandung had been growing stronger with the leadership of Sukarno, who was trained as a civil engineer at the *Technische Hoogeschool* in Bandung between 1921 and 1926. Since 1920s, Bandung as other cities in Java such as Batavia, Yogyakarta, Surabaya, and Semarang had been the centers of politically emerging youth organizations. Needless to say, the native Indonesian students in Bandung were socially from the *priyayi* class. This was in the hope that the Dutch colonial regime would get the cooperative tie and support from the indigenous elites in maintaining their indirect rule and control over the East Indies as their colony. Under the Dutch Ethical Policy, the children of ordinary people did not have access to go into the Dutch school system; in Java, most of them went to Islamic boarding school, *pondok pesatren*. In this traditional school, the learning process is to master the recitation of the Holy Qur'an and the teaching of Islam with less critical thinking.

The sympathy of indigenous people to Islamic teaching has been historically constructed through a fair trade and respectful treatment of Moslem traders since the sixteenth century. The Dutch colonial rule under the Governor-General Van der Capellen ordered an inquiry to integrate the Javanese Islamic school with the Dutch educational system in 1819 from village to urban scale; however, this plan had never actualized and successfully carried out until the implementation of the Ethical Policy in the early of the twentieth century (Steenbrink 2006, 85–6).

The nationalist movement was inspired by the indiscriminative humanism from Islam and Western texts, especially Greek philosophy and European Enlightenment writings, as well as Marxism and Leninism. Sukarno, Hatta, and Syahrir were familiar with such texts; these native students were motivated by the ideas of the national independence of Ki Hadjar Dewantara and the Dutch Edward Douwes Dekker since the movement of *Boedi Oetomo* in 1908 led by Soetomo, Wahidin Soedirohusodo, Goenawan Mangoenkoesoemo, and Soeradji. The trial and arrest of Sukarno and other nationalist leaders in 1929 had put Bandung in the map of a historical national movement toward Indonesia's independence. The outcome of the Ethical Policy was significant to lead the young generations of indigenous populations to literacy. Most young educated people in Dutch East Indies colony experienced the inefficacy of the Ethical Policy for the improvement of indigenous welfare and their health, but they were aware of the importance of literacy for the future development of their history as a nation with one language and similar cultural root. In the 1920 census, 943,000 native populations were literate in Indonesian language and 87,000 in Dutch language (Avonius 2015, 140). The ultimate of the youth movements was the announcement of the Youth Pledge or Soempah Pemuda on October 28, 1928, in Batavia.

The opening of the Dutch educational system for the native people was the beginning of the social interaction between the Western and indigenous values system.

Classrooms and study clubs became the platforms of both sides to learn each other's culture. Despite the patronage of European superiority, such interactions built mutual understanding and unconditionally shared values of meritocracy and fairness. For the native people, the Dutch educational system was a systematic training process for analytical and critical thinking, as well as building a train of thought based on the Dutch language. In the urban context, such social interactions brought about a new process of enculturation in architecture, culinary, literature, and political culture.

5.7 Post-colonial Bandung

After the end of the Pacific War in 1945, Urbanism in Bandung had experienced a period of consolidation for its administration and livability as the place for stay and work. The land ownership in the city and region had been administered with a similar way of the Dutch colonial rule. The administrative status of Bandung had been granted by the President of the Republic of Indonesia as the capital of West Java Province, the seat of Bandung regency, and the municipality. The public offices take place in former Dutch colonial agencies with similar function and service. In private sectors, several plantations in the region had been taken over by the state-owned companies with a similar role and functionality. The consolidation of urbanism in the first decade of the republican era had achieved its culmination in 1955 when Bandung had been chosen as the host of the international conference of 29 Asian–African countries. The conference was held from April 17–24, 1955, and attended by the newly emerging leaders of the world that included Sukarno of Indonesia, Chou En-Lai of China, Gamal Abdul Nasser of Egypt, Jawaharlal Nehru of India, Josip Broz Tito of Yugoslavia, U Nu of Burma, and Kwame Nkrumah of Ghana. Richard Wright, a Western journalist, observed the event of the conference in 1955 as a meeting of almost all human race living in the geopolitical center of gravity of the earth (Wright 1956, 12).

Prior to the event in 1955, Sukarno, the first President of Indonesia, instructed the mayor of Bandung—Raden Enoch—to remodel the great mosque of Bandung; the president initiated to change the design of the roof system from traditional form with the three-layered structure to an onion form with the traditional base structure. With this form, Sukarno wanted to impress the delegates of the Asian–African countries that Islam in Indonesia had been part of the modern world; traditional form had been perceived by Sukarno as a sign of weakness and indifferent statement to the politically transforming world in Asia and Africa from the politically and economically colonialized countries to the independent nation-states. Sukarno had paid less attention to the traditional public square, alun-alun; he just ordered to tidy up the area and put the sidewalk around the square. The onion roof was massively damaged by the severe storms in 1967. In 1979, this mosque was rebuilt again with a traditional roof with a three-tiered structure in 1970.

Fig. 5.2 Urban scene at the corner of alun-alun and Jalan Asia Afrika (former Groote Postweg), Bandung 2015. Photograph by author

The post-colonial era is the historical period when the exercise of power in Indonesian urbanism has been characteristically dealing with the interplay between modernity and tradition. The obvious stage of this interplay is the architecture of the city. Public buildings and open spaces become the object of representation and corporealization how the ruling regimes interpret themselves as the authority of the public realm. Given the political atmosphere of the country in favor of the Army's leadership under Suharto, the major of Bandung between 1966 and 1968, Raden Didi Djukardi, was accused as a politically sympathizing person of Indonesian Communist Party; public opinion accused the Communist as the organization behind the September 30, 1965, movement that was accused of the main initiator and actor of coup d'état against the nation-state of Indonesia (Fig. 5.2).

5.8 The Joglo Mosque and Suharto's New Order

After having fallen in oblivion because of economic struggle since 1965, the idea of restructuring the center area of Bandung was conceptualized with Governor Decree in 1971. The West Java governor—Major General Solichin Gautama Purwanegara or Mang Ihin—took charge to reinforce the commitment of the New Order to Muslim community after the 1965 affairs; he set up a development team and instructed to begin the construction in 1972. The support of Suharto's administration was to actualize this commitment with the remodeling and expansion of the great mosque in Bandung. Under this program, the architecture of the mosque was transformed from its traditional form to modern urban facility. At the same time, the seat of Bandung regent had been moved from the city to Bale Endah in Soreang town in 1973. The old complex of regent's residence was transferred to be the municipal residence of Bandung major.

During 1971, preparation for the remodeling of Bandung city center took place under the supervision of the West Java governorship. The focus of this development was the transformation of the mosque. This was an integrated part of the New Order political movement to embrace Islamic component in the national platform; Suharto, like Sukarno, had been well known as a national leader with strong attachment with

Javanese spirituality—*kejawen*. The total area of the mosque was expanded with upper floor and basement. At the beginning of the 1970s, the New Order consolidated the power in various fields that included arts and architecture in the search for indigenous identity. The design team of the mosque chose and unveiled the Javanese *joglo* form that recalled to the historical belonging of Bandung to the grandeur of Banten and Cirebon Sultanate in the sixteenth century. Even though the 1973 *joglo* form was adopted from the Javanese architecture, its roof form did not architecturally follow the Javanese building type for a sacred edifice; the form of 1973 Bandung mosque was 'limasan' form instead of 'tajug'. Most of mosques in Java used to apply steep pyramidal *tajug* form instead of the *joglo* one; the first roof is characterized by a pointy apex to the sky, while the second is architecturally constructed with a line peak.

On October 1, 1973, the mosque was publicly reopened by the Governor of West Java Province with a ceremony in commemorating the emerging of the New Order under Suharto's leadership. During the time, the popularity of *joglo* form for public buildings was a portion of the favorable architecture of the ruling regime. The massive sponsorship for the use of *joglo* form had been demonstrated in most new public offices and facilities in Java and other islands.

The monumental work for the search of local identity was the Miniature Theme Park of Beautiful Indonesia Project, *Proyek Taman Mini Indonesia Indah*; the park was the idea of Madame Tien Suharto after her visit to Disneyland. Criticism against the project was not about the idea but the timing and fund. The project was to be financed partly by regional governments from who the First Lady solicited voluntarily sponsorships. The strong opposition to the project was ideologically associated with the anti-Sukarno group who was critical to Suharto's New Order regime with its monopolistic capitalism; most of them were the journalists, students, and lecturers, most of them the contributors of the *Journal of Mahasiswa Indonesia*, and anticommunist Army officers from Siliwangi Division of West Java such as HR Dharsono and Sudhandi (Hefner 2011, 75–79). The climax of this opposition was the January 15–16, 1974, affairs or 'Peristiwa Malari' on the occasion of Japanese Prime Minister Tanaka's visit to Jakarta. This was a fierce demonstration after seven years the New Order in power. Nine demonstrators were killed in this event.

The Governor of West Java took over the renovation and remodeling of the great mosque of Bandung as a sign of support to the project of Suharto's wife, Taman Mini Indonesia Indah. The adoption of the *joglo* form for the mosque confirmed the position of West Javanese governor, Major General Solichin, for his loyalty to Suharto's New Order regime; he was not in line with the Army Chief of Siliwangi Division, H. R. Dharsono, who was the dissident of Suharto. Dharsono was critical for the despotic leadership of Suharto who used to crush their political opponents with the use of state ideology of Pancasila as a coercive instrument against political contenders (Vatikiotis 2013, 140).

The reason why the 1973 mosque of Bandung had been architecturally profiled with a joglo form was not by accidence and impartial from the fact that during the period between 1971 and 1975, the New Order was politically in crisis and called for support and loyalty from regional governments. The West Java government stood up to this call with a joglo form of the mosque. With this architectural sign, West Java under Solichin G. P. was in line with Suharto who was a Javanese patron with personal preference for the use of the Javanese architectural joglo form as a sign and symbol of Javanese patrimony and dominance; only Javanese dignitaries are traditionally eligible to build their front hall, *pendapa*, for their residence and office with joglo form.

By 1970s, the design of the great mosque of Bandung was distinguished from the previous ones, the traditional and the onion roof, because of its administrative status that was not under the municipality but within the provincial administration. From an urban design perspective, the design embraced the public square as its spatial extension. The connection between the mosque and the square was provided with a bridge structure crossing over Dalem Kaum street underneath. The monumentality of the mosque architecture was visually created by the colossal stairs that led people from the square to the upper floor of the mosque. A 36 m high and circular minaret was erected at the northern side of the staircase that enhanced this grandeur.

From the front upper floor of the mosque, people would have a theatric view of the square; the stairs provided spectators with the row of staging seats. In the following years after the completion of the mosque, the improvement of the square had been made that transformed the square from its traditional emptiness to a modern urban park with amenities, benches, ornamental lighting poles, fountains, and flower boxes. The image of the 1970s public square of Bandung became a destination of casual social interactions of people with local characteristics. The presence of street vendors and petty cash food stalls—*kakilima*—characterized the livability of the square with vibrant indigenous resources. This transformation had swept away the symbolic patrimonial relationship between the *kabupaten*, local indigenous residence, and the mosque. In other words, the alteration of public square design underscored the urban secularity of the public square and decreasing the symbolic authority of the traditional and religious institution.

5.9 Bandung and Post-Suharto Development

The end of New Order administration in Bandung was signified with the redevelopment of its city center area in 2001. The mosque was considered as the focus of this redevelopment with a radical transformation from the New Order's joglo form. The design involved the team of architects with native backgrounds. The main figures of this team included Ahmad Noe'man, Slamet Wirasonjaya, Keulman, and Arie Atmadibrata. With a total area of 8750 m^2, the mosque is designed to accommodate more than 12,000 prayers inside the building. The main idea of form is to integrate the architecture of the mosque in the global community of Muslims.

The mosques in the Middle East countries have been addressed as the referential image for this redevelopment. Three dome structures are composed in a symmetrical arrangement. The 30-m diameter dome stands in the centerline while both 25-m diameter domes at the left and right. Two 81-m-high minarets of the mosque are constructed in front of the mosque; the building stands with two dominant turrets on the public square. Architecturally speaking, the post-New Order mosque is historical without indigenous traces. Although the minarets of Bandung great mosque remind people to the turret of old Banten's mosque, as a whole composition, this West Java provincial great building adopts a Middle Eastern architecture with massive solid domes and flat roof system. The visual relationship between the mosque and the square is provided with front arcades that create a transparent façade with rhythmic opening arches and shades.

The further development of Bandung city center has been transformed to the next level with the improvement of its public square at the end of 2014. The surface of the square is covered by synthetic grass. The whole area of the square is made free of any structure that creates an open green surface. With this openness, people have ample open space for social interactions and leisure. Family with small children enjoys the artificially covered open space with safe and free from automobiles. This open space recalls the ancient idea of a traditional city square, *alun-alun*, with a different purpose. The post-New Order public square is intended to accommodate all people.

The square represents the center of a democratic openness for casual and social communications. What is missing from the design of this public realm is the historical connectivity between the residence of the local authority, *dalemkabupaten*, and the mosque. The mosque, in this sense, stands visually dominating and directing so that the axial orientation to the seat of the traditional authority has been faded away from the public square. Nevertheless, the presence of traditional authority is always there. The open space provides the traditional patron to exercise their patrimonial order and hegemony through rituals and ceremonies that take place in the square for special national and Islamic occasions.

Why the dome had been chosen for the great mosque of Bandung? The question leads us to understand Islam in the second millennium in Indonesia. Historically speaking, Islam had been understood as an integrated part of Indonesian culture since the sixteenth century. Mosque architecture in the archipelago had syncretized the teaching of Islam with local and indigenous resources. The mosques of Demak, Banten, and Cirebon have been the exemplary models of this syncretism in architectural form. The syncretic tradition in understanding and practicing the teaching of Islam has been surviving until today. After the fall of Suharto's New Order in 1989, various political ideologies have flourished within the framework of the democratic and tolerant platform under the sovereignty of law and order. From this historical setting, ideological radicalism and extremist find their way to grow at any level and area of activities. Islamic radicalism and fanaticism become extremely attractive for those who are spiritually insecure in dealing with modern global capitalism, material-hedonism, and consumerism.

The search for authenticity in the field of spirituality leads some Muslims in the country to Islam in the sixth century when the Prophet Muhammad S. A. W. was living and teaching this religion. Searching this authenticity moves more and more Muslims in Indonesia to adopt cultural traditions and customs of the land and people where Islam was originally taught, established, and developed. In short, the need to transfer of knowledge and practice of religious teaching from the first hand has been leading to adopt everything Arabic that includes outfits, manners, communication, comportments, and architecture.

The inclination to adopt the Middle Eastern culture and tradition has been understood by some Muslims in Indonesia as the conspicuous identity of being good Islamic followers. This identity provides people with a sense of community and security in the globally exposed world. Despite incompatible practicality here and there because of different background, history, climate, and situation, the adoption of Arabic culture and tradition superficially is considered as an easy way than cultural syncretism. The last is necessarily studied, tested, and experimented with a profound philosophical framework. The adoption of the dome architecture of the mosque in Bandung is not by accidence; this choice reflects the contemporary situation of Islam in Indonesia. To a certain extent, regarding the fact that Indonesia is a country with populous Muslims on the globe, keeping the hegemonic status quo of a patrimonial tradition for the elites is unavoidable from their affiliation with Islamic mainstream. The dynamic power relation between the moderate and fanatic Muslims has been always the main political business since the pre-colonial times. Until today, the syncretic culture has been able to overcome the extremist and radical elements of ideology.

5.10 Concluding Remarks

Dutch colonialism had utilized the locally rooted tradition of patrimonial authority. This practice is well known as an indirect rule. In doing so, the Dutch did not need to have a dirty hand in dealing with the exploitation of resources. Architecturally speaking, the urban form of Bandung shows how the Dutch employ the traditional layout of the urban center as a symbolic representation of local hegemony. However, the construction of the Groote Postweg or interregional way between West Java Anyer and East Java Panarukan crossing the city center of Bandung speaks loudly about the who is the boss in this country. Moreover, at the corner of the city center, the Dutch put and erected the prison building for the colonial dissidents and criminals.

The socioculturally constructed reality of urbanism in Bandung has been experienced as a historical journey of power relationships within the dualistic system of populations. This dualistic grouping consists of the patrons and the clients. The patron controls access to resources and supervises the process of productions and services. On the other hand, the clients who are traders, workers, peasants, or other ordinary people take charge to run the activities in society in terms of labors and services. This pattern of power relation has been culturally endured since pre-colonial times until today.

Democratic processes and transformations in the country since the Republican era have not yet totally changed the patron–client relationship in Indonesian society. Political and business elites have been expected more like the fathers of people than as the leaders of an egalitarian society. Rituals and ceremonies have been being used as a mechanism and practice for enduring patrimonial hegemony. Even though democratization and modern education enable the social mobilization of people from the clientele to the elites, these processes do not make any radical transformation of the patron–client system in populations. In subtlety, the elites and the political ruling group, *priyayi*, maintain their traditional role as trustee of populations, *panutan*, while the majority of populations is in the function of followers, *anak buah* or *pengikut*.

References

Anderson B (2006) Imagined communities, reflection on the origin and spread of nationalism. Verso, London, New York

Ardjo ID (2012) New Sundanese dance for new stages. In: Lindsay J (ed) Heirs to world culture: being Indonesian, 1950–1965. Brill, Leiden, pp 397–420

Avonius L (2015) Building nation and society in the 1920s Dutch East Indies. In: Schulz-Forberg H (ed) A global conceptual history of Asia, 1860–1940. Routledge, London, pp 129–148

Bosma U (2016) Franz Junghuhn's three dimensional and transcendental Java. In: Arnez M, Sarnowsky J (eds) The role of religions in the European perception of insular and mainland Southeast Asia. Cambridge Scholar Publishing, Newcastle upon Tyne, pp 175–206

Bourdieu P (1984) Distinction: a social critique of the judgement of taste (trans: Nice R). Harvard University Press, Cambridge, Massachusetts

Bourdieu P (1986) The form of capital. In: Richardson J (ed) Handbook of theory and research for the sociology of education (trans: Nice R). Greenwood, New York, pp 241–258

Bourdieu P (1987) In other words: essays towards a reflexive sociology (trans: Adamson M). Stanford University Press, Stanford, CA

Bourdieu P, Wacquant LJD (1992) An invitation to reflexive sociology. University of Chicago Press, Chicago IL

Burns P (2007) Custom, that is all before all law. In: Davidson J, Henley D (eds) The revival of tradition in Indonesian politics: the deployment of adat from colonialism to indigenism. Routledge, London, pp 68–86

Casey E (2011) From remembering: a phenomenological study. In: Olick JK, Vinitzky-Seroussi V, Levy D (eds) The collective memory reader. Oxford University Press, Oxford, UK, pp 184–187

Colombijn F (2013) Under construction: the politics of urban space and housing during the decolonization of Indonesia, 1930–1960. Brill, Leiden

Ekajati ES, Hardjasaputra S, Mardiana I (1985) Sejarah Kota Bandung, 1945–1979. Departemen Pendidikan dan Kebudayaan, Direktorat Sejarah dan Nilai Tradisional, Proyek Inventarisasi dan Dokumentasi Sejarah Nasional, Jakarta

Eliade M (1959) The sacred and the profane: the nature of religion (trans: Trask WR). A Harvest Book, Harcourt, Orlando, Austin, New York

Fasseur RA (1992/1994) The politics of colonial exploitation: Java, the Dutch and the cultivation system (trans: Elson RE, Krall A). Cornel SEAP, Ithaca, NY

Furnivall JS (1956/2014) Colonial policy and practice. Cambridge University Press, Cambridge

Halbwachs M (1992) Maurice Halbwachs on collective memory (trans: Coser LA). University of Chicago Press, Chicago, IL

Hauser-Schäublin B (2004) The politics of sacred space; Using conceptual models of space for socio-political transformations. Bijdragen tot de Taal-, Land- en Volkenkunde (KITLV) 160(2/3):283–314

Hefner RW (2011) Civil Islam: Muslims and democratization in Indonesia. Princeton University Press, Princeton

Heidegger M (1971) Poetry language thought (trans: Hosfstadter A). Harper, New York

Houben VJH, Thomas Lindblad J (1999) Coolie labour in colonial Indonesia: a study of labour relations in the outer islands, C. 1900–1940. Otto Harrassowitz Verla, Gottingen

Ingleson J (1930s) Workers, unions and politics: Indonesia in the 1920s and 1930s. Brill, Leiden
Kat Angelino ADA (1931/2012) Colonial policy: volume II the Dutch East Indies (trans: Renier GJ). Springer, Berlin
Klaveren NA (2013) The Dutch colonial system in the East Indies. Springer, Dordrecht
Mrazek R (1999) From darkness to light, optic of policing in late colonial Netherlands East Indies. In: Rafael VL (ed) Figures of criminality in Indonesia, the Philippines, and Colonial Vietnam. SEAP Cornell University, Ithaca NY, pp 23–46
Mrazek R (2002) Engineers of happy land: technology and nationalism in a colony. Princeton University Press, New York, Princeton, NJ
Otto JM, Pompe S (1989) The legal oriental connection. In: Otterspeer W (ed) Leiden oriental connections: 1850–1940. Brill, Leiden, pp 230–249
Pemerintah Kabupaten Bandung (2013) Sejarah Berdirinya Kabupaten Bandung. Dec 19. Accessed 12 Sept 2016. https://www.bandungkab.go.id/arsip/16/sejarah-berdirinya-kabupaten-bandung
Ricklefs MC (1991/2008) A history of modern Indonesia since 1200, 2nd edn. Palgrave MacMillan, London
Robequain C (2001) Plantations and scientific agriculture. In: Kratoska P (ed) South East Asia, colonial history: high imperialism (1890s–1930s). Taylor and Francis, London, pp 25–44
Schmutzer EJM (1977) Dutch colonial policy and the search for identity in Indonesia. Brill Archive, Leiden
Schrader H (1997) Changing financial landscapes in India and Indonesia: sociological aspects of monetization and market integration. LIT Verlag, Münster
Schulte-Nordholt H (1999) Temple and authority in South Bali, 1900–1980. In: Geertz H (ed) State and society in Bali; Historical, textual and anthropological approaches. KITLV Press, Leiden, pp 37–63
Steenbrink KA (2006) Dutch colonialism and Indonesian Islam: contacts and conflicts, 1596–1950 (trans: Steenbrink J, Jansen H). Rodopi, Amsterdam, New York
Taylor JG (2004) Batavia (Sunda Kelapa, Jacatra, Batavia, and Djakarta/Jakarta). In: Ooi KG (ed) Southeast Asia: a historical encyclopedia, from Angkor Wat to East Timor, vol 1. ABC-CLO, Santa Barbara, CA, pp 226–229
van Ossenbrurggen FDE (1977) Java's monca-pat; Origins of a primitive classification system. In: Josselin de Jong PE (ed) Structural anthropology in the Netherlands; A reader. Nijhoff KITLV translation series 17, The Hague, pp 32–60
Vatikiotis MRJ (2013) Indonesian politics under Suharto: the rise and fall of the new order. Routledge, London, New York
Volkstelling (1936) Volkstelling 1930, vol 18. Landsdrukkeri, Batavia
Wallace J (2008) Indonesian land law and administration. In: Lindsey T (ed) Indonesia: law and society. The Federation Press, ISEAS, Singapore, Hong Kong, pp 191–223
Wright R (1956) The color curtain: a report on the Bandung conference. World Publishing, New York
Yong MC (1973) Conflicts within the Prijaji world of the Parahyangan in West Java 1914 1927. Institute of Southeast Asian Studies, Singapore

Chapter 6
Urbanism and Ethnic Minority: Chinese Quarters and Society in Jakarta-Indonesia 1600–Present

Abstract The role and function of Chinese culture and society in the archipelago are apparently and undeniably important in terms of architecture, socioeconomic, and cultural activities, as well as culinary and leisure. This study argues that the significance of Chinese contribution to urbanism in Indonesia is crucially reliant upon its patrimonial culture and tradition. The aim of this study is to investigate the aspects and practices of patrimonial urbanism concerning the contribution of Chinese urban tradition and culture to urbanism in Indonesia today. This study is intended to divulge and unfold the contribution of Chinese urban tradition and culture to the development of Jakarta. Historically, the existence of the Chinese quarter has been part of the reality of urbanism in Indonesia under the patrimonial governance based on racially segregated politics. Despite its hardship of the process, the disappearance of the China town is a sign that the Chinese populations have been well integrated into the Indonesian society. However, it does not mean that patrimonial hegemony has been dissolved, but it has been transformed from divisive grouping to collusive exclusiveness.

Keywords Chinese quarters · Batavia · Dutch V.O.C · Colonial urbanism

6.1 Urban Quarters of Minority

This study is a historical exploration with critical analysis concerning the dynamics of power relationships between patrimonial regimes and groups that have shaped and developed the Chinese quarters and urbanism in Batavia from the seventeenth century until today. The research question is to find out why patrimonial culture has not able to deal with multicultural urbanism throughout the historical process. The question leads this study to investigate the issues, constraints, opportunities, and challenges of multicultural urbanism that unveil and promote social justice and civil liberty for all in Indonesia.

In this study, urbanism is understood as a system of ideas, forces, processes, and activities of a demographically concentrated settlement with paternalistic ordering body of governance; the core of is urban economy relies on the patron–client relationship between the ruling regime and the clientele. The practice of patrimonialism in

B. Wiryomartono, *Traditions and Transformations of Habitation in Indonesia*,
https://doi.org/10.1007/978-981-15-3405-8_6

the Indonesian archipelago had deeply rooted in the ancient social structure of Austronesian village system that consists of the landlords and the commoners. However, the dualistic social class system in Austronesian communities did not work within the agricultural production system but for sociopolitical activities within subsistent economy. In this sense, landlords were the oligarch elites of the community with their privilege in traditional knowledge and in the collective ownership of resources.

The practice of wet rice cultivation since circa the first century had transformed Austronesian villages in Java and Sumatra from subsistent to the surplus economic production of food. The development of this agricultural village system had led to the state organization. Moreover, surplus agricultural production had developed a more complex economic system. One important aspect of this system was the establishment of the go-between economic function that works as traders and distributors. However, the ruling group controls the access to resources while the traders run the businesses for the production and distribution of resources.

The patron–client partnership has been working as an informal relationship that establishes an oligarchic network and system of political economy in Java and Southeast Asia from ancient times until today (Eisenstadt and Roniger 1984, pp. 117–138). Patrimonialism has been a prevalent culture in Southeast Asia for centuries (Fukoaka 2015, pp. 88–89; Crouch 1979; Scott 1972; Brown 2003, p. 209; Bertrand 2013, pp. 28–29); in this region, the patrimonial culture historically has worked, organized, and developed urban societies and their settlements based on hierarchical top-down organization and patron–client ties between their leaders with mutuality of beneficial exchanges. The arrival of the Chinese traders and settlers since early thirteenth century (Gungwu 2004, pp. 331–334) and the European explorers, traders, and colonialists since the sixteenth century (Manguin 2000, p. 409) had developed a new constellation of power relation within the patrimonial culture in Indonesia that involved tensions, conflicts, negotiations, and collaborations in a multicultural context.

6.2 Patrimonialism and Chinese Settlers in Pre-colonial Java

The existence of Chinese settlers in Sunda Kelapa of West Java had begun with the search for the site of sugarcane cultivation since the fifteenth century. The area around the port of Sunda Kelapa had been chosen by the Chinese manufacturer. The sugarcane production did not grow swiftly until the European came to Java. Like many other Chinese settlers in Southeast Asia who came from mainland South China between fifteenth and the seventeenth century, the Chinese settlers in Sunda Kelapa had been traditionally organized in socioeconomically units or patrimonial guilds or simply gathered along the principle of brotherhood; they were mostly single men without protection of lineage, village, or state prompted organization along the principle of brotherhood (Ownby and Heidhues 1993, pp. 4–5). The nature of the organization is patrimonial that relies on top-down leadership and collectively oriented work and interest, fewer options for individual liberty, and loyalty to leader and organization In terms of comradeship. To a certain extent, such organizations seem to be surreptitious societies because of the exclusiveness of their customs, traditions, and activities, such as for rotating capitals, funeral rituals, and religious festivals, social and political movements.

Regarding their patrimonial tradition, dealing with Chinese settlers was made easy through their leaders. Local rulers used to work together with Chinese leaders for international trade matters; most indigenous chiefs did not care that much on the economy, but indigenous women did. Domestic exchange and marketplace had been traditionally part of indigenous women's domain, while international trading activities were a room for Chinese settlers. The native rulers used to provide land and location for Chinese settlers concerning their trading activities. Areas at the traditional market or port used to be the locations of Chinese settlement in Java called *pecinan*.

Most Chinese settlers used to practice their own customs, traditions, and rituals based on their culture, including gambling, prostitution, and entertainment. Predominant single male settlers commonly married with indigenous women, and built Chinese families and societies in their newly adopted land for generations. The patrimonial culture sustained Chinese customs and traditions with unique characteristics, manners, fashions, and styles in terms of architecture, culinary, literature, performing and visual arts, poetry, and philosophy.

Urban Chinese societies are distinguished from rural communities because of their awareness of exchange for goods and foods with added values. Trade for urban populations is not only skills for the exchange of things, but it involves and deals with the refinement, enhancement, improvements, and elaborations of products and services. The Chinese culture and tradition had established the necessity for trade and exchange as their integrated part of urbanism since more than centuries years ago prior to the arrival of the Europeans to Southeast Asia.

North coastal areas of Java were the friendly vicinities of landed immigrants and traders from China, India, and the Middle East. Even though multicultural people characterized seaports and estuaries of Javanese sea, Chinese populations stood out in number and concentration since early of the Ming Dynasty. From the late fourteenth century into the seventeenth century, the Southeast Asian trade for spices and goods from China expanded very swiftly. Important vicinities of Chinese migrants in Central Java are the coastal areas such as Bantam, Sunda Kelapa, Cirebon, Tegal, Semarang, Lasem, Tuban, and Surabaya.

The Chinese settlers arrived in Sunda Kelapa—the former and original name for Jakarta's port— in the early sixteenth century. At those times, Sunda Kelapa was the entrepot of Hindu–Buddhist kingdom of Pajajaran with its capital in Pakuan—now Bogor. According to Chinese records, the port had been well known as one of the important trading posts for spices and rice since the twelfth century. Unlike its neighboring seaports of Bantam and Cirebon, Sunda Kelapa was politically not under Islamic control until the arrival of the Portuguese in 1513. Sunda Kelapa's port authority—*shahbandar*—had been always in the hand of Chinese persons. The presence of Chinese populations and settlement had been part of a trading partner of ceramics, silk, and gems with the Pajajaran Kingdom. Opium was another important commodity from China to Java.

The Portuguese geographer: Tome Pires describes Sunda Kelapa in his *Summa Oriental* from his visit to the seaport in 1513. According to Pires' account, Sunda Kelapa was a Hindu–Buddhist Kingdom part of Pajajaran. The King of Sunda Kelapa was Prabu Sang Hyang Surawisesa, the Prince of the Pajajaran Kingdom or Prabu Anom. Pires's noted the indigenous authority with structural hierarchy; the dignitaries of the Sunda Kelapa Kingdom from Mantri Dalam, Tumenggung Sang Adipati, and Shahbandar. The follow up of the visit, the Portuguese sea power made a treaty of trade concession with Sunda Kelapa in 1522 (Gunn 2011, p. 165).

According to the treaty, the Portuguese were allowed to establish its lodge in Sunda Kelapa. The treaty was signed by Prabu Surawisesa of Sunda Kelapa Pajajaran and Captain Henrique Leme' and Captain Jorge de Albuquerque of Portuguese Seapower. In the treaty, the Kingdom of Sunda Kelapa and Portuguese sealed the deal of trade of spices. In the following year in 1523, the Portuguese had established a fortress at the estuary of Ciliwung River. The commemoration for the treaty was inscribed on the memorial stone—*padrao*—at the edge of the Ciliwung riverbank. The stone is now in the collection of National Museum of Indonesia in Jakarta.

The visit of Tome Pires opened the possibility of Sunda Kelapa to be part of the Portuguese route of the trade from Malacca to The East archipelago: Halmahera and Molucca. The presence of Portuguese in Sunda Kelapa made uncomfortable the sultanate of Cirebon Sharif Hidayatullah or Fatahillah—an Egyptian-born scholar and son in law of Pajajaran King Siliwangi. The free-trade status of Sunda Kelapa for Portuguese did not last long. Fatahillah founded his provincial kingdom in Bantam in 1522. He entrusted his son Hasanuddin to hold his reign in Bantam. Fatahillah asked the Sultan of Demak for help by sending his forces to Bantam in 1524–1525. In 1557, Fatahillah, his son Hasanuddin, and Demak's forces attacked and seized Sunda Kelapa and the Portuguese sea power in Sunda Kelapa; he declared the seaport and as an Islamic kingdom with a new name: Jayakarta meaning the place of victory. Fatahiilah named Surosowan as the vassal King of Bantam in Sunda Kelapa.

As the vassal of the Bantam's Sultanate, the kingship of Jayakarta maintained Sunda Kelapa as the seaport of West Javanese hinterland. Fatahillah restored the town and established Sunda Kelapa as the free-trade international port under Islamic control. The settlement at the estuary of the Ciliwung River became the main supporting habitation of the seaport. As the previous ruler of Pajajaran Sunda Kelapa—Surawisesa—Fatahillah arranged a customary agreement with various ethnic groups to live and work together in the port town. He left the trade matters in the port to Chinese *shahbandar* who was in charge of taxes and tolls of exchanges in the international market of Jayakarta.

The Chinese quarter had been part of the seaport activities concerning the supply of silk, ceramics, and opium for the native since the twelfth century. The Chinese group was granted by the King of Sunda Kelapa for their authority controlling and managing internal trades at the seaport. The title of the port authority was well known as *shahbandar* or lord of the seaport. The number of *shahbandar* was not only one, depending on the load and intensity of the port. In Malacca for instance—prior to the arrival of the Europeans—there were five *shahbandar*s working in different sections of the city. During the visit of the Portuguese navy to Sunda Kelapa, Tome Pires only reported one *shahbandar* supervising and managing the port activities. Originally, the concept of *shahbandar* was commonly practiced by the Persian ports prior to the arrival of Europeans in the Indian Ocean region (Hall 2010, p. 316). The main task of *shahbandar* was in charge of international trades with specific currency or gold, and customs duty for all export–import commodities. Periodically, the *shahbandar* gave gifts, tributes, and revenues to the King of Sunda Kelapa.

6.3 The Dutch V.O.C and Chinese Settlers

Like many other European countries, the Dutch expedition was in the search for the source of spices—pepper and cinnamon, and three highly desired trinity—clove, nutmeg, and mace from the Southeast Asian archipelago (Tarling 1999, p. 10). The Dutch fleet arrived in Sunda Kelapa on November 13, 1596. Cornelis de Houtman led the first Dutch expedition to Southeast Asia. De Houtman described the town of Jayakarta was inhabited by predominantly Sundanese-speaking populations who lived within a bamboo enclosure at the Ciliwung River (Taylor 2003, p. 3). Some years later, the Dutch fleet sailed back and forth from the Netherlands to the archipelago. At the end of 1610, the Jakarta Ruler Wijaya Krama and the V.O.C agent Jacques l'Hermite sealed the deal; the native chief allowed the Dutch to build a stone house within a walled compound on the land where the Chinese quarter at the southeast bank of Ciliwung River.

After several unsuccessful attempts to capture Malacca between 1606 and 1618, the V.O.C Jan Pieterzoon Coen decided to establish their headquarters in Java after they were able to seize Sunda Kelapa in 1619. Coen was well known as a strong patrimonial leader with pragmatic business interest; he had Chinese friends and relations during his business in Southeast Asia from 1606 to 1619. Unlike the Portuguese forces that only established their fortification in the area of the port and hired local people for their mercenaries and workers, the Dutch V.O.C captured the port and burnt down the indigenous town center of Jayakarta. The indigenous elites fled out of the town and several were killed. After the fall of Jayakarta to the Ditch V.O.C, there was nothing left in Jayakarta but a small number of Chinese populations, Mardijkers, and slaves.

Jan Pieterzoon Coen was the V.O.C first Commander who was in charge of the capture of Jayakarta in 1619. After his successful attack and capture of Jayakarta, the Herren XVII of the V.O.C directors in the Netherlands confirmed Jan Pieterzoon Coen as the first Governor-General in East Indies colony. Coen planned and designed the general idea of the capital town of the V.O.C in East Indies. He laid out the position and location of buildings and facilities as well as the infrastructure of the town with the help of his V.O.C engineers and officers. His dream was to project his hometown as a model of the town. He renamed the town from Jayakarta to Batavia. After having examined from various aspects, Jan Pieterzoon Coen decided to model the town of Batavia after the Low Countries' town such as Leiden. The most important work for Batavia urban settlement was to establish its fortification and canal system. The Castle of Batavia became the first priority of the construction. The V.O.C garrison was planned to be in the castle.

The problem of the town development was obvious that the V.O.C did not have sufficient skills workers. He could not expect such resources from the indigenous people because the V.O.C had flattened their settlement. Most native populations fled out of town after the capture of Jayakarta. Even though for the labors for the canal construction, the V.O.C had numerous slaves and Mardijkers of former Portuguese militia and workers, building the town fortifications and facilities needed special expertise and experience in carpentry, stonemasonry, and bricklayer works. Regarding their long-standing urban tradition and well-organized teamwork, the Chinese societies in town were approached by the V.O.C officers for cooperation. The V.O.C arranged and formalized the Chinese organizations and societies with a legitimate status for trade contracts and construction works. The leaders of Chinese societies were given names showing their rank and status like in the Company, such as *major*, *kapitan*, *lieutenant*, and secretary.

One important Chinese connection of Jan Pieterzoon Coen was So Bin Kong or Su Ming Kang, a Chinese entrepreneur and community leader; he was known by the Dutch people in Batavia as Ben Kong (Parmer 2012, p. 29). Coen appointed Ben Kong as the Chinese *kapitan* in Batavia a few days after the fall of Jayakarta in 1619. The V.O.C entrusted Chinese matters and affairs in town to Ben Kong; he was in charge for settling down disputes among Chinese populations, a collection of taxes and tolls, peace and order within the Chinese quarters. In return, Ben Kong was granted by the V.O.C the monopoly of gambling, opium traffics and distributions, prostitution, and trades with Chinese merchants from mainland China. For the construction work, Jan Pieterzoon Coen connection was Jan Kon who was by profession a carpenter (Parmer 2012, p. 30).

As a chief carpenter, Jan Kon was the man of connection for skilled workers from mainland South China; Chinese stonemasons, bricklayer workers, and carpenters were brought to Batavia during the period between 1619 and 1626. The Chinese contractors accommodated the workers in the barracks in the west area of the Batavian *bastide*. A few years later, they married indigenous women and slaves. In most cases, Chinese migrants worked as a coolie or *kuli* for their Chinese boss—*tauke*. Most of them did not survive too long because of various diseases.

Consuming opium was common among Chinese coolies that worked as a drug for their anguish and sickness. Only a few survived and were able to manage their flight; some of them were gradually integrated with the indigenous people and lived in the local settlements around Batavia. The construction of Batavia as the V.O.C town and headquarters in the East Indies was likely impossible without the support of Chinese populations. They were not only skillfully for commercial activities, but also well experienced in various trades and team works. Their patrimonial tradition was significant leverage to the V.O.C for working with the Chinese societies.

Chinese societies had representative elders sitting on the board of an association house called Kong Koan or Gongguan or Gongtang. Persons of the board are only male persons who are highly respected by their members because of their experience and expertise or just of comradeship. Profession, trade, and hobby are the common platforms of Chinese settlers for establishing their clubs, guilds, and associations. The purpose of the establishment of such organizations is to regulate their activities and to control the quality of their products and services. The councils met regularly either weekly, monthly, or annually according to their own statutes, codes, and internal needs.

Unlike the traditional way of building the urban settlement, the construction of Batavia was necessary to transform the landscape of area at the estuary of Ciliwung River from its natural environment to clear and open terrain; the Dutch approach to landscape was like to deal with a blank page of the paper. Site clearing means providing space for human possibilities based on mastery and territorial control. By clearing site from any tree, bush, and structure that is not in compliance with the white sheet or empty space, the site are conceived by the Dutch free from unwanted things and events in terms of savage nature. Historically speaking, the artificial environment had characterized the western approach to landscape design and development. It was since the ancient Greek times. Accordingly, the establishment and construction of the built environment were geometrically ordered and organized free from the wilderness of vegetations and scrubs. The fifteenth-century Renaissance architecture in Italy renewed the ancient Greek-Roman building design approach with geometrical proportion. The seventeenth-century European approach to landscape design and development was strongly influenced by Cartesian coordination of points, lines, and shapes. The purity of Platonic geometric form and mathematical spatial coordination lead western architectural approach to master the landscape with artificial design and construction. The peak of an artistic approach to landscape design and development was achieved in Europe and well known as Baroque art and architecture.

The capture of nature under human control is undoubtedly the reason for site clearing by flattening and cutting down all existing elements and structure on the landscape but the ground surface. In doing so, the environment is not at its wild and raw condition, but environmentally ready and open for human habitation. In contrast to the western way of building the site, indigenous approach to site development departs from the perception that nature and culture are necessary to synergize for complementary growth and development. Any transformation by humankind in dealing with nature needs a ritual and initiation that prevents and shortens shocks, abruptions, jeopardies, and gaps of cohabitation. In doing so, there is the necessity for communications and negotiations as well as adjustments and assimilations between humanity and landscape.

Sacrificing domesticated animals for a communal meal and feast belongs to the ritual and initiation of site development and building construction. According to native beliefs, site development is an act of changing the elements, structure, and function of habitation that includes tangible and intangible forces. The spirit of the place for indigenous beliefs is not simply a credo of site development and experience. Rather, the spirit of the location is concrete energy and atmosphere that plays an important role in building a position, orientation, and habitation. Rituals and feast along the building process are to signify regard, acknowledgment, and recognition of native people to the transcendence.

For the Dutch culture, the site clearing is a rational process of making a dominion—*herrschapij*— that includes inserting plans and designs according to the rational geometry. Canal and wall system for the town of Batavia rectifies the wilderness of nature with floods and the doubt of rational men against their fear of uncertainty in a foreign land. The origin of *bastide* design was not impartial from historic sites of Greek towns and the development of European firearms and cannons in the twelfth century. The form of the township is based on the *bastide* design of the medieval town architecture in the thirteenth century French (Hanlon 2009: 50). Such *bastide* design is commonly found in the plans of towns in Dutch Lowland Countries such as Leiden, Delft, and Gouda. Ideologically speaking, the establishment of *bastide* Batavia was more about a military sea trading post of the Dutch V.O.C frontier instead of a depiction of a geometrical ideal city in the middle of a huge landscape, such as Citta Fortezza Palmanova designed by Vicenzo Scamozzi from circa 1593. The buildings are architecturally constructed ass perimeter blocks that shape and reinforce the open structure of public space where street and pedestrian are. All buildings are arranged on the frontage street. Building inside the *bastide* is mostly two or three stories units with attic under gable roof; some are provided with go downs for storage.

By 1730s, the construction of Batavia had put most institutions in place (Abayesekere 1987, p. 38). The multiethnicity of Batavia characterized the populations either inside or outside the *bastide*. Only Dutch—other European—Japanese, and Christian Chinese lived inside the *bastide.* The area of outside the fortified town was the settlement for other populations, such as Chinese, Gujarati, Arab, Persian, and indigenous populations from Java, Makassar, Ambon, Sumatra, and other islands, including slaves and Mardikejrs—freed slaves and former Portuguese militia. The area of Batavia *bastide* was a rectangular form along with the flow of Ciliwung River from the South hinterland to North Java Sea. The river divides the area of the *bastide* on the west and east side with almost equal areas. The plan of the *bastide* is 1200 m across and 2250 m in length. Unlike its original intention of a *bastide*, the Batavian walled town was not a farming community with diverse populations' ethnicity. Architecturally speaking, the *bastide* of Batavia was provided with a fortified island called Castle of Batavia which was situated at the northwestern side of the *bastide*. The core area of the *bastide* was at the intersection of major streets on the east side of the *bastide*. The core was mainly characterized by town square, town hall, and main Church. The best documentation of the scene of Batavia's town square in 1770 was engraved by Danish painter and V.O.C officers: Johannes Rach (1720–1783).

Town development with canal system and canal dredging work had been familiar for Chinese builders since the Ming Dynasty (1368–1644). When Jan Pieterzoon Coen appointed Chinese contractors to do the canal system, their problem was not only to bring the skilled workers from mainland South China to Batavia but also to provide them the space for their settlement; most of them were young and single men. They were provided by the Chinese contractor barracks in the southwestern and western areas outside the *bastide*. Gradually, the Chinese men left their barracks after they were married to indigenous women. Some of them moved to the Chinese sugar plantation areas which were located in the southern area of the *bastide*. The development of the Chinese quarters followed their patrimonial tradition.

Unlike the irregular settlements outside the *bastide,* the habitation inside the Batavia's wall was a well-organized territory with a well-structured layout of streets and blocks within the gridiron system. The grid system enabled the V.O.C security guards overseeing the activities on the street and mobility for social and trading activities along the canals and streets. The best areas for such socialization and exchanges were at the Ciliwung Riverbanks and Tigers Gracht. The scene is to be seen in the engraving by Johannes Rach 1682—Atlas van der Hagen Deel 4, inv. nr. 1049B13_022, a collection of National Library of Indonesia. Tigers Gracht was one of the inner canals where prominent members of Dutch Batavian society lived. The canal and both side streets established a promenade of Batavia. Not far and away from Tigers Gracht, by 1740 there was a residential block of 2500 Chinese houses (Ricklefs 2008, p. 112); they lived at the west area of the *bastide* behind the Grand River blocks. During the 1740 incidence of Chinese massacre, most of their houses were burnt down.

The patrimonial relationship between the Dutch V.O.C officers and the Chinese leaders had been never always smoothly going because of many reasons. The vibrant urbanism of Batavia was achieved between 1700 and 1730. One important fact is that the Chinese society in Batavia outnumbered the Dutch and European populations in 1730s; in 1733, the Chinese settlers were about 80,000 whilst the Westerners were only less than 3000. By 1790, three dominant groups in Batavia consisted of Chinese, Indigenous, and slaves population with a total of 107,400 from total 135,000 (Hoerder 2002, p. 183), whilst the Dutch/European and Eurasian populations were only less than 6000 populations.

The massacre of not less than 10,000 Chinese populations in 1740 was probably the anticlimax of the vibrant metropolitan Batavia (Cribb 2000, p. 92). The livability of Batavia was reliant upon exchanges and exporting activities of commodities from East Indies to Europe, China, India, and other parts of Asia. The decline of the East Indies jewel port town was not only circumscribed by the socioeconomic and political situation in Europe and Asia, but also by natural disasters, epidemic diseases, and internal power struggles within the Dutch Indies Company. All these had led the V.O.C to the end of its trading and military hegemony in Southeast Asia in 1799. The consequence of the V.O.C's bankruptcy was crucial for the well-being of Batavia. Travelers and visitors of Batavia at the end of the nineteenth century, such as Weitzel, Couperus, and Tavernier depicted the town in gradual demise (Blusse 2012, p. 66), from well-being to illness and death.

The Chinese quarters outside the *bastide* were not planned as the supporting natural environment that upheld and prevented the town from flood and the uncontrollable influx of migrants from hinterlands or other islands. From the onset, the *ommenlanden* or the surrounding areas of the town of Batavia was left out to the Chinese populations, slaves, former Portuguese militia—*mardijkers*—indigenous Javanese, and Balinese populations. The surrounding areas were not planned as farming land such as it was commonly found in French *bastide* or other medieval towns in the

Low Land countries. Jan Pieterzoon Coen did not plan and design the surrounding areas of the Batavian *bastide* but left it to the Chinese *kapitan* for sugar plantations and settlements.

Between 1619 and 1740, the surrounding areas of the *bastide* were gradually developed as a highly concentrated population of Chinese migrants and indigenous settlers from hinterlands and other islands. The way of Chinese populations built, developed, and inhabited the surrounding areas were indivisible from their urban tradition in the mainland. Even though the area outside the bastide was architecturally not well organized, the fate of Batavia's economy during the half until the end of the seventeenth century was sealed by the sugarcane cultivation in these outer districts (Ward 2009, p. 98).

As mentioned earlier, the Chinese settlers had involved in various urban development, productions, and services. One important economic activity of the Chinese in Batavia was in the sugarcane cultivation and production. The V.O.C was interested in the export of sugar to the Netherlands. Java, especially Batavia was one of the important producers of sugar in Southeast Asia. The V.O.C was reliant on the *kapitan* of Chinese settlers for sugar business. The collateral manufacturing product of sugar factory was the alcoholic beverage—*arak*. During the establishment of Batavia, the demand of *arak* increased due to the need for export and the V.O.C sailors and soldiers as well as visitors and Chinese people in town. The growth of populations between 1619 and 1740 increased the demand and production of sugar and *arak* in Batavia. By 1730, there were at least 89 factories producing sugar in Batavia; 84 of them belonged and were run by Chinese businessmen (Matsuyama 2013, pp. 259–60).

The township of Batavia was not like other townships in the Netherlands. The town hall of Batavia was not the center of administration and public services. The town hall was mostly for the function of the courthouse. The administration of Batavia was run by the V.O.C senior officers in the Castle of Batavia; the councilor of the East Indies was the chairperson of the aldermen (Taylor 2003, p. 10). The township of Batavia was administratively managed by the patrimonial leadership of the Company in the Castle of Batavia; its statutes were only applicable for European, Eurasian, and limited Asian populations such as Christian Chinese and Japanese inhabitants. The Statutes of Batavia—*plakatten*—were issued by the V.O.C Governor-General and adopted by the High Council of V.O.C in 1642 neither to protect and serve the political right and privilege of all Batavian populations nor to reinforce the free citizens. Rather, the statutes were initially more about dress codes, ordinances, and regulatory uses of status symbols according to the Company's official rank and racial identity.

After the Chinese upheaval in 1740, the V.O.C Governor-General Adriaan Valckenier issued restriction order with the *wijkenstelsel* or district system for Chinese populations in Batavia. According to this system, the Chinese settlers were not allowed to move outside their district, later known as *pecinan*. The V.O.C restricted Chinese populations in contact with indigenous people. Based on the experience of Chinese rebellions in Java (1740–1742), the Chinese–Javanese alliance was obviously powerful and endangered the V.O.C position in maintaining peace and order in most

Javanese towns and ports. The necessity for segregating the Chinese populations from indigenous people was not only a preventive measure but also a strategic practice of 'divide and rule' of the V.O.C for maintaining their economic and political hegemony above Chinese and indigenous populations.

6.4 Chinese Settlers Under the Dutch Colonial Rule 1740–1830

The historical period between 1780 and 1830 was undoubtedly not a favorable era for Dutch interest and business in the East Indies. During that period, the V.O.C had exercised experimentation in socioeconomic policies. In 1747, the Company abolished tollgates for several ports of Java and followed with the abolishment of farm revenue in some regions in 1753 (Gie 1992, p. 11). After the Chinese uprising incidence, the Chinese populations in Batavia declined; at least, 8000 were murdered, some flew out of town to find places for refuge in Java and other islands.

Economically speaking, the Batavia's main commodity—sugarcane—came into the lowest level of production. The V.O.C tried to attract the Chinese populations back to town. By 1764, Chinese settlers were compulsory to obtain permits from the V.O.C headquarters for settling down in Priangan region; the Company required them to promote the cultivation of tobacco and peanut there (Gie 1992, p. 12). Later in 1870, the permits for agricultural cultivation outside urban areas were withdrawn by the Dutch East Indies administration. Meanwhile, several former Dutch V.O.C officers and Eurasian employees moved out of the bastide. They became the free *burghers* and bought lands in the outer districts of Batavia. Daendels (1806–1810) liberated the ownership and right of land in Java that enabled the Chinese populations to buy the lands.

From the Chinese massacre in 1740 to the collapse of the V.O.C in 1799, the livability of Batavia slowed down from various aspects. Demographically, the populations of the Europeans and Eurasians did not make any significant growth (See Tables 6.1 and 6.2). On the other hand, the increasing number of Chinese and Indigenous populations was obvious from the influx of immigrants from China and migrants from hinterlands and other islands. From this demographic figure, the decline of the European and Eurasian populations was because of various reasons, such as wars against the indigenous rulers, dysentery pandemics, malaria, and typhoid. Despite the Anglo-Dutch wars in 1780–1784 had led the V.O.C into serious troubles, the most crucially undermining forces that brought down the Company sinking into bankruptcy was the corruptive behaviors of the V.O.C officers, from high to low level.

In spite of the Chinese settlers had been an asset and productive agent for the revenue of the V.O.C—especially from plantation and sugar mills—the Company did not allow them to engage in the Dutch organization. The highest rank of a Chinese person in Batavia was *kapitan*. The title of *major* was not granted by the V.O.C

Table 6.1 Demographic figure of the Dutch colonial town of Batavia 1673

Description	Populations	%
European and Eurasian	2750	8.6
Chinese	2747	8.6
Mardijkers	5362	16.7
Javanese	6339	19.8
Balinese	981	3.1
Malay	611	1.9
Slaves	13,278	41.4
Total	32,068	100

Sources Salabrini Migration Centre (2006: 58) from Dagh Register (2011); Batavia (1902: 27–30) as quoted in Castles (1967)

Table 6.2 Populations of Batavia in 1720

Origin of Populations	The year 1720	The year 1790
Dutch/European	2000	
Eurasian	1100	
Chinese	11,700	37,700
Arab	600	
Mestizo/Mardijkers	9000	
Indigenous Malay	3500	
Indigenous from other islands	5600	
Indigenous Javanese and Balinese	27,600	44,000
Slaves from various ethnicities	29,000	25,000
Total of populations	90,100	135,000

Source Derk (2012: 183)

ruling system, but by the Dutch East Indies colonial rule that begin after 1837; every important city in Indonesia had a Chinese *major* (Heidhues 2006, pp. 77–78). The patrimonial relationship between the Dutch and the Chinese leaders was continued after the death of Jan Pieterzoon Coen in 1629. However, Coen's successors did not have a personal relationship with the Chinese *kapitan*s. Meanwhile, at lower officers' level, the patrimonial relationship between the V.O.C and Chinese traders was strongly developed but for personal fortune through illicit businesses, smugglings, illegal tolls, and briberies. Some V.O.C officers resigned from their positions and became private citizens to pursue their private businesses; they bought lands, married indigenous women, and built their houses in the hinterland of Batavia.

For centuries prior to the massacre of Chinese populations in Batavia 1740, Southeast Asia has been a refuge land for political dissidents of the Qing Dynasty, traders, pirates, adventurers, and criminals (Parmer 2012, p. 33). Whilst at the end of the eighteenth century, the V.O.C was in the hardship and at the edge of financial and military collapse, the Chinese merchants dominated the local and international trades. In the surrounding areas of Batavia, the Chinese settlers were back in town and consolidated their sugar cultivation and production until 1810. During the Dutch East Indies regime, the Chinese populations in Batavia still maintained their status as migrants who terminated their formal link with China after their arrival in Java or other islands; so they could not assume any diplomatic or military protection from their homeland.

The patronage of the Dutch V.O.C was continued by the Dutch East Indies colonial regime. Jan Pieterzoon Coen commended the place of Chinese settlers in Batavia as an integrated part of the Company's interest and business for gathering cloves, nutmeg, and mace; he wrote the letter to the V.O.C board of directors on October 5, 1618, one year before he seized Jayakarta (Gie in Fernando and Bulbeck (eds) 1992: 8). Since the decline of the sugar market in circa 1740 in Batavia, mass unemployment and poverty increased criminal activities. Social unrests and crimes urged the V.O.C to capture several Chinese activists. Some of them were transported to the Cape of Good Hole in Ceylon; rumors among the Chinese were that the dissidents were thrown to the sea on the way to the destination. The rumors aggravated the Chinese resistance against their masters, *kapitan*s, and the V.O.C. As the massacre of Banda 1621 led by J. P. Coen, the Dutch V.O.C used to hire several Japanese *ronin* or *samurai* without Japanese lord for the execution of the Chinese massacre in Batavia. It is certainly the first and largest Dutch colonial act of *Sinophobic* violence (Derk 2012, p. 256). A more detailed account on the 1740 Chinese massacre in Batavia is exposed by J. T. Vermeulen in his dissertation (Vermeulen 1938; Heidhues 2009).

Under the Governor-General Hermann Willem Daendels, the Dutch East Indies regime still maintained the Chinese populations as Dutch trading partner. The mission of the Dutch regime in 1806 was to hold back and defend the East Indies from the attack of the English forces. The hand of Daendels had a lot of works for defensive constructions. However, the public works were less contracted to Chinese contractors. Instead, the Dutch Batavian Republic under Daendels preferred to mobilize and oppress the local rulers with forced labors.

To a certain extent, the Dutch administration in Batavia moved their headquarters from Sunda Kelapa area to Gambir and Lapangan Banteng area. Gradually, the *bastide* and the Castle of Batavia were demolished and left out, the building materials were used for the construction of New Batavia and a new line of defense in today Jatinegara. For the defense of Java, Daendels built the Post-Weg from Anyer in West Java to Panarukan in East Java almost 1200 km in length. Daendels centralized and rationalized the administration of the Dutch East Indies in such a way so that his government worked efficiently from Batavia to districts in Java and some outer islands.

Administratively, the Dutch colonial government under Governor-General Wilhelm Daendels established the interior administration—*Binnenlandsch Restuur*—and divided Java into prefectures. During the British interregnum 1812–1816, Thomas Stamford Raffles modified the administrative structure from the prefecture to the residency system—*regentschappen*. Each *residentie* consisted of two or more *afdeeling—kabupaten*. The Dutch colonial administration established dual management: *Europeesch Bestuur* and *Inlandsch Bestuur* (Claver 2014, p. 16). The Dutch resident was the chief official of the European administration whilst the indigenous ruler was responsible for the native affairs. Both had their seat mostly in the same place: the capital town of the traditional kingdom. However, the Dutch resident indirectly supervised the local ruler.

The patronage of Dutch colonial administration to the native one was hierarchically derived from indigenous Javanese concept of center: from the regional state (*negara*), regency (*kabupaten*), region (*kadipaten*), district (*kawedanan*), and village (*desa*). Their patron–client relationship was observable on the spatial planning and architecture of the town. The presence of the Dutch authority in most of the traditional town center area was represented by a fortress or prison nearby the native ruler's seat. This phenomenon is apparent in Yogyakarta, Surakarta, Bandung, Cirebon, and Malang. The seat of Dutch residence was commonly at a distance from the traditional ruler's palace. Each residence was usually to oversee two or more regencies, *kabupaten*. From top to bottom, the Dutch colonial administration was run by Governor-General for the whole Dutch East Indies, resident for the region, assistant resident for regency, and a controller for the district.

Dual administrative representation characterized the towns of Java under the Dutch colonial system. This was not the case for Batavia. The traditional center of Jayakarta had been burnt down from the beginning of the establishment of the city on May 30, 1619, by the Dutch V.O.C—*Vereenigde Oostindische Compagnie*. In order to maintain the political and economic hegemony, the Dutch colonial authority practiced indirect rule with the principle of 'divide and rule.' The local rulers and their vassal chiefs became the important clients of the Dutch patron for tax collection and tributes from indigenous commodities. The Dutch colonial exploitation of agricultural and industrial production in Java had reached its peak during the compulsory cultivation system, *cultuurstellsel* between 1830 and 1870 (Fasseur 26–55. Bloembergen 2006: 28–30). During the compulsory cultivation system, one-fifth of the land in each Javanese village had to produce crops for European commodities.

Under the Governor-General Johannes van den Bosch (1830–1833), Java became a huge private estate of coffee, tobacco, tea, indigo, sugar, quinine, and cotton. For the distribution and delivery of the commodities, the Dutch authority consolidated their urban areas as the hubs of commodities from rural areas. Some of them were developed as international trade seaports with a direct line to the Netherlands. The exploitative practice of compulsory cultivation system did not go without criticism in the Dutch Indies as well as in the Netherlands. After the abolition of the cultivation system in 1870, the trade line between Rotterdam/Amsterdam and Batavia was wide, open after the opening of the Suez Canal in 1869.

During the Dutch liberal period between 1870 and 1900, private capitalism came to employ a preponderant influence upon colonial policy in the Dutch Parliament (Ricklefs 2008, p. 193). The development of modern infrastructure for Dutch colonial towns began at the end of the nineteenth century. During those periods, the Dutch colonial rule invested heavily in the railways, roads, bridges, ports, irrigations, and sanitation. The Dutch laid the systems of urban infrastructure and built an environment of most Javanese cities and towns. Its main commodities and services for the European market characteristically determined the structure and function of urbanized areas. The surplus of revenue from the compulsory cultivation system enabled the Dutch East Indies to modernize the towns and cities with modern infrastructure and communication system such as telephone, automobile, and steam engine train.

In 1870, the Dutch East Indies regime issued the Agriculture Act that abolished the compulsory cultivation in Java and Sumatra. Afterward, the Dutch East Indies was open for private enterprises and international investors. One thing that the entrepreneurs, liberal politicians, and the liberal cultural elites did not pun in question was the public accountability and legitimacy of the Dutch East Indies concerning the colonialized land property (Bloembergen 2006, p. 27).

Since 1820, the Dutch East Indies administration had issued the promulgation concerning the zoning and pass systems that restricted the Chinese populations to stay outside the urban areas of Java and other outer islands. The restriction of mobility was applied until 1870. By excluding the Chinese populations in rural areas and agricultural sectors, they were forced by law to consolidate their role and function in trading activities in urban areas. The *Staatblad* 1835 no. 37 and 1866 no. 57 regulated the Chinese populations on their status of stay and their quarters in urban areas. Accordingly, the Chinese settlers were by law to live and work in certain areas of towns or cities that did not mix up with the European and Eurasian populations as well as with indigenous people. During the Ethical Policy, the Dutch East Indies administration lifted up the protection for native peasants in rural areas of Java from the Chinese money lenders. In 1920, the Dutch colonial rule deprived the Chinese role and function in the monopoly of gambling, prostitutes, the opium trade, and pawn house. Since then, the Chinese quarters were full of people with unemployment. Only few Chinese populations were selected by the Dutch to fill the official positions, who were able to the elites with other Asian populations. Most of them were converted into Christianity. Their children were educated in the Christian private schools and Dutch school.

Morphologically, the Chinese settlement was constructed according to the ward system or based on clan clustering system. Traditionally, the Chinese settlements were organized into a hierarchical group of social units from the city, district, sub-district, neighborhood, block, and household. Urban blocks were arranged to accommodate lots of 12–20 households in a row shophouses system. Commonly, the live–work units consisted of two stories or two parts: front and back area. Gridiron system had been applied to integrated blocks and streets as a compact organized settlement with gates and locally adapted orientation. The inhabitants of Chinese-*kamp* were mostly small business people, shopkeepers, and workers while well to-do Chinese businessmen were granted to have their private houses outside the *pecinan*; they had been in a historical patron–client relationship with the Dutch officials since the V.O.C times. Most Chinese businessmen and their family had been members of the Catholic or Protestant Churches.

On Batavia's map of Wagner & Debes Geog Establ Leipzig 1914, Chinese settlers developed the *Chinese-kamp* in today Glodok and Pancoran area. Close to the area was Chinese hospital on the west and the V.O.C prison on the east. Krukut River flows into the area dividing the *Chinese-kamp* into two areas at the west and at the east. Under the Dutch colonial rule (1806–1942), the Chines-*kamp* did not find the way out to participate and contribute to the civic services and developments beyond their Chinese district, *pecinan*. According to the Dutch *regeriingsreglement* 1854, the Chinese populations in East Indies fell into the category of *vreemde Oosterlingen*, together with the Arabs, Persians, and Indians; this regulation addressed the Chinese as foreigners or people with racial, class, and customary identity from outside of Southeast Asian archipelago.

In daily life, the identity of people in the public sphere was compulsorily demonstrated with dress codes—as an integrated part of the Batavia's Statutes—that each racial group had to wear their own traditional outfits. Indigenous and Chinese people were forbidden by law dressing like the Europeans, except the elites and those who had been educated in Dutch schools or recruited as Dutch officials, soldiers, or policemen. The Dutch colonial rule established the schools for native Javanese elites in 1816, but not until 1914, the Dutch elementary schools were open for low-class native populations. Meanwhile, the Chinese well-to-do populations used to educate their children with private tutors, depending on their social milieu. Some of them sent their children to Catholic or Protestant missionary schools.

In spite of the Chinese populations in Indonesia came from various places in China—with different traditions, beliefs, and languages—from the Dutch perspective, the Chinese were a socioeconomic and political entity. In a matter of fact, the Chinese, other Asian, indigenous, and European populations were not necessarily segregated by the racial category or background. While during the Dutch colonial times, other urban institutions—such as school, office, company, hospital or Church—were not open for all inhabitants, marketplace was probably one important place where all urban dwellers—regardless of their race, background, and social class—had the chance for interaction and communication as persons.

The Dutch policy of racial segregation was vulnerable to interracial and intercultural conflicts and confrontations. The ethnic tension came out from the fact of fewer interactions and communications. The Dutch authority practiced this apartheid mechanism of '*divide and rule*' to maintain their hegemony among the potential players of socioeconomic urbanism (Dorward 2003, p. 129). The Buddhist and Taoist Chinese populations and the indigenous Muslims in various towns of Java were often to deal with open clashes because of misunderstandings and misperceptions of each other's sociocultural values; such clashes happened during the Chinese Lunar New Year's celebration in Kudus and Pati 1920.

6.5 Post-colonial Chinese Populations in Indonesia

The Japanese authority treated the Chinese differently. The overseas Chinese and Indian populations in Indonesia were perceived by the Japanese Imperial power as potential Japanese nationals (Touwen-Bouwsma 2005, p. 55). The Japanese military rule handled the Chinese populations in Java with care concerning their economic assets, such as sugar factories, textile productions, pharmacies, manufacturing facilities, and construction industries. To mobilize and collect Chinese resources, the Japanese military rule dissolved all Chinese organizations and associations. Then, the Japanese authority established the overseas Chinese organization—Hwa Chiao Tsung Hui—as an official agency of Chinese population in Java from the capital city to regional and local level (Willmott 2009: 35). The mission of the overseas Chinese agency was also to monitor and control the socioeconomic and political activities of the Chinese populations—either Chinese *peranakan* or Chinese *totok*. In 1944, the Japanese military rule realized the importance of harmonious cooperation among all populations of Java regardless of their ethnic and cultural identity for the Japanese defense system.

The Japanese authority established several organizations for self-defense—*kebondan, keibota*—economic—*kumiai*—and social organizations—*Djawa hokokai*—that allowed Chinese and indigenous people to participate in the organizations. Such interracial platforms enabled diverse populations in Java and other islands to find and establish a common ground for multicultural urbanism. In most cases, the patrimonial authority of the Japanese military power did not allow other patrimonial establishments outside their system. Unlike the Dutch patrimonial practice that built and maintained power relations between racial and cultural groups, the Japanese patrimonial rule integrated, organized, and mobilized all sociocultural, economic, and political potentials into a fascistic military system.

The end of the Japanese occupation of Indonesia and the independence of Indonesia were an opportunity to establish a multicultural platform and nation-state of all people in the former Dutch East Indies territory. The road toward an interracially unified system of the nation-state was still far and away from peaceful and harmonious cooperation between the newly established Republic and all political streams of the whole Chinese populations. After the end of the Japanese occupation in 1946, the Chinese populations and their properties were the targets of racial hatred and rage such as in Tangerang, Surabaya and other towns and cities in Java.

The violence against Chinese populations has been repeated for several times since the Chinese uprising and massacre in Batavia in 1740. The difference was clear that in 1946 the newly established Republic of Indonesia was not capable of managing public safety and security because of lack of resources. While the violence against the Chinese populations happened, the Dutch colonial power came back to the archipelago to reclaim their East Indies colony.

During the struggle for independence of Indonesia against the Dutch colonial power (1946–1949), the Chinese communities were divided in their position; some groups were in collaboration with the Dutch firms, some other organizations worked together with the Republican forces (Willmott 2009, p. 37). One important reason for this was the fact that the Chinese populations were in uncertainty concerning their citizenship status; under the Indonesian Citizenship Act 1946, most Indonesian born Chinese—*peranakan*—were free to choose either for Indonesian citizenship or foreigner status. In most cases, the Chinese *peranakan* decided to adopt Indonesian citizenship. Uncertainty concerning the status of overseas Chinese nationality came to the moment of decision 1955 when the People's Republic of China and the Taiwanese Republic of China offered citizenship for overseas Chinese populations; acculturation with locality was an option (Qiu 1997, p. 209). All over again, the formality of Chinese integration into Indonesia's nation-state was easy on paper, but the reality of daily lifeworld did not smoothly dissolve the prejudice and resentment against the appearance of Chinese identity.

Under the patrimonial regime of Suharto's New Order, the enforcement of Chinese integration into Indonesian culture was enacted by law; discriminatory treatments for public services and facilities were implemented in various fields and cases (Winarta 2008, pp. 62–63); the government closed Chinese schools, banned Chinese publications in mandarin, forbade Chinese cultural performances and dissolved Chinese associations and organizations. Under the Suharto's New Order, the Chinese populations were oppressed and compulsorily obliged to change their names from Chinese to Indonesian names. As with any other populations, the Chinese people were requested to adopt one of five religious affiliations for their personal identity card. The discriminations and mistreatments of Chinese populations culminated and were blown out by the mass action of violence against Chinese populations on May 13–14, 1998.

For over two and half centuries after the Chinese pogrom in Batavia, the Chinese settlers in Glodok area had never been so heavily devastating like in mid of May 1998. Witnesses on site saw how a group of well-trained men arriving in the location, provoking people with racial slurs to Chinese there, then, systematically throwing stones and burning down their shops and stores, and looting their properties, and raping the Chinese women. During the times of riots, 297 of 600 kiosks in Glodok market and *Orion Plaza* were destroyed, burned, and looted. The incidence was hardly said as a spontaneous attack or an accidence because it was region-wide and possibly national-wide movement against the Chinese populations.

Even though the violence against the Chinese populations was inseparable from the student protests and political upheaval against the New Order, the scenario behind the scene was clear that the Chinese matters were the scapegoat and distraction of power struggles among political elites in Indonesia. For the umpteenth times, the violence against Chinese ethnicity and their architectural heritage in *pecinan* Glodok; most Chinese architectural heritage and structures in the area were burnt down and left by their owners.

Although the administration of Abdurrahman Wahid in 2000 had recovered and reinstated the Chinese culture and heritage by Chinese populations in Indonesia, some sites in *pecinan* have not been rebuilt and redeveloped until the beginning of 2004. Wahid issued the Presidential Decree No 6/ 2000 concerning the freedom of sociocultural expression of Chinese populations that was automatically to denounce and redrawn the Presidential Decree of 127/U/Kep/12/1967—on the compulsory change of name from Chinese to Indonesian—the decree of 49/U/8/1967—about the abolition of media in Chinese characters and mandarin language—and the decree of 14/1967 concerning the obliteration of Chinese tradition, culture, and education. Since January 2000, the Chinese populations of Glodok and other parts of Indonesia

Fig. 6.1 Kiosks and street vendors in the Glodok area 2014, photograph by the author

have begun to reinstate the Chinese customs, traditions, and culture. The first celebration of Chinese Moon Festival took place at the Vihara Dharma Bhakti in Glodok area on February 16, 2000 (Fig. 6.1).

However, most of the former store owners have not restored their old buildings for several years after the May riots; only some of them rebuilt their shops with architecture without traces and clues of Chinese features. Some former inhabitants of Glodok quartiers do not come back to their home place. The trauma and horror of the May 1998 riots still remain unforgettable for most Chinese populations in Glodok area; they were afraid of building their shops and stores with Chinese identity. Nevertheless, there are few buildings with Chinese architectural characteristics in Jakarta that are to be found in the area on Jl. KH. Mohammad Mansour, Jalan Toko Tiga, Jalan Petak Sembilan, and Jl. Blandongan. At Jl. KH. Mohammad Mansour building No. 102 s/d 106, No. 254, No. 238. Jl. Pekojan 38, 45, 46, 47, 54, 55, 60, 61, 67, 71, 86, and 87, Jalan Tambora VI and VI. Most of the old Chinese quarters had been burnt down in May 1998 riots; the Chinese populations were looted, murdered, burnt, and harassed while several Chinese women were raped and humiliated. The violence against the Chinese populations was mostly with the demolition and looting of their property, assets, and investments. The tragedy of May 1998 is likely the event that has disposed toward the absence of Chinese architecture and town in Indonesia (Fig. 6.2).

Since the end of May 1998, the Chinese populations in the sub-district of Glodok have mostly moved out from their home; the populations of this 36 ha sub-district in 2004 were 11,148 or 3281 households and 1179 families in 2010 (Pemprov DKI, Dinas Informasi, Komunikasi dan Kehumasan 1995–2010). Since the Indonesian independence in 1945, the Chinese Glodok populations have been moving out from the old town of Jakarta to other parts of the city. West and North Jakarta are the home of Chinese populations. Almost 78% of populations of West Jakarta municipality are Chinese descendants. The migration of the Chinese populations from Glodok had begun since the first twentieth century when the Dutch colonial authority lifted up their regulation on Chinese matters.

Statistically speaking, the Chinese inhabitants in Indonesia were 2832 million in 2010 or 1.2% of Indonesia's 237 million populations (BPS (Badan Pusat Statistik)

Fig. 6.2 Abandoned buildings after May 13–14, 1998 riots in Glodok area, photographs by the author

Indonesia 2010, p. 8). The number is confusing because most Chinese populations do not feel comfortable to identify themselves as *keturunan cina* or Chinese descendants; they rather see themselves as Indonesian because socially and culturally they are not attached and affiliated with Mainland Chinese tradition and culture. In many ways, acculturation and assimilation of Chinese descendants into local Indonesian culture and tradition have been well worked out through formal education, mixed marriage, sports, and other sociocultural activities and organizations.

Prior to the fall of Suharto's New Order in 1998, the political integration of Chinese descendants in Indonesian political parties has been a challenging issue for several decades. Even though the amendment of Citizenship Law No. 12 2006 allows the Chinese ethnic group to express their sociocultural tradition and culture, they do not have courage and intention to participate in rebuilding the Glodok sub-district as a Chinese quarter as it was in the past; many former inhabitants of these historic quarters never want to go back but move on with their new life in other parts of the city or in other countries.

6.6 Concluding Remarks

The patrimonial practice of authority is demonstrated with the exclusivity of the ruling group. In order to stabilize and maintain its hegemony, patrimonial regime develops the structure of hierarchy for a line of command and domains. In urbanism, the practice of patrimonialism is demonstrated with the ghettoization based upon racially grouping settlement. In doing so, spatially speaking, urbanism is the socioeconomic and political platform to show who is the privilege and who is the less and the marginal.

As a socioculturally and politically organizing system, patrimonialism has been practiced by the Dutch, Chinese, and indigenous communities in Indonesia. Dutch colonialism was patrimonial domination over others: the Chinese and native groups. The Chinese groups were in the role and function of intermediaries between the Dutch colonial power and indigenous communities. The relationship between the

Dutch V.O.C and Chinese populations was historically based on patron–client ties that involved the personal relationship between both leaders; without such personal relationship, the patron–client ties between them were vulnerable that had led to distrustful and suspicious business. The pogrom of the Chinese populations by the Dutch V.O.C in 1740 was not only because of the insecurity and anxiety of the Dutch colonialism in dealing with the Chinese power, but also that of the absence of a personal relationship between the patrimonial leaders of the Dutch and Chinese part.

During the Dutch East Indies colonial rule, the patrimonial hegemony was exercised with various experiments, from absolute control for sociopolitical and economic exploitation to decent patronage for the cultural hegemony of Western civilization. However, this patronage did not intend to dissolve the segregation of population groups based on race, class, and beliefs system. The Dutch colonialism in the twentieth century was eventually incapable to handle its own values system based on civil liberty and universal humanism; it is probably an unfinished project of Dutch patronage over others, the Chinese and Indigenous culture.

Characteristically, ghettoization and marginalization are indivisible from the way of patrimonial thinking so that the hierarchy of domains comes into play in urbanism. The Dutch and European occupy the most privileged location and are followed with other Asian foreigners, the indigenous elites, and the native commoners.

The Japanese occupation (1942–1945) was another exercise of patrimonial power over the Chinese and indigenous people. Blessing in this guise, the Japanese fascistic patrimonialism was able to unite various groups and factions together to be a patrimonial nation-state of Indonesia. The necessity for unity in diversity among all populations was challenged by the comeback of old-fashioned colonial patrimonialism without any success but surrendering the stupidity of presumption for cultural superiority in 1949.

During the Republican administration, the road toward a national integration is still far and away. The Republican vision concerning universal humanism, social justice, and civil liberty is the work in progress. Despite it's a long-standing historical presence, Chinese customs, traditions, and heritage in Indonesia have not yet been officially acculturated and embraced as an integrated part of Indonesian culture and heritage. Resentment and racial hatred still linger in the mind of some indigenous populations for many reasons. The most obvious is socioeconomic jealousy because most Chinese people are successful entrepreneurs either with or without a patrimonial relationship with the ruling political elites and state apparatus. The sense of otherness from the indigenous perspective, especially the elites, has not easily and totally dissolved within the process of growing up together as an Indonesian nation-state from 1945 until today. As far as the multiculturalism is concerned, the assimilation, acculturation, coalition, and cooperation of all beliefs, traditions, and customs are necessarily worked out as a national agenda of Indonesia.

The May 13–14, 1998 riots were indivisible from the struggle of patrimonial hegemony at the national leadership. The unrests were just the ignitions of momentum for greater movements of power transition. Despite the initiators and intellectual actors behind the riots were not publicly announced, the involvement of armed forces' elites—especially the high-ranking army and police officers—was possible because most of them shared something in common for the fall of Suharto. The actual issue was that the elites were frustrated by Suharto's determination to stay in power despite the serious economic crises 1997–1998. The riots were intentionally designed with the indirect support of students' movement for socioeconomic and political reformation. By killing the students as martyrs, the uprising against the New Order's status quo became a massive and national-wide movement. The Chinese populations and their settlements were the scapegoats and easy target of the riots; the patrimonial elites of the nation needed a common enemy for the diversionary strategy of their internal contestation of power.

References

Abayesekere S (1987) Jakarta: a history. Oxford University Press, Singapore

Bertrand J (2013) Political change in Southeast Asia. Cambridge University Press, Cambridge

Bloembergen M (2006) Colonial spectacles: the Netherlands and the Dutch East Indies at the world exhibitions, 1880–1931. National University of Singapore Press, Singapore

Blusse L (2012) An insane administration and insanitary town of the Dutch East Indies. In: Ross RJ, Telkamp GJ (eds) Colonial cities, essays of urbanism in colonial context. Martinus Nijhof, Dordrecht, pp 65–85

BPS (Badan Pusat Statistik) Indonesia (2010) Sensus: Kewarganegaraan, Suku Bangsa, Agama dan Bahasa sehari-hari Penduduk Indonesia. BPS Pusat, Jakarta

Brown D (2003) The state and ethnic politics in Southeast Asia. Routledge, London

Claver A (2014) Dutch commerce and Chinese merchants in Java: colonial relationships in trade and finance 1800–1942. Brill, Leiden

Cribb R (2000) Historical atlas of Indonesia. Curzon Press, Surrey

Crouch H (1979) Patrimonialism and military rule in Indonesia. World Polit 31(4)

Derk H (2012) History of the opium problem: the assault on the East, ca. 1600–1950. Brill, Leiden

Dorward D (2003) Colonial administration. In: Page ME, Sonnenburg PM (eds) Colonialism: an international, social, cultural, and political encyclopedia, A-M. vols 1, 2. ABC CLIO, Santa Barbara, CA, pp 128–130

Eisenstadt S, Roniger L (1984) Patrons, clients and friends: interpersonal relations and the structure of trust in society. Cambridge University Press, Cambridge

Fukoaka Y (2015) Demystifying "people power": an elite interpretation of democratization in Southeast Asia. In: Case W (ed) Routledge handbook of Southeast Asian democratization. Routledge, London New York, pp 85–101

Gie PL (1992) The changing economic position of the Chinese in the Netherlands India. In: Fernando MR, Bubeck D (eds) Chinese economic activity in the Netherlands India: selected translations from the Dutch. Institute of Southea Asian Studies, Singapore, pp 5–18

Gungwu W (2004) China imperial. In: Ooi KG (ed) Southeast Asia: a historical encyclopedia, from Angkor Wat to East Timor, vol 1. ABC CLIO, Santa Barbara CA, pp 331–334

Gunn G (2011) History without borders: the making of an Asian world region, 1000–1800. Hongkong University Press, Hongkong

Hall K (2010) A history of early Southeast Asia: maritime trade and societal development, 100–1500. Rowman & Littlefield, Lanham

Hanlon D (2009) Composition in architecture. Willey & Sons, New York

Heidhues MS (2006) Chinese voluntary and involuntary associations in Indonesia. In: Kuag-Pearce KE, Hu-Dehart E (eds) Voluntary organizations in Chinese diaspora. University of Hongkong Press, Hongkong, pp 77–120

Heidhues MS (2009) 1740 and the Chinese massacre in Batavia: some German eyewitness accounts. Archipel 77:117–147. Retrieved from https://www.persee.fr/doc/arch_0044-8613_2009_num_77_1_4127
Hoerder D (2002) Cultures in contact: world migrations in the second millennium. Duke University Press, Durham
Manguin P-Y (2000) City states and city state culture in the pre 15 century Southeast Asia. In: Hansen MH (ed) A comparative study of thirty city-state cultures: an investigation, vol 21. Kgl. Danske Videnskabernes Selskab, Copenhagen, pp 409–430
Matsuyama (2013) Traditional dietary culture in Southeast Asia, its formation and pedigree. Routledge, London
Ownby D, Heidhues MS (1993) Secret societies reconsidered: perspectives on the social history of modern South China and Southeast Asia. Sharpe, New York
Parmer NJ (2012) Historical perspectives of Chinese populations in Southeast Asia. In: Amstrong MJ, Amstrong RW, Mulliner K, Amstrong A (eds) Chinese populations in contemporary Southeast Asian societies: identities, interdependence and international influence. Routledge, London, pp 18–54
Pemprov DKI, Dinas Informasi, Komunikasi dan Kehumasan (1995–2010) Kecamatan Taman Sari. Retrieved 9 April 2015, from https://www.jakarta.go.id/web/encyclopedia/detail/3177/Taman-Sari-Kecamatan
Qiu Z (1997) The overseas Chinese problem and solution. In: Suryadinata L (ed) Political thinking of the Indonesian Chinese 1900–1995: a sourcebook. NUS Press, Singapore, pp 204–206
Register D (2011) Dagh-register gehouden int Casteel Batavia van passerende daer plaetse als over geheel. Brill, Leiden
Ricklefs MC (2008) A history of modern Indonesia since c. 1200. Palgrave Macmillan, New York
Scott J (1972) Patron-client politics and political change in Southeast Asia. Am Polit Sci Rev 66(1)
Tarling N (1999) The Cambridge history of Southeast Asia. Cambridge University Press, Cambridge
Taylor JG (2003) Indonesia: peoples and histories. Yale University Press, New Haven Connecticut
Touwen-Bouwsma E (2005) The Japanese policy towards Chinese on Java 1942–1945 in. (Kratoska P, ed). National University of Singapore, Singapore
Vermeulen JT (1938) De Chineezen te Batavia en de Troublenen van 1740. Eduard Ijdo, Leiden
Ward K (2009) Networks of empire: forced migration in the Dutch East India Company. Cambridge University Press, Cambridge
Willmott D (2009) The national status of the Chinese in Indonesia 1900–1958. Equinox, London
Winarta F (2008) No more discrimination against the Chinese. In: Suryadinata L (ed) Ethnic Chinese in contemporary Indonesia. Institute of Southeast Asian Studies, Singapore, pp 57–74

Chapter 7
Masculine Cult and Territorial Control: "Premanisme" in Javanese Urban Areas, Indonesia

Abstract Masculine potentiality is human nature. What is the relationship between the practice of territorial control and masculine cult from this perspective? The question leads this study to unfold and divulge the social behavior of aggression and struggle for hegemony. This study argues that the necessity for territorial control is human nature, but its practice is necessarily within the framework of the socially ordering system in terms of law and order. In the Javanese context, the practice of masculine cult has been likely acute and hinders the supremacy of law and order in urban public spaces. As far as masculine supremacy is concerned, territorial control becomes the stake of struggle for such a hegemony that involves violence and extortion. The aim of this study is to examine and dismantle the phenomenon of premanisme from the perspective of human nature.

Keywords Masculine cult · Premanisme · Territorial control · Urban crime

7.1 Masculine Cult

Recently, there are scholarly works concerning the relationship between masculine cult and various gangs of criminals (Ford 2012; Bowker 1998; Blok 2001; Messerschmidt 1993; Archer 1994). This study is to find out the relationship between the phenomenon of territorial control of public domains and masculine cult in Indonesia, especially in Javanese towns and cities (Blok 2001, 21, Wilson 2012, 122). Territorial control is sociologically considered as the validation and justification for masculine supremacy and hegemony. This relationship is not only challenging for the question of status quo concerning public safety from a legal perspective but also about the cult of violence in Indonesian society. Theoretically speaking, from Max Weber, we learn that the state is the only legitimate and monopolistic agency for exercising violence. This is unfortunately not the case for Indonesia today.

Evidently, in Javanese urban areas, the public domain is never empty of a claim for land lordship. In this case, conquering and controlling areas and resources are not impartial from the cult of masculinity; in traditional Javanese context, this supremacy is understood under the notion of *digdaya* meaning masculine potentiality.

B. Wiryomartono, *Traditions and Transformations of Habitation in Indonesia*,
https://doi.org/10.1007/978-981-15-3405-8_7

Accordingly, land and its resources need conquering and controlling that is not simply for the benefit of material gain. This tradition was well known as *babat-alas* or forest clearing for settlement and domestication. Meanwhile, the general concept for conquering and controlling area is known as *ngrambah* that is to demonstrate the sense of growth and expansion.

Most importantly, dominating resources is showing masculine potentiality. Since today the traditional need for this masculine supremacy has been taken over by the supremacy of rule of law, the traditional practice of masculine domination becomes part of underworld business. Such a business is known as *premanisme*. The essential drive of this practice is to hold the status quo of masculine supremacy; a proprietor of this status quo is called *preman*. The best illustration of what and who *preman*s are and their activities in Indonesia is lively presented in the Joshua Oppenheimer movie: 'the Act of Killing' (Final Cut for Real ApS 2013).

The aim of the study is to divulge and unmask the practice of domination and control territory beyond the statutory supremacy of law and order in public places. Moreover, this study is also an attempt to explore the local tradition why does the practice of *premanisme* hold and maintain the status quo. Why does the state not take any measure and action against the violence and extortion of the hoodlum? In order to deal with the questions, this study will be presented in three sections.

The first is the search for the historical accounts of the practice of hoodlum in traditional Javanese society from pre-colonial, colonial, and modern Indonesia. Historical analysis from literary sources belongs to the main part of this first section. The second section will discuss the incidences of the *premanisme* that examines the actors and matters of violence and extortions in the public domains of Jakarta and other Javanese urban areas. The third is a summarizing exposition of finding and theoretical construction on the practice of *premanisme* in the context of the politics of domain.

A number of scholars have explored studies on criminals and thugs in Java (Kartodirjo 1966; Barker 2001; Cribb 2008; Lindsey 2005; Siegel 1998; Henk and van Till 1999). From their studies, reveal the necessity for an in-depth study on the roots of crime and violence in the Indonesian context from the perspective of gender complex concerning patrimonialism. This study argues that *premanisme* in Indonesia is the masculine struggle of men for their patrimonial hegemony in society. Since the struggle is considered human need, its energy should be managed wisely through sports or politics or other activities where public accountability and fairness apply. As far as the practice for masculine supremacy in society is concerned, violence and brutality are potentially exercised. In order to prevent these, society should give space for managing masculine potentialities under the supremacy of law such as extreme sports and other thrilling racings and competitions.

7.2 Historical Accounts and Roots of Territorial Control

Historically, the hegemony of public and private domains in Java is an open question; it is not easy to draw a strict line between legal and illegal matter in the context of social order because of the ambiguity on the meaning of law and order in Indonesia since pre-colonial times. The Dutch colonial rule did not introduce to people in the archipelago the united regime of law and order for crimes, but the plural regimes of control over land and populations (Lev 2000, 99). The legacy of the Dutch colonial rule on the apartheid politics of space and populations has been likely overshadowing the status quo over the public domains of today Java.

Even though in modern Indonesia there is no segregation of land ownership based on racial background, the legal status of land ownership is still plural. Because of the pluralism of legal system on land tenure, the idea of the legitimacy of social order, from the V.O.C Dutch colonial times to the end of Suharto's New Order regime, has been problematic in Indonesia. Moreover, the legitimacy of a social order based on the supremacy of legal system does not only bring any economic benefit to them but also that kind of supremacy is not favorable for their masculine cult to those who are in power.

In the Post-Suharto era, the perception of ordinary people on law and order as well as legal and illegal remains unchanged; the indicators for this are obvious in the piling-up of the unsolved legal cases for crimes of corruption, collusion, nepotism, abuse of power, and violence against humanity. All of these pertain to the fact that state officials and criminals did the same things: violence and extortion for their own interest; probably the establishment of modern Indonesia has not significantly changed the perception that the establishment of law and order system is another word for domination, *penguasaan*, instead of for social justice, *keadilan sosial*.

Neither Sukarno nor Suharto, their patrimonial regimes did not make any solid foundation for the justice system and civil society based on the system of law and order. The obvious reason is that both regimes were, from gender perspectives, the same; they upheld and practiced patrimonial hegemony (Jenkins 2010). What the ordinary people experience today is the fact that the state has a monopoly and legitimacy in the exercise of violence. The supremacy of law becomes uncertain as to the source of fairness; law, *hokum*, is not for social justice. Rather, the law is punishment, *hukuman*, and sentence, *putusan*. Why is the ambiguity on the concepts of the legal system in effect?

The impact of the ambiguity on the concepts of the legal system is apparent in various cases of daily life in public space. One important consequence of these cases is the patrimonial control of public domains by private persons, *premans*. Patrimonial territorial control by tough men has been inseparable from daily urban street life and rural areas in Java; in this sense, the control means the might of doing a collection for those who use the areas, which are legally defined as public property. The main issue is not simply about the practice of collection. Rather, the problem of this lies in territorial status and control of the area, which is not comprehensible under any applicable legal system. Claiming territoriality under private patrimonialism is likely more about to show who is the boss of the area?

In most cases of parking cars in public spaces, one has to pay a parking ticket or cash to a person; we never know either the payment is for the municipality or for somebody else. The practice of violence and extortion is unsurprisingly the part of the daily business of the public realm. From a legal perspective, the state fails to protect people from informal fees, briberies, and inducement. What is confusing for those who are not familiar with the Indonesian and Javanese way of life in the public realm is that the collection agency could be on behalf of the local municipality or the business is just run by private persons.

The root of daily violence and extortion in some public transits and spaces is not simply for the illegal tolls. Rather, the cause is more about the necessity of local tough men for acknowledgment and respect concerning their masculine hegemony in their area. Commonly accepted, that ordinary people or citizens do not expect too much that the collection will go back to them as part of the municipal services and welfares. Such expectation has been out of the question from the beginning because the concept of what government is about. Accordingly, the government in the Indonesian context is identical with *pemerintah* meaning literally instructor, supervisor, and headmaster; *pemerintah* incorporates the idea of authority and politically ordering body for public matters. In doing so, it hopeless to expect the idea of government as a public servant, this concept is considerably humiliating the authority of the public institution. Until today, despite *reformasi* and Joko Widodo–Jusuf Kalla's administration, the concept of what government is does not dramatically change from dominator and supervisor to facilitator and administrator.

Traditionally, the local ruler controls the use of public areas in Java for rituals and common utilities. Territorial control over settlement area in pre-colonial Jakarta, as well as Java, was indivisible from feudal Javanese society. Even though there was no exact date of the beginning of *premanisme*, the root of the practice was probably rooted in the ancient Javanese feudal states. The phenomenon of *premanisme* could have been dated back to ancient Java with the concept of *rakai* for the lord or master of a territorial unit of tax collection called *wateg* (Bellwood et al. 1995, 313).

Ancient Javanese village called *wanua* was an independent political institution of agricultural-based settlement. The generation of village founders became the aristocrats or elites called *rama* in several ancient Javanese inscriptions of the seventh to the tenth century. The elites held and maintained their status as the proprietor and authority of fee collection for various economic activities in their areas. A *rakai* could control several villages, *wanua*s within his *wateg*, depending on his consolidation of power in various ways. Esoteric power and martial arts were part of the power consolidation for a *rakai*.

In order to establish and maintain his power, which was constantly challenged by other aristocrats, *rama*s, in his area, the village chief and the elites would have collaborated with religious leaders and authorities. The elites welcomed the visits of Hindu Brahmins and Buddhist priests because their ideological doctrines, rituals, and ceremonies were able to strengthen their ruling position, with secrecy and privileges: literacy and knowledge. In return, the priests enjoyed the protection and high status in the Javanese society as advisors and scholars. The conspiracy of local rulers and religious priests in ancient Javanese culture reached its peak with the establishment of a territorial *wateg* as a state/kingdom, *negara*, and the masculine cult of the ruler.

The other concept of Javanese village governmentality was *demang* and *kademangan* for village settlement; *demang* means somebody who controls, taking care, and has power over. *Ki demang* is the title of the village head. Most of the *demang*s were originally the descendants from village founders who opened the forests and settled down there as pioneers. Meanwhile, their descendants were those who successfully took a lead in achieving his masculine hegemony over his feuding leaders. Even though the village chief in ancient Javanese tradition was the hereditary position, he or she had the absolute rights of agricultural land disposal remaining unredeemed from debts for paying tributes or taxes (van Stten van der Meer 1979). In many cases, such *demangs* were impossible in power to dominate the collection of taxes and fees from peasants and traders without overpowering other criminals in his territory. *Demang*'s masculine supremacy has been the way to control resources.

Historically speaking, consolidating the power in pre-colonial Javanese villages was impossible without violence and bloodshed. The great Javanese leaders such as Ken Arok of Singasari, Jaka Tingkir of Pajang, and Senapati of Mataram were not impartial from the practice of the masculine cult of hegemony. The root of *premanisme* in Java lies in the fact that the practice has been traditionally exercised since ancient times. In a matter of fact, the ancient Javanese society with their absolute ruler did not develop a legal system in the search for justice, but for punishing those who evicted and rejected to pay taxes, tributes, and fees to the authority. It is probably the reason why in Indonesia today, the search for punishment, *hukuman*, is practically more important rather than that of justice, *keadilan*.

Schulte-Nordholt and Till's study shows the life and role of *jago* in the Dutch colonial times (Henk and Till 1999). The conclusion of the study is probably surprising because the Dutch administration was not able to penetrate the world of *jago*. The reason is simply that the whole population where the *jago* lives are not able to overcome the fear of his charisma and power which associated with mysterious and occult forces. The Dutch colonial rule did not have something to do directly with the *jagos*. However, during the colonial cultivation times in the nineteenth century, the Dutch needs to make sure that the flow of commodities from plantations to the ports run well without ambush from indigenous bandits, *rampok* and *begal*. The colonial power never touched but acknowledged the existence of *jugos* to secure their interest.

The colonial administration acknowledged the practice of Javanese taxation under the traditional kingdoms. The Javanese King controlled the land beyond the capital city with the help of landlords called *bekels*; their areas called *perbekelan*. For the taxation matter, landlords were assisted by tax collectors called *patuhs*. Since *bekel*s were mostly only interested in the outcome, they allowed their collectors to operate independently for collection purposes. The abuse of power by collectors often caused collective resistance among the land cultivators and traders. In colonial Batavia, such collectors were known as *centeng*s or *jawara*s were originally security guards for warehouses and houses of the rich, the Chinese merchants and local rulers.

In the broadest sense of the word, *centeng* is a territorial lord who collects tributes and fees from the users of public domains. Batavia in the nineteenth century was famous with several territorial lords or tough men such as Si Pitung, Abang Jampang, Si Gantang, and Sakam. Unlike in the rural areas, urban tough men in Batavia were not in collusion with the local authority. People called them with the Dutch word *vrijman* meaning literally independent person who was not part of the colonial authority. *Vrijman* is another word for colonial officers who are off duty without uniform. Indigenous people called Si Pitung, Sakam, and Abang Jampang as freemen because people felt the latent fear of their actions at the same feeling as the fear they got from the colonial authority. Of course, freemen were not bandits or outlaws. Colonial *Vrijman* was actually *jago* in urban areas. They were not living apart from their community. To a certain extent, some of the freemen like Si Pitung did burglary and robbery in Robin Hood-style. The colonial authority was likely to tolerate the existence of *vrijman* as a local ruler.

However, in many cases, freemen did also robbery, burglary, rape, and murder against the rich people that included Chinese and the Dutch. In this case, Si Pitung, Oenoes, and Sakam were on the list of Dutch colonial law enforcement. The indigenous people treated freemen in an ambivalent way. On the one side, they were afraid of the freemen's bullying power. On the other side, they sympathized with freemen's rebellious gut against the Dutch authority. Unsurprisingly, freemen like Si Pitung gained sympathetic support and acknowledgment as their heroes. Today, indigenous populations of former colonialized Batavia adore Si Pitung as an independent hero against the Dutch colonial power; indeed, it is probably an exaggerated way to fill the absence of opponent of colonialism in the nineteenth century Batavia. The life and activities of Si Pitung were probably away from reality.

Although the interest of the freemen and the Dutch was the same, collecting fees and taxes, the colonial power never directly did in touch with indigenous populations. The Dutch power used the local rulers for such collection. During colonial times, the Dutch did not care about how to do the collection. To manage the administration of collection, the Dutch maintained, installed, and protected local and traditional rulers to deal with their own people. Thus, the indigenous people in colonial times had to deal their payment to formal and informal dominators: the Dutch/local ruler and the freemen: *centeng*, *jago, jawara*, *bramacorah, jagabaya,* etc. From indigenous people's point of view, colonial authority and freemen were not different in their nature; both are economic dominator with their expertise in bullying, harassing, and oppressing people; they need to do so because they have to make sure to themselves that their masculine supremacy is acknowledged and respected.

During the Dutch rule until the beginning of the Pacific War in 1942, control over public domains in the colonial towns and cities of Java was ethnically divided. The Dutch authority controlled their area where the Europeans lived, played, and worked while the local rulers took care of the territory of indigenous settlements. During the Dutch colonial period, other foreign populations, such as Chinese, Arabs, and Indians, lived in their compounds. The Dutch left their territorial control over land to each ethnic group based on their own rules and tradition.

Under the Dutch colonial rule, indigenous criminals were not prosecuted with the Dutch legal system. The Dutch colonial legal practice did not exercise a normative law for all populations but valued justice system as cultural norms and codes (Cribb 2005, 53). Consequently, local people lived in their own legal system under the notion of *hukum adat*. The only main concern of the Dutch colonial interest was not about to establish the justice system but the collection of land and property taxation; the Dutch colonial rule did not establish and develop the system and management of taxation on an individual basis or personal account but on grouping collection base. The Dutch colonial regime did not hand over a legal practice system to the natives, but the bullies, watchdogs, and spies of local rulers and 'tough men,' *jago*, *jawara*, and *centeng*. What crime for the native populations was fully the business of local rulers? The Dutch colonial rule did not give the native populations the sense of justice based on law and order for all, but fractured regimes of distinct social order.

In his study on brigand in Java (1950–1942), Suhartono (1997) identifies various titles for criminals with a single Javanese word: *bajingan*. The word for the Javanese people is a concept for furious and vicious men with no connotation of criminals or thugs in terms of the normative legal system. People fear such men because they can kill and extort them for fees, tributes, taxes, and other contributions. The way to deal with them is to accept their rule and order. The origin and existence of tough men in Javanese society is probably ancient history. From West to East Java, the men are known with different titles: *benggol kecu*, *brandal*, *durjana*, *warok, centeng*, *bromocorah*, *jawara, jago,* and *gali*. Rural stories are full of chronicles of tough men who are often depicted with extraordinary skills of martial art and esoteric knowledge, *elmu kebathinan*.

Traditionally, the feeling of ordinary people to tough men and to the local authority is not quite different; both are socially in distance and do not care about people's well-being. To a certain extent and until recently, local authority and tough men are interchangeably in their role and function as the dominator of public space. However, what makes them different was because of their status in the cosmological system; prior to Indonesian state's independence, the local authority had a specific relationship with the religious rituals and ceremonies while the tough men were associated with black magic power. In a matter of fact, official and private persons likely share something in common with their necessity for masculine hegemony. Today, public domains remain open for the contestations and likely collaborations between formal authority and tough men; as far as the patron–client ties between them clear, there will be no clashes between them in public spaces.

The ambiguity of the relationship between the state and tough men never stands without mutual interest. Since 1945, the date of Indonesia's independence as a nation and state, the relationship has been part of the history of a power struggle. Unsurprisingly, most riots, social upheavals, and demonstrations in the era of *reformasi* are not free from gangsters' participation. To a certain extent, police officers and *premans* share something in common in the occult powers that bring about crimes as power struggles among men for their male pride; the crueler and sadistic their way to defeat others is the more pride and respect they get from others. Unsurprisingly, the public domains become male territory for the contest of the masculinity between the formal authority and the freemen.

What kind of state is that if its police culture works within the framework of *premanisme*? This is a simple question but the answer becomes sociologically complicated because the establishment of a social order based on justice system does not work. The officers and agencies of public safety do not get a direct benefit from the users of public space and their customers. In his notable study, Barker (2001, 20–53) observes and shows how the culture of Indonesian Police has been evolving since the 1980s within the occult powers of *jago* or *jawara*. From a gender perspective and the culture, it is unsurprisingly to know that the policemen and *preman*s have something in common concerning their desire for masculine cult and patrimonial hegemony.

7.3 *Premanism* in Modern Cities

Conflicts between the users of public spaces and local authorities have been observable during the New Order in power between the 1970s and 1990s. Notorious conflicts between ordinary people and the state happened during Jakarta city's restructuring development under the governorship of Ali Sadikin between 1966 and 1977 (Silver 2008, 15). He introduced modern and rational urban planning in Jakarta. Grouping areas into distinct land-use category, restructuring connectivity and a network of communication, and improving infrastructure for healthy sanitation system and flood control belonged to the main agenda of Sadikin's Jakarta development.

Governor Sadikin or Bang Ali as former marine Lieutenant General sees the urban world of Jakarta as his battlefield (Warren and Ross 2007, 237). His masculine adrenaline is challenged by a disorderly urban form of kampungs and its poor sanitation. Restructuring Jakarta spatial structure and form is nothing but to manage its populations and their activities under a rational system of modern production. Surprisingly, the controversial thing of his action for modernizing Jakarta metropolitan was not about the legal issue on the status of the land. Rather, the controversy was about the fund for his restructuring project, which was made available from National Lottery. In the country with the majority of Muslim populations, money from gambling is considered not *halal*.

Even though the cost of the development was the displacement of people from their homes and the disappearance of the features of traditional settlements, Ali Sadikin as the strong leader got nationwide support to move forward with his vision and action. The restructuring of the Jakarta metropolitan area from the 1970s to 1990s did not go without violence and extortion. However, there were surprisingly no significant cases of legal issues against the Jakarta government. Sadikin, who was well known as a strong rational leader and populist, coupled with the rational development of modern urbanism with populist Kampung Improvement Program such as Husni Thamrin project. In doing so, urban development was not only a top-down program but also an attempt to embrace local participation for coercively self-healing action from poor sanitation, unhealthy environment, and illiteracy. Sadikin in Jakarta and Indonesia is a controversial and respectful leader of development. However, from the civil society's perspective, his way of doing things is a kind of *premanisme*.

Since the restructuring of urban areas in Jakarta, the use of violence by the state/local government for urban development is also a common practice in other Javanese towns and cities. The crucial matter for the potential conflicts of interest between state and ordinary people is the value and status of the land. In dealing with this matter, the state/local government recruits tough men as their task forces for the negotiations, reimbursements, and compensations of land and property value.

Indeed, the legal status of land in most areas of Indonesia is problematic because the land law had just enacted in 1960 and the Indonesian government had not initiated to conduct a nationwide land registration until 1981, under the program of National Operation Agrarian Project (Reerink 2010, 532). Prior to the establishment of land status based on formal registration, the land ownership status was apparently in question. Indeed, most urban traditional settlements, *kampungs*, do not have legal status.

Briefly, the justice system in Indonesia does not provide a rigorous platform for seeking fairness; all this is because of the fact that the practice of *premanisme* has deeply penetrated into various state institutions, from the government to legal system (Agusyanto 1995). Accordingly, issues and problems, including legal matters, are manageable beyond the court system. *Premanisme* will be able to solve problems, accusations, corruptions, and other crimes with bribery, inducement, violence, and extortions. In other words, a complete and solid legal system in this country is still the work in progress and far cry from reality (Vos 2001, 184). The consequence of the ambiguity is apparent in the relationship between the state and tough men, as well as state and ordinary citizens throughout various eras from pre-colonial to modern Indonesia.

The phenomenon of a brigand in terms of *warok, jago, jegger*, and *jawara* has been recorded with a long history of violence in Javanese society since pre-colonial times. The quintessence of *jago* in Javanese tradition is a masculine contest of dominators based on their guts, nerves, and skills in dealing with cruelty, sadism, violence, and the impossible for ordinary men. The currency for the contest is valor for male pride and respect in terms of phallic superiority. Controlling territory in Javanese tradition is not simply about dominating resources and populations for economic advantages. Rather, it is about achieving the highest pride and respect for masculinity by creating a sense of fear, horror, and dread. In order to be a super male, *premans* or tough men learn and train themselves to gain specific knowledge and skills from martial arts to occult powers. In doing so, they become confident in their male pride. However, they need to demonstrate their masculinity with their extraordinary guts in dealing with sexuality, violence, and murder. In short, a *jago* or *preman* is a vicious male person without mercy for their enemies.

Under the administration of the Republic of Indonesia, control over land use and property taxation had not well administrated and managed with proper procedure and mechanism until the enactment of Law 12/1985 and 1994 on land and property tax. Even though public domains—according to Government Regulation 15/2010—remain under the control of state or municipalities/local governments for common use, in daily practice it is not so; there is no any further direction of the management of public use. The public domains in the Jakarta metropolitan area (Jakarta Bogor Tangerang, and Bekasi) are not fully under the control of their respective municipalities; illegal parking guards dominate the parking fee collection.

Indeed, the local authorities issue parking tickets and other local fees but they do not charge the users of the public domain on an annual basis. They collect the fee on a daily basis in terms of local retribution for facility maintenance, such as garbage collection and cleaning routine. In many cases, the municipal authority and local *preman* share parking and spot fee; *preman*s usually get more share for such collection. Behind *premans* stand local military or police offers; in colloquial Indonesian words, people call such officers as *beking* from the English word backing. The practice of *backing* is commonly found in various public facilities such as bus or transit terminals, local traditional markets, and entertainment districts such as nightclubs, discothèques, bars and restaurants (Collin 2007, 290).

Formally, respective municipalities or local authorities legally control public domains of towns and cities, for public safety and health reason. Theoretically speaking, public areas are common property and free for being used by all. In Javanese tradition, it is not always the case, public places need guards, such as *lurah pasar* for market and *kuncen* for the cemetery and sacred places; *kuncen* means literally key. Only certain public areas such as the street in the village are not guarded because they use it as the daily outdoor living areas of community life. However, guarding and maintaining public domains belong to community voluntarily job; all this is done based on collective awareness for common good. Traditionally, such stewardship is subject to community voluntarily work called *bersih desa* in terms of *gotong-royong*.

Most neighborhoods in Java exercised and experienced such a voluntarily stewardship for centuries. Where does the practice of *premanism* take place? In the healthy and socially established community, *premanism* does not have room to grow. The practice takes place mostly in anonymous areas, beyond the stewardship of local community or neighborhood. In a socially well-established neighborhood, male community members, *warga*, voluntarily guard their area in turn; the head of the neighborhood organizes and coordinates the rotation of the job among the households on the weekly or monthly basis. The *warga* do their guarding job, *ronda*, as part of traditional service of community security called *siskamling*, *system keamanan lingkungan*, and neighborhood security system. A strong neighborhood with a well-established *siskamling* is usually free from the practice of *premanism*; *premans* will recoil to live in such a strong neighborhood.

Premanisme in Indonesia is a modern concept for those who claim territorial control for illegal fees because they find their habitat, which is traditionally out of neighborhood watch areas. The essence of *premanisme* is deeply rooted in the fact *premans* are people in the underworld because they are not able to survive in the ordinary world, but they have the need, gut, and drive to fight for proofing their masculine vigor. In this sense, identifying and defining *preman* practice as the act of hoodlum is likely incorrect because its *modus operandi* is not characterized by the geography of the street, but by the exercise of domination with violence and extortions. Sadism and cruelty belong to their valuable measure for their rank and status. To a certain extent, *premanism* is about the way of life in a group of male persons where might is right.

The agency of *premanisme* could be a person, a group of people, and an agent who is independent or a part of the organized body of the private or public sector. Today, the concept has been widely understood as any action for fee collection or obtaining goal without respect to normative procedure and process; to a certain extent, *premanisme* is ironically double standard because it is done by state clerks, officers, and agencies. The practice varies from the collection of a fee for applying: driving license, development proposal, building permit, passport, to winning state project bids and high state positions. In short, obtaining something from state or government is likely impossible on schedule without paying extra and illegal fees or commissions. If it is no extra fee or commission, it must be an odd or uncommon process that takes a longer time than it is supposed to be.

On the street, in order to achieve their goal, fees, commissions, and contributions, *preman*s used to exercise violence and extortion. In daily life, the real *preman*s are hardly recognized; they work and perform their duty as ordinary persons at the state institutions. They are distinguished from street *preman*s not simply by their scale of the fee they get, but also by the way they dress and lifestyle as well as their connection to high profile politicians, *ulama*s, and businessmen. In contrast, street *preman*s dress like gangsters; their body is mostly decorated with furious or macho tattoo and long hair or bald. Some have been self-promoted from the street—*jalanan*—to creamy—*gedongan*—*preman*s; the most notorious figures for this are Hercules of Tanah Abang until 1996, Bang Ucu Kambing of Tanah Abang, Basri Sangaji of Pulo Mas, John Key of Senen, Logo Vallenberg of Bumi Serpong Damai.

Hercules is one of the controversial figures of *preman* who was supportive of New Order's special military operation in East Timor during the 1980s. In a matter of fact, most notorious *preman*s must have been loyalists of the New Order. After the fall of Suharto's New Order, several people with *premans* background work as politicians, such as Yoris Raweyai form Golkar party, Ongen Sangaji from Hanura party, and Lulung from Partai Persatuan Pembangunan (Badan Pertahanan Nasional Indonesia 2010). Obviously, the pride of ethnic origin plays an important role in the *preman* world. Makasar, Bugis, Flores, Manado, Timor, Ambon-Maluku, and Banten are common references for the *preman* figures. Such regions are well known for their male cult and traditional manhood contest through martial arts.

Today, the struggle for domination among *premans* is part of the underworld and nightlife of Jakarta. The main business of *premangedongan* includes snack and drink supply, drug trafficking, debt collection, and security of casinos, hotels, nightlife places. To a certain extent, *preman*s have some civil associations and groups, such as *Pemuda Pancasila, Laskar Jayakarta, Forum Betawi Rembug*, and *Pemuda Pancawarga*, which are seemingly supportive of civil society. Even though several organizations, such as *Forum Rembug Betawi, Front Pembela Islam,* and *Angkutan Muda Pembaharuan Indonesia*, deny falling into any group of *premans*, their actions and moves do not comply norms and principles of ethics in civil society; their actions are mostly associated with violence.

The practice of *premans* looms after the fall of Suharto's regime. The high rate of unemployment and poverty in rural areas is probably one important cause of the increasing number of *premans* in towns and cities; they move to urban areas looking for jobs. Male migrants who are not willing to go for labor work but prefer to have a shortcut for money establish a brotherhood group of men. They had to survive in urban areas as temporary parking guards or other part-time jobs. Most of them are not able to compete with other migrants for many reasons, such as dropping out of schools and too many problems with violence, drugs, and sexes. What they do most of their time is just hanging around in public domains asking street vendors for changes.

In many cases, the presence of unemployed young people in public domains is annoying; they usually fight with each other for no reason but harassing each other. In order to minimize the crime rate among the young urban poor, municipal authority recruits them as their community police task force known as *polisi pamong praja*, PolPP. In the private sector, security companies hire some of them for parking business and debt collection. The municipality and the security agencies need them because of their familiarity with violence. After the fall of Suharto from power, the status and existence of the PolPP are confirmed and established as a municipal task force by the Laws: 22/1999, 25/1999, 32/2004, and 8/2005 concerning regional autonomy and local government for establishing the law and order in public domains.

So far, the use of potential *premans* for security forces in terms of municipal police had brought about clashes and violence. The practice of *premanisme* in Indonesia is probably a phenomenon of the exercise of power in daily life. Fatah describes the practice as the child of New Order system from the period of 1970–1989 (Fatah 2004, 77–9). In his observation as a political analyst, *premanism* is not simply the practice of illegal fee collection from the street based on the use of public domains. Rather, the practice has penetrated and exercised in the state bureaucracy and political mechanism of Post-Suharto Indonesia.

Even though he does not mention any further on the concept of *premanism*, it is obvious that state bureaucrats, officers, and politicians had exercised such practice for any form of decision making and approval. Paying extra fees to officers for any business of obtaining legal approvals, permits, and documents becomes a common practice in the daily life Indonesian government. Despite the law of anticorruption, no 31/1999 had been in the enactment, there is no indication that the practice of collecting illegal fees in various levels of public service decreases.

Even though since ancient times *premanisme* in Javanese towns and cities had been known dormant in relation to the state and formal authority, the phenomenon was likely different in the 1980s. In dealing with Suharto's political opponents, the use of *premans*' service was obvious (Rytter 2001, 128; Lindsey 2005, 290; Honna 2010, 149), because the regime could indirectly control the antigovernment movements without legal procedures. Kidnappings, killings, torturing, and extortions the antigovernment activists would have been with no legal consequences because all these were done by hitmen or by tough men on the street.

Indeed, the use of *premans* for the political interest of the New Order was not without danger and risk in the future. In order to protect the New Order's interest,

Suharto's administration responded the practice of *premanism* with a death squad called *Penembakan Misterius* or mysterious killings and known with abbreviation *petrus*, which was in action between 1983 and 1985. The authority of public safety made official statements concerning mysterious killings, *petrus*, as the consequence of intergangs power struggle. Indeed, the victims were mostly male persons known as *gali*, the abbreviation of *gabungan anak liar* or union of untamed youths. The target of the silent operation of *petrus* was the criminals or gangsters who were mostly male between 17 and 60 years; their body is mostly decorated with tattoo on their chest, back, hands, and feet; all this is a sign of ex-convict. Moreover, for *preman* tattoo signifies specific class and rank of their masculine bravado for sadistic and vicious domination.

During the New Order in power, such actions were practically common under the interest of national stability and unity. Army officers at any chain of commands, from district area to national level, used to conduct special operations through the practice of *premanisme* against separatist movements and antigovernment activists such as in East Timor and Aceh 1980–1989. One important reason why was the practice of *premanisme* was looming again in the era of *Reformasi* probably because of the high rate of unemployment in urban areas.

Youths in the urban area become *preman*s because they are barely to find jobs in the formal sectors of productions and services. Lack of training and education is mostly the common case of being *preman*. Before having become *preman*s, boys spend their early years in public domains as parking guards, beggars, or fee-collectors of street vendors. Fighting and winning for territorial claim becomes part of *preman* job. Pemuda Pancasila and Forum Betawi Rempug, as well as Front Pembela Islam, are associations that recruit such unemployed people for their patrimonial struggle for economic and political hegemony.

Public space in Javanese towns and cities is never free from the claim and interest of the underworld people for their platform of masculine competition. *Premanism* in Javanese towns and cities is a long-standing tradition of masculine culture of *jago*. The root of the *preman* tradition lies deeply in the cult of masculine superiority. To a certain extent, the masculine practice of *jago* is not restricted for tough men and criminals on the street and underworld, but male persons of the security forces and state bureaucracy are potentially member and supporters of the tradition. Even though at the superficial level, the business of *premanisme* is terrorizing, harassing, and extorting ordinary people for fees and tributes, deep inside the practice is the practice of sexual domination with the practice of sadism and masochism. Most public spaces in Javanese urban settlements are under such patrimonial domination with various agencies, either formal or informal. From this situation, a social order based on the system of law and order in public domains becomes problematic.

The main purpose of patrimonial claim on the territory for *preman*s is not simply to control the resources through the collection of illegal fees and tributes. Rather, the claim is the show-off their masculine practice for patrimonial sovereignty. Since the sovereignty is meaningless without public acknowledgment, public domains become the platform for those who believe in the cult and contest of phallic supremacy within *jago* or *jawara* tradition. The contestants of the cult and competition are men with a special need for their masculine supremacy. The collection is considered the reward for their supremacy. The cult and contest are probably in need because most of the *premans* are not able to compete and gain influence in civil society with their formal qualifications and skills. Unsurprisingly to realize that frustrated state officers could be part of the men in this category.

7.4 Concluding Remarks

The practice of patrimonialism is not only exercised at the seemingly civilized realm but also in the underworld. The core of patrimonial practice cannot afford to ignore the masculine cult for domination. The practice of *premanism* is not the excess or consequence of income inequality. Rather, it is a socioeconomic part of the patrimonial tradition at the raw and 'primitive' level and realm. Therefore, it is unsurprising to observe that several political riots and social upheavals in the country have involved several *preman*s as agitators and provocateurs of actions. The patrimonial politicians and premans share something in common concerning the necessity for masculine power with violence and extortion for control and domination.

In many incidences involving *premanisme*, there is never free from bloodshed and mayhem; all these are signs of weakness because they are historically lack of care of the home. The Oppenheimer's movie: 'the Act of Killing' is the best documentary exposition on how violence and extortion are exercised by the private groups of gangsters while the state agency for public safety does not care to prevent for such actions. Most of the *premans* are spiritually homeless and insecure; they never find a peaceful state of mind because of their inability to work out the fear of otherness. Instead of negotiation, adjustment, and accommodation, *premans* prefer to fight against otherness for their masculine supremacy. Some try to live and adopt an ordinary lifestyle but they mostly fail into street and nightlife trap.

An ordinary and normal home place for *preman*s is inconvenient habitation because their masculinity is not challenged to exercise violence and viciousness for domination. The normal course of life for *preman*s is boring and not interesting. *Preman* way of life is on the contrary to normal cycle because they are spiritually insecure and in latent anxiety for social identity and emotional stability. Do violence and extortion reflect the male anxiety and insecurity in dealing with feminine counterpart? *Preman*s are petrified by the feminine aspect of home and wholeness; most active *preman*s do not have a family life. Masculine occult and contest are to overcome their weakness in this matter that they are not able to manage and handle the feminine aspect of life with respect and dignity; they are too proud of themselves as separate entities. At the end of the day, they are tired to compete, because they are actually impotence and not the competence to deal with women as equal partners. They are not able to deal with the reality that is toward wholeness and perfection.

Unemployment among youth engenders the continuity of *premanisme* as the practice of masculine domination in the underworld of Indonesia. This practice is considered the inevitable way out to deal with poverty and their masculine potentiality.

In other words, creating and providing employment among youth are not adequate. Other activities should be managed and provided so that the human nature of masculine potentiality is able to be transformed into economic, social, and cultural activities. This potentiality needs to be seen as a productive capacity for society and humanity as a whole. The problem is how this masculine potentiality becomes an integrated part of public policy for job creation and opportunity of youth in the country.

References

Agusyanto R (1995) Premanisme. Majalah Gatra, Indonesia, March 25: 4

Archer J (1994) Male violence. Routledge, London & New York

Badan Pertahanan Nasional Indonesia (2010) Wordpress, September 12. https://bpn16.wordpress.com/2010/09/12/hercules-preman-tanah-abang/

Barker J (2001) State of fear, controlling the criminal contagion in Suharto's new order. In: Anderson B (ed) Violence and the state in Suharto's Indonesia. SEAP Publication, Ithaca NY, pp 20–53

Bellwood PS, Fox JJ, Tryon DT (1995) The Austronesians: historical and comparative perspectives, comparative Austronesian series. Australian National University Press, Canberra

Blok A (2001) Honour and violence. Polity, Oxford & Cambridge, UK

Bowker L (1998) Masculinities and violence. Sage, Thousand Oaks, CA

Collin EF (2007) Indonesia betrayed: how development fails. University of Hawaii Press, Honolulu, HI

Cribb R (2008) Gangsters and the revolutionaries, the Jakarta people's Militia and the Indonesian revolution 1945–1949. Equinox, Jakarta

Cribb R (2005) Legal pluralism decentralization, and the roots of violence in Indonesia. In: Adam AW, Anwar DF (eds) Violent internal conflicts in Asia Pacific: histories, political economies, and policies. Yayasan Obor Indonesia, Jakarta, pp 41–57

Fatah EEPS (2004) Mencintai Indonesia dengan Amal, Refleksi atas Fase. Penerbit Republika, Jakarta

Final Cut for Real ApS (2013) The act of killing press release. Accessed 23 Aug 2016. https://theactofkilling.com/wp-content/uploads/2013/09/THE_ACT_OF_KILLING_press_notes_sept2013.pdf

Ford M (2012) Men and masculinities in Southeast Asia. Routledge, London & New York

Henk SN, van Till M (1999) Colonial criminals in Java 1870–1910. In: Rafael VL, Mra'zek R (eds) Figure of criminality in Indonesia, the Philippines, and colonial Vietnam, vol 25 of studies on Southeast Asia. SEAP, Itacha, NY, pp 47–69

Honna J (2010) The legacy of the new order military in local politics: west. In: Aspinal E, Fealy G (eds) Suharto's new order and its legacy: essays in honour of Harold Crough. ANU Press, Canberra, pp 135–149

Jenkins D (2010) Suharto and his generals: Indonesian military politics, 1975–1983. Equinox, Singapore

Kartodirjo S (1966) Peasant revolt in Banten in 1888. KILTV, Leiden

Lev DL (2000) Legal evolution and political authority in Indonesia: selected essays. Martinus Nijhoff, Dordrecht

Lindsey T (2005) The criminal state: Premanism and the New Indonesia. In: Lloyd GJ, Smith SL (eds) Indonesia today: challenges of history. Institute of Southeast Asian Studies, Singapore, pp 283–297

Messerschmidt JW (1993) Masculinities and crime: critique and reconceptualization of theory. Rowman & Littlefield, Lanham

Reerink G (2010) Land registration programmes for Indonesia's urban poor: need reach and effect in Kampongs of Bandung. In: Ubink J, Hoekema AJ, Assies WJ (eds) Egalizing land rights: local practices, state responses and tenure security in Africa, Asia and Latin America. Amsterdam University Press, Amsterdam, pp 102–178
Rytter L (2001) Pemuda Pancasila: the last loyalist free men of Suharto's new order. In: Anderson B (ed) Violence and state in Suharto's Indonesia. SEAP, Ithaca NY, pp 124–155
Siegel JT (1998) A new criminal type in Jakarta, counter-revolution today. Duke University Press, Durhan NC
Silver C (2008) Planning the megacity: Jakarta in the twentieth century. Routledge, London
Suhartono (1997) Bandit-bandit di Pedesaan Jawa: Studi Historis. Aditya Media, Yogyakarta
van Setten van der Meer NC (1979) Sawah cultivation in ancient Java, aspects of development during the Indo-Javanese period 5th century (ANU Faculty of Asian Studies). Oriental Monograph Series (22)
Vos ML (2001) International cooperation between politics and practice: how Dutch Indonesia cooperation changed remarkably little after a diplomatic rupture. Het Spinhuis, Antwerpen
Warren R, Ross JA (2007) The global family planning revolution, three decades of populations policies and programs. The World Bank, Washington DC
Wilson I (2012) The biggest cock: territoriality, invulnerability, and honour amongst Jakarta's gangster. In: Ford M (ed) Men and masculinities in Southeast Asia. Routledge, London & New York, pp 121–138

Chapter 8
Masculine Aesthetics and the Works of Frederick Silaban: Post-colonial and Patrimonial Architecture Indonesia

Abstract This study concerns the relationship between masculine aesthetics and a built environment that examines and scrutinizes Silaban's architecture as a case study. Given that Silaban—by his sociocultural background and characters—grew up in the patrimonial tradition of Batak Toba, Indonesia, this study argues that his masculine aesthetics is useful for Sukarno's nation-building in Indonesia. In other words, Sukarno and Silaban find the common ground of self-actualization, visualization, representation, and justification of monumental architecture for the imagined Indonesian nation-state. Sukarno and Silaban share the common ideology of aesthetics that conformity and integrity are crucially important to represent the nation-building of Indonesia.

Keywords Masculine aesthetics · Silaban · Sukarno · Suharto · Architecture · Indonesia

8.1 Masculine Aesthetics

This study concerns the aspects of the relationship between gender and architectural design in the context of masculine aesthetics. Masculine aesthetics in this study is understood as an experience that arises from masculine corporeality in the sense of muscular strength; being a male person means having muscles and physical bulk, in opposition to inert simulacrum (Carden-Coyne 2005, 223). To what extent does masculine aesthetics play an important role in the patrimonial political culture and architecture? In order to deal with the question, this study investigates and collects architectural and literary data on Silaban's works from his archive in Bogor, West Java, as well as his buildings in various locations of Indonesia. This study is intended to unfold and unmask the masculine aesthetics for political purpose and interest under the concepts of totalitarianism, nationalism, and patriotism.

Friedrich Silaban (1912–1984) is an eminent architect in Sukarno's era with his renowned works for the post-colonial architecture in Indonesia. His architectural works mostly become iconic buildings and enact as landmarks of its area and location. The contribution of Silaban to the historical geography of Indonesian capital is ostensible. Studies on Salaban's works have been done by scholars concerning

B. Wiryomartono, *Traditions and Transformations of Habitation in Indonesia*,
https://doi.org/10.1007/978-981-15-3405-8_8

various topics and interests (Sopandi 2009; Ondang 1992; Ardhianti 2005). Even though their studies are significant regarding his contribution to Indonesia's nation-building, a comprehensive and profound study on his works is still not available in the publication.

Until recently, the ideas, conditions, and circumstances behind his works are still less publicly unfolded and less theoretically explored. This study is an attempt to fill a room of discussion between the previous findings and the available data of Silaban's works. The purpose of this inquiry is to provide an explorative exposition of the relationship between the architect and his works in the context of masculine aesthetics of patrimonial political culture. Most recently, the American academy of political and social science has published studies on patrimonialism from various perspectives (Adams and Charrad 2011); the discourse on patrimonialism has attracted many scholars from various fields of social sciences and humanities because it has something to do with the concept of authority and the practice of power that matter for public interest concerning social justice and freedom of expression.

Methodologically, this study departs its investigation from the context of Silaban's personality and his works as an architect, then followed with reading the traces of his works from Silaban's archive at his family residence in Bogor. Cross-checking of data and confirmation of findings are conducted with interviews and discussions with former colleagues, friends, and family members, especially with his son: Panogu Silaban. Since the relationship between the architect and his works is indispensable personal, this study argues that the underlying structure of Silaban's works lies in his personality and sociocultural background.

Under the notion of patrimonialism, this study understands an ideological system that works in society for cultural production and deployment of its resources based on paternalistic leadership and authority. Patrimonial hegemony has been part of the political culture in Southeast Asia (Scott 1972; Bertrand 2013; Case 2014; Brown 2003; Crough 1978). In daily life, patrimonial culture is reliant on the patron–client relationship. The politically ruling groups play their role and function as the patron that controls the access to resources: labors, materials, money, and information. On the other hand, the politically ruled group is in the function of clients that serve the patron as followers and supporters.

Despite democratization, patrimonialism in Indonesia and other countries in South East Asia has been practiced as an integrated part of their tradition, especially under the Sukarno's Guided Democratic era and Suharto's New Order period (Brown 2003, 90). Under the patrimonial regimes of Sukarno and Suharto, architecture in Indonesia has played its significant role at the service of nation-state's ideology (Kusno 2014, 69; Lukito 2015, 123–5; Kusumawijaya 2001). The most important thing about patrimonialism lies in the need of society for strong leadership that unites various sociopolitical streams and differences. However, this need is actually not an only social and political condition, but also spiritual requisite for the sense of unity. The state of being politically united is perceived as the state of being healthy and prosperous. Even though the region has been changing in terms of the political system from less to more democratic administration, the society still holds and maintains patrimonial power relation at informal interactions; they still uphold and exercise respect to person based on his patriarchal seniority and paternal reputation.

This study concerns the aspects of the relationship between sociocultural values system and architectural design. To what extent does masculine aesthetics play an important role in the works of Silaban? In order to deal with the question, this study investigates and collects architectural and literary data on Silaban's works from his archive in Bogor, West Java, as well as his buildings in various locations of Indonesia. The analysis of the data is focused on two areas of concern: historical setting and theoretical development on patrimonialism. Studying the architecture of Silaban in this paper is neither to judge its presence for democratic nor for an authoritarian regime. Rather, this study is intended only to unfold the process and production of architecture based on patrimonial context and condition. In doing so, the relationship between patrimonialism and architecture is recognizable as masculine aesthetics.

Indeed, in most cases, it is complex and difficult to draw the line between the masculine and feminine identity of the works of architecture because there is the necessity of harmony between contradictory elements and aspects in any composition. The necessity for the structural logic of building construction entails masculine components while the attractiveness of building appearance requires feminine characteristics; most works of architecture involve and demonstrate the interplay and embodiment between both complementary aspects in a composition. In other words, only a few works show a dominant expression either masculinity or femininity.

Cross-culturally, the most common traits and features of patrimonial practice are associated with strength order, and power through muscular, directing, and controlling expression; all these are closely related to and associated with the muscular power of man or masculinity. In the Indonesian context, masculine and feminine traits and features have been inseparable entities of symbolism, as shown in most ancient temples, such as the temple of Penulisan in the village of Sukawana, the district of Kintamani, Bali and the temple of Sukuh in East Java. The ancient architecture of Java and Bali demonstrates male, *linga*, and female, *yoni*, symbolism as the Supreme Unity and Totality of all beings (Kempers 1959: 19) that signifies the safe, healthy, and harmonious world.

Masculinity and femininity in ancient Indonesian arts and architecture are complementary elements of duality. Hindu–Buddhism in Java and Bali integrated and underscored the ancient duality of antagonistic elements for the sign of Siva–Durga divine power of growth and sustainability. In other words, great monuments, from Candi Prambanan, and Candi Borobudur do not show a preference for patrimonial aesthetics with dominant masculine traits and features; the aesthetics of antagonistic duality plays an important role. The phallic and curvy element comes into play in the composition with elegant unity, such as shown by Borobudur's stupa and Prambanan's Trimurti buildings.

Even though Batak, Toraja, and South Nias communities are culturally patriarchal, their traditional houses and village settlements show the complementary masculine–feminine aesthetics. Batak and Toraja houses share a similar way of architecture; their substructure shows the muscular feature of orderly columns and beams while their superstructure is characterized by expressively curvy gable roof form made of bamboo. The South Nias houses do not complement their masculine substructure with the feminine superstructure but with their structural beam called *sikholi* representing the go-between the upper and underworld.

The traditional houses of patriarchal communities in another Indonesian archipelago, such as in Sumba, Flores, and Roti islands, masculine–feminine elements work together as a complementary system for their architecture and settlement. In most cases of Indonesia, masculine and feminine elements have been complementary in an architectural composition that consists of substructure and upper structure. The substructure is to establish the foundation and platform of the upper structure in which the domestic space for working, living, and playing is provided. In other words, human space is constructed between the substructure and upper structure. The space of substructure is commonly provided for domesticated animals while the superstructure is dedicated to the spirits of an ancestor.

In contrast to other ethnic groups in Indonesia, the Batak settlement is not organized that is based on the village association but that on the clan system. The patrimonial patronage of clan signifies the Batak settlement. Clans and its kindred establish and build their settlements in the region of Batak land based on customarily law and traditional concession from generations. The compound of Batak Toba clans is known as *huta* consisting of a row of houses, *ruma*, and a row of granaries, *sopo*. The rows of buildings stand face-to-face in a compound square. Thick bamboo clumps are planted to encircle the compound that is surrounded by rice fields. Traditionally, Batak settlements in the form of *huta* spread out in the landscape of open green and rice fields.

The masculinity of Batak Toba house, *ruma*, and granary, *sopo*, is demonstrated by its rational organization and construction of its substructure. The joints of the columns and beams at the substructure are simple, clear, and rigid. In contrast, the superstructure demonstrates a sophisticated, flexible, and accommodative gable system of roof planes construction that is made of wooden and bamboo elements; the flexibility of the upper structure is provided by its joints and systems that work for tension and compression, especially in response to wind forces, monsoon storms, and to earthquake.

As a whole system, the upper and substructure are integrated and tied up by the tensile element made of rattan and bamboo ropes. One important element of this roof construction is called *tali benguet*; this element works as wind bracing structure for the gable roof planes that joins the roof ridge and the roof beams into a stable but flexible roof structure against wind forces. Architecturally speaking, the roof construction system of Batak Toba buildings is mostly established by the joints of bundling with rattan and bamboo tope. All this reminds people of the weaving and knitting activities which are predominantly occupied by female persons. On the contrary to the superstructure of the house, the substructure is chiefly reinforced by simple peg construction. Is the superstructure of ruma and sopo to represent the role and function of a woman while the substructure for a man in Batak Toba society? The architecture of traditional Batak Toba values the integration of masculine and feminine way of construction into a livable system of space for family and community activities (Fig. 8.1).

Fig. 8.1 Rumah Bolon, in Ambarita at Toba Lake, North Sumatra, drawn by the author

8.2 Silaban (1912–1984) and Patrimonial Tradition

Silaban was born on December 16, 1912, in the village of Bonandolok, Tapanuli in North Sumatra; he is a son of Protestant priest, Jonas Silaban, and his mother is Noria boru Simamora. Friedrich is the fifth child of the Silaban's family with a strong patrimonial tradition of the Christian Batak family. By tradition and education, Silaban is no way egalitarian and liberal person. His Batak sociocultural background plays an important role in his perception and preference for the relationship between people and architecture; gender category is an issue for Silaban. Accordingly, function, role, and position of person have been traditionally arranged by Batak tradition for their habitation and socialization based on patriarchal culture.

The educational background that trains and guides Silaban is the Batak mission school and the colonial Dutch educational system at the time (1883–1940). By experience and circumstantial, the school is a patrimonial institution. The missionaries worked in Batak land were mostly Germans; they shared something in common concerning patrimonial tradition in the family and sociocultural life. The Dutch colonial policy for the education of native people targeted the children of nobility (Aritonang 1994, 13–4); its intake students are boys from privileged parents.

Friedrich Silaban is the son of a Batak priest who has the privilege among other children in his village to attend the mission school. The patrimonial line of thinking of Silaban is timely appropriate for the project of nation-building under Sukarno's leadership (1959–1965). Unsurprisingly, Silaban's architecture finds its historical momentum in the project of Indonesia's nation-building of Sukarno because they share something in common under the notion of patrimonialism for public aesthetics.

As any son in the Batak family and kinship, young Silaban has to behave and follow the traditional rules and manners that fit and proper for the respect and dignity of fatherhood in terms of provider, protector, and leader. Since his boyhood, Friedrich has to conform to Protestant ethics and Batak patrimonial family life. After his Dutch elementary school education at the Narumonda in Tapanuli in 1927, he was sent by his parents to Konigen Wilhelmina School—a vocational school for building design and construction—in Batavia—now Jakarta—and finished in 1931.

The engagement of Silaban with architecture was introduced in the practical high school with technical drawing drills and construction workshops. The interest in architectural design has been developed after he graduated from school. He worked for a Dutch modern architect J. H. Antonisse. This Dutch architect came to Indonesia at the age of 26 from the Netherlands in 1914. He was trained as illustrator and designer; he worked as the chief of technical department for the public work of Batavia municipality since 1920. One important work by Antonisse was the design of the pavilion of Pasar Gambir for the annual festival and exhibition of Batavia from 1923 to 1932.

Antonisse was well known as a creative architect with a passion for an experiment in dealing with local and non-Western heritage. His works for the annual exhibition of Pasar Gambir demonstrated and explored the variety of forms and constructions of tropical Asian architecture; his exploration was an attempt to assimilate various architectural traditions from India, Thailand, China, Japan, and Indonesia. Antonisse opens Silaban's eyes for the fascination on the variety of traditional forms and constructions. Silaban draws, paints, and sketches traditional buildings from various regions of Indonesia. Some of his collections are displayed on the walls of his home studio in Bogor.

Under the mentorship of Antonisse, Friedrich Silaban developed his skills and knowledge on architecture. This was developed through practical experience as a draftsman at the office of military service. He prepared drawings of the public works in Batavia from 1931 to 1937. From his work experience in this office, Silaban learned and adopted modern architectural design and construction. Modern design inspired him to apply the clarity and efficacy of form, material, and construction. Based on his achievement, the Dutch military authority installed him as a development officer in Pontianak, West Kalimantan from 1937 to 1939; his notable work in this city is the Equator Monument built in 1938.

In his career as a government officer for public work, Silaban finds his passion for tropical building design. His fascination with the simplicity of modernity is not something apart from his practical experience under his mentor—Antonisse. The works in the Dutch public works in various cities led him to get in touch with the diversity of indigenous tropical architecture; from the *uma lepo*" of Dayak Iban populations at Kapuas river to Sundanese *imah julang ngapak*. The important characteristics of such building are its adaptation to a tropical environment. This was achieved with cross air ventilation and long roof overhang of shade. For him, tropical design of such traditional buildings is architecturally integrated whole as a building system that uses local organic materials. The improvement of design in modern times is thought by Silaban as necessary that makes such buildings adaptive for modern activities.

Lesson learned from the Dutch colonial architecture in the archipelago is that masonry, metalwork, and carpentry are important components of a modern building that need to be managed for its long-lasting use. In his letter to the United Nations (Muramatsu 2008: 26), he emphasizes the importance of building material regarding its durability, affordable maintenance, and manageable operation. During the Pacific War, the Dutch East Indies was controlled by the Japanese forces between 1942 and 1945. Like many other Dutch East Indies employee and populations, Silaban and his fiancée were sent to a Japanese detention camp for a few months.

After the Pacific War, Silaban was back to work for the public work in Bogor in 1947. He was appointed as the director of the office there until 1965. His main job was to design several public buildings, in Bogor and Jakarta area. Between 1949 and 1950, Silaban was invited to attend the Akademie voor Bouwkunst in Amsterdam. From his international excursions to 30 countries of Asia, Europe, and America, Silaban experiences modernity alive with its iconic buildings, such as Arch de Triumph and Eifel Tower in Paris, stock exchange building by Henrick Berlage in Amsterdam, Parliamentary buildings in Brasilia, Empire Building in New York City; such buildings caught Silaban's attention to monumental and classic aesthetics of modernism.

As a person, young Silaban is an assertive person with self-confidence on his professional competence as an architect. His friends and relatives regard and respect him as an insistent and passionate person. Han Awal—a senior Indonesian architect and Silaban colleague who was born in 1930—has encountered with Silaban for several occasions and gets the impression of Silaban's personality concerning originality in his thought and strong self-confidence; he is the man of principle (Han Awal 2007). As his son, Panogu Silaban describes his father, Fredrick Silaban, as somebody who was desirable and enjoyable to work together in terms of a senior–junior relationship (Silaban in Muramatsu ed. 2008: 8–9). In the house, the relationship of father and son does not give Panogu Silaban a room for exchange of mind with his father concerning architectural issues or other related matters to design and construction. The passion for the classical music of Mozart, Beethoven, and Wagner, is shown by his collection in his office at his home.

Notes from his family circle tell us that Silaban is an obedient Christian. His father who is a priest teaches him to follow the Christian way of life as a servant of the Almighty, therefore for his work and prayer are undividable (Sindhunata 2007: 46). Being a Christian Batak for Silaban is by no means falling into a fanatical stream of ideology. He is spiritually affiliated with Christianity in the spirit of universal humanism. On the contrary, Western education brings about his outlook to the world to be a universal humanist with deep Christian spirituality. The patrimonial tradition for Silaban is not about good or bad, but it is the way to the grace, harmony, and beauty of God. Patrimonial obedience is conceived by Silaban as respect and thankfulness to his ancestors. Like any Batak male person, Silaban is responsible and in the role to establish rule and order in his family.

Based on the author's interview with family members, friends, and colleagues of Silaban, they feel his radiant eagerness and sound judgment. He is aptly to dress in a conventional and proper manner with regard to practicality, durability, and worth. To a certain extent, he does not care too much about his appearance. However, proper attire for a specific occasion is part of how he presents himself in public and at home, even though Silaban loves to be judged on the basis of his work rather than on his appearance. In the family circle, Silaban leads his family by example for correctness and respectability. Dark and earthy color is his common choice for his attire. As a father, his family members and relatives find him as a reliable and good protector and provider. He motivates and encourages his children to pursue their dream without forgetting to pray and hardworking.

Beyond his private life, Silaban was a well-known person of social activism with an interest in universal humanism, arts, and culture. His sociocultural engagement reached its peak when he signed a petition for the cultural conception of Indonesia. The petition was published by the magazine of *Lentera* on March 16, 1962. The magazine was part of the *Bintang Timur* daily—a communication medium of *Lembaga Kebudayaan Rakyat—Lekra* or organization of culture for people—of Indonesian Communist Party. Silaban was never a member of any political party or its organizational wing such as *Lekra*. The signature of Silaban for the support of cultural conception was purely professional in order to support the government's program and commitment to the indigenous arts and culture.

Silaban was familiar and a friend of several Lekra's activists such as painter Henk Ngantung and painter Hendra Gunawan. Especially with Henk Ngantung who is a prominent leader of *Lekra* (Jones 2013: 106), Silaban has intensive discussions while working on several design competitions such as Monument of Selamat Datang, National Monument in Merdeka Square, and the monument in Banteng Square. Henk Hermanus Joel Ngantung was the deputy of Jakarta governor from 1960 to 1964. Sukarno installed Henk Ngantung as governor of the capital city on August 27, 1964, and honorably discharged on July 15, 1965. Ngantung and Silaban were close to Sukarno and worked together as a team for a vision of the president in shaping Jakarta as a cultural city. For Sukarno, Silaban was a man to consult concerning urban design quality of Jakarta.

As the first generation of the native architect, Silaban worked together with other native graduates of the School of Architecture at the former *Technische Hoogeschool te Bandoeng* to set up *Ikatan Arsitek Indonesia*—Indonesian Institute of Architects. They met together in Jakarta 1959 in response to the initiative of the Ministry of Public Work for instituting planners/designers and builders/developers under a single roof. He and Suhartono Susilo did not agree with the idea and gathered newly graduated people with an architectural degree to establish a professional association of architects. Indeed, the persistence of Silaban and the conscience of Susilo led the way to the formation of the Indonesian Association of Architects. Silaban, Suhartono Susilo, Lim Bwan Tjie, and 18 other graduates of the architectural program held a meeting in Lim's residence at Jalan Wastukencana Bandung, then followed with a decisive meeting at Theehuis in Dago area of Bandung on September 17, 1959; from this meeting they drafted and conceptualized the charter of the organization known as IAI—*Ikatan Arsitek Indonesia* (Aristiana and Murtiyoso 2005) (Fig. 8.2).

Fig. 8.2 Monument of West Papua Liberation, Banteng square Jakarta, photo by the author

Since its conception, Silaban was an active founder and member of IAI. He participated in most periodic meetings and conferences of the organization. One notable contribution was his paper: *Idealisme Arsitektur dan Kenyataannya di Indonesia*, presented at a conference in Yogyakarta in 1982. In his paper, Silaban carefully outlines his long-standing experience as a practicing architect in dealing with a tropical climate (Silaban 1984, 75–89). Between the lines of his paper, he is in the hope and appeals that architects for their trustworthy judgment are based on best practices experience. He unveils the necessity for honesty, modesty, durability, and simplicity of architectural design and construction in the search for graceful beauty.

8.3 Reading Silaban's Mind from His House

Silaban designed and built his family residence in Bogor between 1958 and 1960. The plan of the ground plan of the house covers 432 m^2 with a mezzanine plan of 72 m^2. The house stands in a setback position on the site of 1004 m^2 at the 19 Gedong Sawah street. In this house, Silaban, his wife, Lefty Kievits, and their children used to live and grow together as a family. His wife is a daughter of a Dutch East Indies plantation enterprise holder in the village of Cibatok, district of Ciampea, Bogor region; the village is located 17 km away from the town. The Kievits family sent their children for school in the town of Bogor, Lefty went to Catholic school—now Regina Facis—in Bogor. In 1939, Silaban was engaged with Letty Kievits; they got married on October 18, 1946.

From his marriage with Lefty Kievits, Silaban had grown to ten children: two daughters and eight sons. Two maids usually help the family and living in the house. The most important ritual in the house takes place in the dining room. Silaban sits at the head of the dining table and leads to praying before they have a meal together, mostly for dinner. During dinner, conversations and talks are mostly dominated by the father. No one is allowed to leave the table before the father says so. The commencement and closing ritual of the meal are made by the father of the family.

The house is juxtaposed deep into the site facing the street on the south that creates an open area in front of the house under shade in the afternoon. Children of the house play and spend their time in the late afternoon. The front yard of 18 m deep from the right of way provides a good visual control from the living room, master bedroom, and office to the street and driveway. In his conference paper for the Institute of Indonesian Architects in Yogyakarta 1982, Silaban points out that the sitting building designates something concerning the personality of its owner (Silaban 1984). Indeed, the deep setback of his residence designates his considerate presence in a neighborhood that values respect and privacy.

The rooms and domains in the ground plan are arranged with a central corridor and divided into two main areas: front and rear wing. The front/south wing comprises one master bedroom of 4.5×6 m^2, office/studio of 4.5×9 m^2, one bathroom, two toilets, and living area of 6×6 m^2 while the rear/north wing consists of four bedrooms of 6×3 m^2, dining room of 6×6 m^2, kitchen and service area, maid room, and car garage. The main corridor of 2 m width enacts as the backbone of the room layout connecting the garage and living/dining area; stairs to the mezzanine are provided at the edges of the main corridor. The mezzanine of the house is mostly not for daily use, but for spare rooms. Visual control to rooms and domains of the house is made possible from the corridor of this mezzanine. Ritually, Silaban walks along the corridors at the ground floor and mezzanine in the night and early morning.

Besides the dining/living area, the office/studio in the house occupies a dominant position and function; Silaban's workplace is the core of the house in terms of control and location. Inside the studio, Silaban puts his books, drawings, drawing table, office desktop, music instruments, memorabilia, music records, gramophone, and his favorite ceramic collections. Silaban collects not least than 600 serious books—mostly architectural and engineering subjects in Dutch, German, and English languages—and reads them mostly in the evening. Like many other men in his generation, Silaban smokes cigar and tobacco with a pipe. He collects several pipes from various designs and materials. Silaban enjoys smoking after having a meal and after his work is done. At home, Silaban spends most of his time in the studio and on the front terrace. In his paper mentioned earlier, he underscores the significance of the long overhang for the terrace—*emper*—in the tropical building. He reveals the importance of such a terrace that is not only for tropical comfort but also for the daily lifestyle of Indonesian people (Fig. 8.3).

The construction of the house is developed from the rectangular ground plan and elaborated to optimize the site. The structure of the house is basically concrete slender bearing walls with steel beams' structure. Simplicity and practicality characterize the whole structural system that consists of main steel beams, purlins, and rafters. In order to manage the heat of the day on the wooden tile gabled roof, Silaban provides wood insulation. The use of steel for the roof system of house construction is a considerably new experiment of the post-colonial building in the 1960s. Silaban leads the residential design of architecture in the country by integrating masonry, carpentry, and steelwork into a light tropical construction system; the construction has survived without any damage from several earthquakes (Fig. 8.4).

In many occasions and talks within the circle of professional architects, Silaban values modernity in terms of simplicity, durability, practicality, and cost affordability. Material and construction should be straightforwardly presented in accordance with its innate visual quality. For his family house, the materials he chooses are stone,

Fig. 8.3 Silaban's studio and office at Jalam Gedong Sawah II/17 Bogor, photo by the author

Fig. 8.4 Front terrace of Silaban's house 2014, photo by the author

wood, steel, glass, wooden roof tile, and ceramic floor tile. People find the resonance of Adolf Loos's dictum for replacing the highly decorated surface with a visually natural look of the material. In many ways of construction, Silaban tries to avoid highly sophisticated and complicated details at his building. He believes in simplicity and durability as the source of beauty and dignity. In addition, the search for beauty for him is correspondingly to reveal pure colors at certain components of his building, such as a door, stairs, kitchen, and steel beam for the roof structure. As a whole, the wall system of the house and its windows is dominated by white color. Some parts of the wall and outdoor floor are covered by stone tiles in their natural appearance, especially in the entrance area and terraces. Architecturally speaking, Silaban treats his family house as a habitually ordering system. Consistency, order, and repetition of elements become an integrated part of his architectural characteristics. Silaban believes in the sense of order for efficient use and cost. He values the durability of materials in dealing with tropical humidity and simplicity for composition and construction. Openly, he rejects the way the Dutch East Indies architects do in Indonesia (1800–1942) that commonly build their tropical building with thick masonry walls, large rooms, and European styles.

8.4 Silaban and Sukarno (1955–1965)

Prior to his encounter with President Sukarno, Silaban was the director and practicing architect at the public work division in the municipality of Bogor. Sukarno heard about Silaban. He prepared the designs of several public buildings and utilities. Some significant design works between 1949 and 1959 are secondary vocational high-school building in 1951, office for fishery affairs in 1951, and the official residence of Bogor municipality's major 1952 in Bogor. Silaban was commissioned by the government of Indonesia to design the following buildings in Jakarta: the gate of Kalibata National Cemetery 1953, the first scheme of Mosque of Istiqlal design competition 1955, National Museum design proposal 1955, Bank of Indonesia headquarters 1960, and BNI 1946 building. Moreover, Silaban in his home studio worked and participated in various design competitions after his office hours in the public work office; the government office at the time opened from 8.00 am to 14.00 pm.

Public buildings and projects that he won from design competition between 1949 and 1970 are third prize for Faculty of Agriculture, University of Indonesia in Bogor 1949, second prize for National Monument in Jakarta 1954, first prize for Bank of Indonesia Headquarters in Jakarta 1954, first prize for the gate of National Cemetery in Jakarta 1954, first prize for Mosque Istiqlal in Jakarta 1954, first prize for BNI 1945 headquarters in Jakarta 1959, first prize for Gedung Pola in Jakarta 1961, and first prize for the Lapangan Banteng Monument of West Papua liberation from colonialism. The aforementioned built and unbuilt works show Silaban's design traits and qualities concerning the tropical monumental building (Fig. 8.5).

At the end of 1949, Indonesia as a newly independent nation-state survived from the comeback of the Dutch colonialism. The roundtable agreement at The Hague led to the transfer of power from the Dutch to the Republic of the United States of Indonesia with Sukarno and Hatta at the center of the political stage. The Indonesian society at the end of 1949 as described by Herbert Feith was beset by the cleavage and obstacle to reach consensus and legitimate authority as the outcome of ethnic and religious diversity (Feith 1962/2007, 27). Internal conflicts and disagreements

Fig. 8.5 Bank of Indonesia Headquarters, Jalan Thamrin Jakarta, photo by the author

within the multipolitical parties endanger the consolidation of the republic from onset since 1945. However, few political leaders and observers at the time spoke of the significance of democracy as a means for nation-building. On December 18, 1949, Sukarno was applauded as the president of the Republic of the United States of Indonesia; he was in the role of the great leader of Indonesian Revolution while Hatta was to undertake his function as the prime minister of the administration. In such a situation, Sukarno was able to play the role as the father of Indonesian nation; he did not need to do day-to-day government business. He could concentrate to bring about Indonesia's Revolution to its triumphal end. His vision was outlined with the architecture and urban design for '*Proyek Mercusuar Nusantara.*'

Silaban mentions his encounter with Sukarno at the first time in an interview with Solichin Salam (1979: 63); Silaban met Sukarno during the Japanese occupation in 1942 in the house of Dutch painter Ernest Dezentje at Jalan Bodongan, Bogor. Both persons involved in a serious discussion on arts and architecture. The first impression Silaban got from Sukarno was that he met a great personality with a broad-minded outlook on universal humanism. Personally, Sukarno knows Silaban because the president asks him as a person and his family; in Silaban's notes, Sukarno wants to visit his home and family. After 1942, Silaban did not have a chance to meet him in person again for several years until 1959. However, Silaban was always there attending and listening to Sukarno's speeches in the theater of maxim or town square when he visited Bogor during the Japanese occupation. Surprisingly, on Sunday morning of August 1959, Sukarno called Silaban to come over to Bogor Presidential Palace. In his notes on the meeting, Silaban was not aware of why he was invited to meet the president. Sukarno asked Silaban for his professional judgment concerning the master plan of Jakarta for hosting the fourth Asian Games in 1962.

The most impressive engagement with Sukarno was when the president conceded the judgment of Silaban on architectural matters for the site planning of the newly proposed stadium and other sports venues for the games. Sukarno listened to Silaban's professional advice on the urban design considerations of the sports venues complex. For Silaban—regarding his background in patrimonial Batak society—such acknowledgment for his competence and expertise was hardly imagined to come from a great leader like Sukarno.

Since his first encounter with Sukarno in 1942, the president accepted Silaban who spoke straightforwardly from his heart. Sukarno regards Silaban's touch for architecture that expresses and articulates energetic expression of form and appearance. Rosihan Anwar notes in his memoirs (2009: 121) how Sukarno was angry with the Prime Minister Ali Sosroamidjojo when the Concordia Building was not well renovated according to Silaban's design treatments; the conference hall was gloomy, and the façade was not visually appealing. All this for Sukarno was the sign of weakness and incompetence that should be not the image Sukarno wanted for hosting the delegates of the Asia-Africa Non-Alignment Conference in 1955 in Bandung.

Prior to the presidential decree on July 5, 1959, that ended the constitutional democracy in Indonesia, Sukarno was not busy with the administrative governance; the parliamentary administration allowed him to focus on 'nation-building' through culture and architecture. For Sukarno, as the capital city of the newly established

Republic of Indonesia, Jakarta should represent a newly constructed image of diverse political, ethnic, and religious groups. Accordingly, the architecture of the new capital should utilize contemporary design, material, and technology. Sukarno believes in the modernity of architecture as the starting point of nation-building; modernity offers the equal opportunity of form, structure, construction, and style to represent new ideas and needs. In its essential terms, the modern design enables new buildings in Indonesia to overcome local/regional differences and to liberate their appearances from colonial traces, idioms, and vocabularies.

The vision of Sukarno on the urban design of Jakarta is to establish an avenue of freedom as performing stage that runs from the Merdeka Square at North to the gate of Kebayoran Baru new town at South; along the avenue, people are expected to experience the new world of Indonesia. Silaban learns this vision from Sukarno during his meeting in Bogor 1959 concerning the site selection for the fourth Asian Games complex. At the time, Sukarno finds his momentum to actualize his vision because the post-war reparation fund from the Japanese government to Indonesia was confirmed in 1958: $223 million for restoration over 12 years and $400 million for economic development aid over 20 years (Ishikada 2005, 23). The fund enables Sukarno to prepare and build facilities and infrastructures of the games. For his sports venues complex, Sukarno does not need to worry about because he finds support from the Russian government.

The contribution of Silaban to Sukarno's vision for the sports venues—especially for the stadium—was not limited to its site selection. Regarding his long career as public work director and architect, in 1950 Sukarno installed him to be a chief of the municipal experts for the architecture of Jakarta. His main job was to attend periodical meetings for advising and reviewing the ongoing plans and physical development of the capital city. Silaban led the municipal expert council for architecture until 1955. One national project that potentially transformed the image of the city is the sports venues complex for the fourth Asian Games.

Silaban prepared technical assistance for the preliminary design of the stadium for 100,000 spectators and its typical construction details. The Russian engineers developed the structural system for the tribunes and roof system, while Indonesian army engineers and technicians worked for the construction. Routinely, Silaban supervised and controlled the development of the construction process; he checked on the working drawings page by page and detail by detail before Sukarno puts his initial for his approval.

Architecture and urban design for Sukarno are not only the vehicle of nation-building but also the actualization of Indonesia in its three-dimensional form. He speaks of three-dimensional construction in his speech on August 24, 1961, for the first lying foundation of the mosque of Istiqlal in Jakarta. By reading between the lines of his speech, Sukarno urges the architectural visualization of Indonesia's rebirth after years of Western colonialization that recalls the glory of the past from Sriwijaya, Singasari, and Majapahit. He takes the examples from Hindu–Buddhist temples from Candi Borobudur, Candi Prambanan to Candi Sukuh. The last monumental building catches Sukarno's attention because its form and design are uniquely constructed in the form of ziggurat; the form has been familiarly found in the shrines of ancient

Fig. 8.6 Candi Sukuh, from ca. fifteenth century, in the village of Berjo, district of Ngargoyoso, the regency of Karanganyar, Central Java, photo by the author

Indonesian societies, known as *punden berundak*. Silaban knows well Sukarno's subtle demand from his personal encounter and speeches that the liberation from foreign influences is architecturally necessary, but it is by no means to cost local characteristics and identity (Fig. 8.6).

8.5 Silaban and His Opus Magnum

In his career as a professional architect and Christian, Silaban never imagined that he won the first place for the national design competition of mosque of Istiqlal. His proposal was selected by a committee consisting of politicians, Islamic clerics, engineers, and architects. In his speech for the construction of the mosque in 1962, Sukarno told his audience that the origin of the idea for the urban mosque had been deliberated in 1944 in an informal meeting of Islamic clerics in his residence at Jalan Pegangsaan 56 Jakarta. Kiayi Haji Wahid Hasyim, Agus Salim, HOS Tjokroaminoto, and Anwar Tjokroaminoto were among the initiators of the mosque; both are well known as the proponents of the Indonesian movement for independence since 1930s. Under the Japanese occupation, the discussion is on urban mosque stroke Sukarno's mind concerning its meaning for people. His audience—as told by Sukarno in his speech—was all excited and eager to donate their money for the construction; they were mostly not aware of the cost and process for the construction of an urban mosque for more than 10,000 prayers.

Because of political hardship, Sukarno was not able to deliver his promise to his audience on the construction of the mosque right after the end of the Pacific War. However, he just waited for the right time. Between 1949 and 1959, Sukarno was personally and publicly not busy with government business; under the constitutional democracy, he was not in charge of the administration, but he was the head of the state; he was busy with ideological and moral duty for the unity of diversity in the country (Poesponegoro and Notosusanto 2008, 301–418). Sukarno was well known as a charismatic father figure for Indonesian nation-state; he was skillful and competent for public mobilization.

Indeed, he passionately mobilized people with rhetorical speeches for the artistic and sociocultural struggle toward the unity and integration of political, ethnic, and religious groups. Sukarno believes that the nation-state needs visually appealing embodiment of pluralistic unity and patriotic spirit. The easy way to fulfill this need is learning from the best practice in other countries. Silaban was one important person for this embodiment and its learning process. For the best interest of people, Sukarno asked Silaban to accompany the president visiting cities in various foreign countries with unique architecture and magnificent histories of civilization, such as Paris, Moscow, Rome, Vienna, Washington DC, Brasilia, Tokyo, Brussel, Antwerp, and New Delhi.

After his first visit to the Netherlands in 1950, Silaban had a chance to visit Japan, India, Burma, and the Philippines. The patronage of Sukarno assigned Silaban as his client for grandeur architecture and urban design. In return, Silaban served his patron with an opportunity of actualizing and expressing his architectural passion for his patrimonial aesthetics. Both men had something in common for patrimonial values system and its architectural representation. In his letter to the United Nation's adviser: Professor Alvaro Ortega, in the opening, he writes and describes about himself as an architect but not an ordinary one (Silaban 1967; Muramatsu 2008, 1); Silaban is proud of himself as an autodidact person with passion in architectural design. For Silaban, Sukarno is not only his architectural patron but also a great national leader and a fair politician in dealing with his opponents (Salam 1979, 63).

Concerning architectural culture, the most appealing things for Silaban are how modernity and tradition come to terms with contemporary needs. In his visit to India in 1952, Silaban observed British experiment of Sir Edwin Lutyens and Sir Herbert Barker with modern architecture for the New Delhi capital in 1912. The integration of landscape and architecture is well juxtaposed with grandeur avenues that connect public buildings and monuments with vistas and open green spaces. For Silaban, Lutyens' plan for New Delhi is uncontestably Garden City on earth. From this Indian experience, probably Sukarno and Silaban came to awareness of the necessity of connectivity of landmarks for urban design. The grandeur avenues and open spaces were found in the newly established government center of Brazil: Brasilia. In his travel notes of 1954, Silaban mentions his admiration for Brasilia designed by Lucio Costa and Oscar Niemeyer. Silaban fascinated by the dramatic scenes of grand avenues and iconic buildings in the midst of woodland. Sukarno shared a similar experience with Silaban on his admiration for the architecture and urban design of Brasilia (Sopandi 2009, 56). Jakarta's location and the setting are likely not close to neither Brasilia nor New Delhi. Sukarno and Silaban discussed what they have observed in New Delhi and Brasilia; they shared the common ground on the necessity for monumentality in terms of durability, order, simplicity, and clarity.

The echo of such design principles was experienced by Silaban during his visit to Japan in June 1959. Unlike modern national capitals, Japan—as observed by Silaban—was able to manage modernity without losing their sociocultural roots, traces, and characteristics. Still, Silaban did not too much agree with Sukarno in his dream to build a national capital of Indonesia's Brasilia in the midst of Kalimantan woodland; Sukarno pursued his dream in central Kalimantan with his preliminary plan in 1956 for its government complex (Poesponegoro and Notosusanto 2008, 533), but later on the plan was abandoned in 1959 (Wijanarka 2006). Sukarno puts the first lying stone for the future capital city and gave its name: Palangkaraya meaning sacred and highly regarded place; the ceremony for the initial construction was on July 17, 1957, witnessed by the Russian ambassador: DA Zukov and the US high-ranking diplomat: Hugh Cumming Jr.

While Sukarno was busy conceptualizing the national capital in its three-dimensional form, Silaban was eventful with working on design competitions for several public buildings in Jakarta. Lessons learned from his visits with Sukarno, Silaban put his attention to his design proposals. From his archive in Bogor, there were traces on sketches and drawing studies that he worked hard for bringing about potentially iconic and classical structures by trials and errors. However, Silaban did not exaggerate his designs just simply for becoming the landmarks of its adjacent urban fabric. Rather, what is Silaban to achieve—as described in his conversation with his family circle—is to embody the grace and beauty of the Almighty on earth for humanity. In this spirit, Silaban participated in the design competition for the national mosque in Jakarta in 1954. His submission code is 'divinity'—*ketuhanan*.

The design proposal of Silaban for the mosque of Istiqlal revealed to the competition committee after several times of deliberation. The committee consists of the components of government, religious leadership, professional, and expert—Rooseno, Djuanda, Suwardi, Hamka, Abubakar Aceh, and Oemar Husein Amin; there was no any female person sitting in that commission. Wahid Hasyim as the ministry of religious affairs 1950 sat in the committee and initiated to find an endowment foundation for the construction of the mosque on December 7, 1954. Under Sukarno's order, the foundation was led by Anwar Tjokroaminoto and launched the design competition openly on February 23, 1953. The selection process took place in Jakarta in 1955. The committee received 30 entries; they examined and evaluated the proposals and came up with five nominees. Silaban's design proposal won the first prize.

How could a Christian architect win the first prestigious place for the national mosque design completion? No one could imagine before including Sukarno. The committee confirmed their nomination for Silaban's scheme to be constructed. Sukarno agreed and mobilized donation for the actualization of Silaban's design. What is interesting to note on the plan of the mosque is the fact the 8 m height dome is supported by 12 columns (O'Neil 1993, 156–8; Pringle 2010, 187). Are the pillars the representation of 12 Christian apostles? Silaban does not respond to speculation on the number of columns, but simply answer the number of columns has something to do with the structural load as recommended by Professor Roeseno and Professor Mann—structural engineer and expert of polyhedron dome construction from the University of Darmstadt, Germany. The original scheme by Silaban uses 24 columns. Mann and Roeseno worked on the columns for the dome with more efficient space so that more people are able to use the floor between the pillars for their prayer.

The intriguing question for Silaban was why he did not apply traditional Indonesian roof form for the mosque like the grand mosque of Demak, Yogyakarta, and Cirebon? The question was sent by a student to Silaban by mail from Surabaya on August 18, 1980. Silaban explained his design consideration for the dome and minaret was in response to the terms of reference as requested by Sukarno for the national mosque. Sukarno did not want to experiment with the international image of a mosque. However, the president demanded modern construction system with a familiar feature of a mosque. Thus, he asked for uniquely minaret design and structurally advanced dome construction, and later this structure was known as a concrete polyhedron membrane system. Silaban responded to the request and proposed the dome of 45 m diameter at the central area of the square plan of 87 $\times$ 87 m^2.

For most scholars of architecture, the dome has been familiarly known since ancient time Rome's pantheon. Silaban understood well the historical root of dome construction. What was new for Silaban to work out was how to ventilate the dome from the warm air inside. Silaban provided a ring structure at the top area and put the circular cover with a small gap for roof ventilation. Indeed, in the execution of the scheme, the structural engineer, Roeseno, asked Silaban to work out the membrane.

After the announcement of the winner on July 5, 1955, Sukarno was still not sure about the fund for its construction. The first step he did is to do the site visit on November 14, 1956. The site was former Dutch military citadel of Prince Friederick Hendrick with its unique landscape architecture. Important structures in the park included the bronze statue of Vredes Angels 1822 and the Waterloo monument dedicated to the Dutch soldiers fallen in Aceh war (1873–1904). The citadel and the monument were situated in the English landscape park of the Queen of Wilhelmina. Based on the photographic collections of the Tropenmuseum in Amsterdam, the site was perfect 'Garden of Eden' in Batavia between 1910 and 1942 (Harper-Brook 1911, 21); it was well composed with river flow, alleys, open lawns, isolated groups of trees, and vistas. Initially, the construction of the park was intended by the governor of Dutch East Indies: Johannes van den Bosch, for the Citadel (Fig. 8.7).

The elaboration and development of the park began in 1910s by the Dutch authority and dedicated to the Queen of Wilhelmina Helena Pauline Marie van Orange-Nassau (1880–1962). Since the beginning of Japanese occupation in 1942, the park was gradually neglected and abandoned for almost one decade. At the end of 1950, Sukarno renamed the park to Taman Wijaya Kusuma. Later, Sukarno saw the site as the potential location for the national mosque of Istiqlal. Indeed, he ordered the public work department and the Istiqlal foundation to do the site clearing on his site visit. The Vice President: Mohammad Hatta who ran the business of the government disagreed

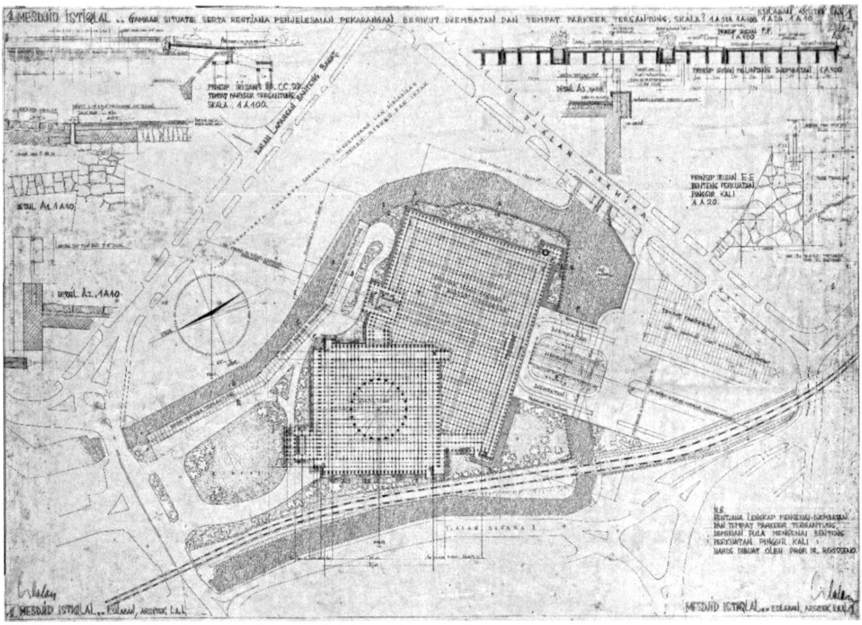

Fig. 8.7 Original ground plan of Silaban's design competition entry for the mosque of Istiqlal 1954, photo by the author

with Sukarno's because the demolition of the citadel would cost the administration too much money (Fachrurazy et al. 1972, 11).

The site of the former park of Wilhelmina occupies an area of 6.4 ha with the ruins of citadel, monuments, and wild bushes. Sukarno did not give up because of the cost and insisted to turn the park as the location of the future national mosque. He was supported by the members of the foundation of Istiqlal. To clear up the site, Sukarno mobilized volunteers from various resources: government officers, Indonesian armies, and citizens of 60,000 people. The most challenging work was to demolish the citadel that was constructed to handle explosives, cannonballs, and bullets. The site clearance was done in a few months after the president's site visit.

In several conversations with friends, colleagues, families, and relatives, Silaban delineates the main idea of the mosque in the search for the experience of grace and beauty of the divinity that is timeless and great. Sukarno's ideals of great architecture from Candi Sukuh have inspired Silaban on the use of durable materials and clarity of form for monumental presence. In order to achieve this, Silaban develops and elaborates the form of the building that is geometrically strong, compositionally consistent, and visually dominant (Fig. 8.8).

The main building consists of two main forms: box and dome with a clear orientation to Mecca's Ka'bah, but is in harmonious visual connection with the adjacent National Monument and other government buildings. The main building covers 5476 m^2 ground floor area and 4800 m^2 mezzanine floor area. The capacity of the

Fig. 8.8 Exterior perspective view of Silaban's design competition entry for the mosque of Istiqlal 1954, photo by the author

mosque was for more than 120,000 prayers, inside and outside of the building. In order to provide more space for special occasions such as Ieds and Friday prayer, Silaban provides and connects the main building with a large open court of 2200 m^2. The arcade structures of 6 m width and 4 m height enclose the open court that gives spaces for other activities related to the function of the mosque. At the basement of the mosque of 22,000 m^2, Silaban provides with rooms for educational purposes, offices, and washrooms (Fig. 8.9).

The monumentality of the mosque is not only managed by the magnificent scale of form and spaciousness, but also by the repetition of columnar elements of the mosque with a monochromatic expression of yellow–gray marble color. Actually,

Fig. 8.9 View from the inner court of the mosque of Istiqlal 2014, photograph by the author

Silaban wanted to modify his design due to its limited area of the site. Sukarno disagreed and asked Sutami as the public work minister for the relocation and realignment of the railways that enlarged the site for almost 50%. Sukarno laid the first stone foundation for the mosque on August 24, 1961. From this occasion, there was nothing materialized from the mosque of Istiqlal's plan to construction; Indonesia was in a confrontation with Malaysia and internal economic and political insecurity. The climax of the political and economic instability was marked with the September 30, 1965, movement by the alleged procommunist army group that murdered six army generals and one lieutenant. A few days later, mass conflicts and riots between Sukarno's supporters and its opponents happened in most towns and cities. Major General Suharto and the army were able to seize the situation by scapegoating the Communist Party of Indonesia; by 1967, Sukarno's power was diminished through the parliamentary process, while at the same time, Suharto was finally confirmed by the Indonesian Parliament as the acting president until 1968. During the uneven transition of power between 1965 and 1967, Silaban was never officially released from his duty for supervising the construction of the mosque of Istiqlal. Although there were no viable options for the execution of his project, he passionately kept working further for its construction details of the mosque, especially for the permeable wall system made of stainless steel.

Despite the mobilization of the fund and other resources, the construction of the mosque did not make any significant progress under Guided Democratic Sukarno's administration until 1967. The end of the light in the tunnel for the construction was delivered by General Suharto, as the New Order President; he invited Silaban to meet him and confirmed his decision with the Presidential Order no. 14 1968 for the restructuring of the committee for the Istiqlal construction; the New Order administration (1967–1998) allocated the fund for the construction in their national budget each year from 1968 to 1972. However, the completion of the construction was also made possible by the public supports and donations from business people and ordinary citizens. Eventually, on February 22, 1978, Suharto announced the official opening of the mosque for public use; on the date, Silaban was invited by Suharto to attend the opening of its public service. Since then, Silaban never visits his opus magnum until his last day on May 14, 1978, at the age of 72 years.

What is interesting to note from a gender perspective is the use of space in the mosque. Under Suharto's New Order, the occupation of the mosque was strictly organized with a clear division of space for male and female prayers based on front and back category. The arrangement of the domains remained the same under Habibie and Abdurrahman Wahid in 2004. The change of categorical occupation happened under the order of Megawati Sukarno Putri. Since then, the male and female domain in Istiqlal mosque and other mosques in Indonesia are arranged equally with the right/left category; the arrangement is sustained until today (Fig. 8.10).

Fig. 8.10 Interior of the mosque of Istiqlal with 12 pillars 2014, photo by the author

8.6 Concluding Remarks

In most cases, patrimonialism presumes masculine gender preference in society that decision making, exploration, and protection belong to masculine world that lead to the necessity for supremacy, domination, and control, while domestication, accommodation, and habitation that lead to the need of maintenance, nourishment, and sustenance; this preference is traditionally maintained as a fixed role and function of people based on gender identity. In the democratically liberated world, patrimonialism will not work because the division of labor is not based on gender preference but the merits of individuals. The need for masculine aesthetics through architectural work reflects the necessity for strong control and direction.

In architectural work, the strongly ordering system of function, structure, and elements is regarded as overcoming the fear of chaos, anarchy, disorder, confusion, and disorientation. All this is to confirm the relationship between masculine aesthetics and patrimonialism that is mostly established within the framework of monumental arts and architecture (Bressani 2014, 192; Handler 1988, 145). On the other side, the monumental expression shows the masculine weakness for dialogs, appeals, and inclusivity; in daily lives, men are not in good measure to the negotiations, entreaties, seductions, and threats and seductions initiated by women (Inhorn 2009, 36).

In modern movement, patrimonial characteristics are obvious in the use of the strong geometrical form with straight lines. The sense of order becomes dominant in building appearance. The amplification of the sense of order is underscored with the repetition of form for a structure that creates the scenes of rhythmic progression and direction. The architectural scheme for Berlin Prachtstrasse of NAZI Germany 1937 by Albert Speer is an exemplary work of patrimonialism in a totalitarian regime of Adolf Hitler (Sereny 1995). The core of patrimonialism is not impartial from the necessity for control and domination. In a matter of fact, there is no practice of patrimonialism without oppression and coercion; the practice commonly utilizes customary, religious, or ideological doctrine.

Friedrich Nietzsche as discussed by Heidegger in his lectures on The Will to Power as Art unfolds the way we deal with the work of arts that masculine aesthetic is in the terms of artist's experience as producer (Heidegger 1991, 77–80); for Nietzsche, the common experience of what is beautiful is, is feminine aesthetics in the sense of submissive perceiver. The essence of association of the work of art and architecture with masculinity lies in its artistic quality in the sense of commanding assertion. At its presence, the work of arts and architecture in terms of Nietzsche's the Will to Power has once the capacity to be masculine or feminine. If it is energetic, muscular, and productive, it is also fickle, mendacious, and seductive (Eagleton 1990, 257).

In many ways, the work is mostly the amalgamation of sexual stereotypes. Masculine aspect of architecture has been familiarly represented and expressed in various monumental buildings; syntactically, such architecture shows strict geometrical structure, dominant repetitive element, and strongly prescribed appearance. Semantically, masculine architectural composition entails the sense of strength, order, direction, and hierarchy. Male preference for aesthetics seems to limit an architectural composition for the options of freedom, informality, and playful experience.

Familiar examples of masculine aesthetics in modern architecture are mostly associated with the works of Adolf Loos, Frank Lloyd Wright, Meis van der Rohe, Aldo Rossi, and Oswald Mathias Ungers. The necessity for masculine aesthetics is to visualize the strength, rigorousness, and clarity of order and authority; architecture becomes the means and medium of the owner's need and wants in terms of personal or ideological preferences. Unsurprisingly, fascism, despotism, totalitarianism, and communism apply architecture with masculine aesthetics that demonstrate not only the contender of femininity but in most cases also the sense of oppression and repression, such as Stalinism in USSR and other totalitarian regimes of North Korea and German Nazi. Besides its essential nature of woman, feminine aesthetics shows the universal qualities and values of affectivity, corporeality, and sensibility (Richard 2004, 29).

The relationship between gender and architecture is apparently inescapable from the association of building and the human body. Ayn Rand's novel: The Fountainhead portrays the protagonist: Howard Roark, architects, as an ideal male figure with the vigorous body, composed of long, straight lines, and angles, each curve broken into planes. Rand's description of the ideal man: Howard Rocks is inspired by Frank Lloyd Wright's the Falling Water. Architecturally speaking, feminine aesthetics is aptly associated with curvy, flexible, and variable lines and shapes while the masculine arts are pertinently expressed with straight, firm, and strong lines and shapes. On the other hand, the architecture of Art Nouveau shows feminine qualities with organic lines and motifs. Rand's Fountainhead unveils the dialectical relationship between architecture and masculinity in mutually reinforcing ideologies that invoke each other to counterbalance and sustain its particular entitlements and purposes (Sanders 1996, 11); the relationship between masculinity and building design takes its hallmark to the multifaceted interdependency of gender and architecture as a cultural construction.

Indeed, in most cases, it is complex and difficult to draw the line between the masculine and feminine identity of the works of architecture because there is the necessity of harmony between contradictory elements and aspects in any composition. The necessity for the structural logic of building construction entails masculine components while the attractiveness of building appearance requires feminine characteristics; most works of architecture involve and demonstrate the interplay and embodiment between both complementary aspects in a composition. In other words, only a few works show a dominant expression either masculinity or femininity.

Cross-culturally, the most common traits and features of patrimonial practice are associated with strength order, and power through muscular, directing, and controlling expression; all these are closely related to and associated with the muscular power of man or masculinity. In the Indonesian context, masculine and feminine traits and features have been inseparable entities of symbolism, as shown in most ancient temples, such as the temple of Penulisan in the village of Sukawana, the district of Kintamani, Bali and the temple of Sukuh in East Java. The ancient architecture of Java and Bali demonstrates male, *linga*, and female, *yoni*, symbolism as the Supreme Unity and Totality of all beings that signifies the safe, healthy, and harmonious world (Kempers 1959, 19).

Historical geography of Indonesia is conspicuously demonstrated with the transformation of landscape urbanism under the patrimonial leadership. Patrimonialism has been playing an important role in the conception and application of virile public aesthetics that had worked for the necessity for the nation-building of Indonesia. Silaban and Sukarno represent prominent figures of the nation-building whose background and way of thinking are paternalistic. They share similar patrimonial values and masculine aesthetics that life is graceful, meaningful, and beautiful because of strength, order, and authority. Artistic work and architecture of Silaban actualize such values and ideals. After the fall of patrimonial regimes under Sukarno and Suharto, the call for patriotic identity and unity becomes less necessary, but the need for consumerism and liberalization of public aesthetics is more apparent through gigantic malls development everywhere in the country. Is this the sign that patrimonialism to be continued through capitalistic private corporations?

References

Adams J, Charrad M (2011) Patrimonial power in the modern world. Ann Am Polit Soc Sci 636

Anwar R (2009) Sejarah Kecil, Petite Histoire, Indonesia Jilid 2. Buku Kompas, Jakarta, p 121

Ardhianti Y (2005) Bung Karno, Sang Arsitek: Kajian Artistik Karya Arsitektur, Tata Ruang Kota, Interior, Kria, Simbol, Mode Busana dan Teks Pidato. Komunitas Bambu, Jakarta

Aritonang JS (1994) Mission schools in Batakland Indonesia 1861–1940. Brill, Leiden

Aristiana, Sutrisno M (2005) Perkembangan Arsitek sebagai Profesi dan Lahirnya Ikatan Arsitek Indonesia. IAI Jawa Barat, Bandung

Awal H (2007) Catatan Kecil Beberapa Pertemuan dengan Silaban (1960–1978), dipersembahkan dalam sesi diskusi workshop tentang. Rumah Silaban, Bogor

Bertrand J (2013) Political change in Southeast Asia. Cambridge University Press, Cambridge

Bressani M (2014) Architecture and the historical imagination: Eugène-Emmanuel Viollet-le-Duc 1814–1879. Ashgate, Farnham Surrey

Brown D (2003) The state and ethnic politics in South East Asia. Routledge, London

Carden-Coyne A (2005) From pieces to whole: the sexualization of muscle in post-war body building. In: Forth CE, Crozier I (eds) Body parts: critical explorations in corporeality. Lexington Books, Lanham, pp 202–227.
Case W (2014) Routledge handbook of Southeast Asian democratization. Routledge, London
Crough H (1978) The army and politics in Indonesia. Cornell University Press, Ithaca
Eagleton T (1990) The ideology of the aesthetic. Blackwell, Oxford
Fachrurazy JH, Nasution S, Bafagih H, Shahab A (1972) Pembangunan Mesjid Istiqlal. Departemen Penerangan RI, Jakarta
Feith H (1962/2007) The decline of constitutional democracy in Indonesia. Equinox, London
Handler R (1988) Nationalism and the politics of culture in Quebec. University of Wisconsin Press, Madison, Wisconsin
Harper-Brook E (1911) Java and its challenges. Jennings & Graham, Cincinnati
Heidegger M (1991) In: Krell DF (ed) Nietzsche, vol I & II (Trans: Krell DF). HarperSanFrancisco, San Francisco
Inhorn MC (2009) Reconceiving the second sex: men, masculinity, and reproduction. Berghahn, London
Ishikada MY (2005) Toward peace: war responsibility, postwar compensation, peace movement, and education in Japan. iUniverse, Lincoln, NE
Jones T (2013) Culture, power, and authoritarianism in the Indonesian State: cultural policy across the Twentieth Century to the Reform Era. The Brill, Leiden, p 106
Kempers AJB (1959) Ancient Indonesian arts. Van der Peet, Amsterdam
Kusno A (2014) Behind in postcolonial: architecture, urban space, and political culture in Indonesia. Routledge, London, New York
Kusumawijaya M (2001) Sukarno, the architecture of the Capital City, June 03. https://www.thejakartapost.com/news/2001/06/03/sukarno-architecture-capital-city.html
Lukito YN (2015) Exhibiting modernity and Indonesian vernacular architecture: hybrid architecture at Pasar Gambir of Batavia, the 1931 Paris international colonial exhibition and Taman Mini Indonesia Indah. Springer, Wiesbaden
Muramatsu S (2008) Silaban's house. mAAn, Jakarta
Ondang SA (1992) Arsitek dan Karya: F. Silaban dalam Konsep dan Karya. Penerbit Nova, Jakarta
O'Neil H (1993) Islamic architecture under new order. In: Hooker VM (ed) Culture and society in new order Indonesia. Oxford University Press, Kuala Lumpur, pp 159–161
Poesponegoro M, Notosusanto N (2008) Sejarah nasional Indonesia: Zaman Jepang dan zaman Republk Indonesia 1942–1998. Balai Pustaka, Jakarta
Pringle R (2010) Understanding Islam in Indonesia: politics and diversity. Didier-Millet, Singapore
Richard N (2004) Masculine/feminine practices of difference(s). Duke University Press, Durham NC
Salam S (1979) Bung Karno di mata Bangsa. DelaRohita, Jakarta
Sanders J (1996) Stud, architecture of masculinity. Princeton Architectural Press, New York
Scott JC (1972) Patron-client politics and political change in Southeast Asia. Am Polit Sci Rev 66(1)
Sereny G (1995) Albert Speer: his battle with truth. Alfred A. Knopf, New York
Silaban F (1967) Letter of Silaban. Letter to Professor Alvaro Ortega. InterRegional Adviser Building Center for Housing, Building and Planning, United Nations, New York, Bogor, West Java
Silaban F (1984) Idealisme Arsitektur dan Kenyataanya di Indonesia. In: Budihardjo E (ed) Menuju Arsitektur Indonesia. Alumni, Bandung, pp 75–89
Silaban P (2008) Tentang Pappie, in Muramatsu, Silaban's House. mAAN Indonesia Publishing, Jakarta, p 8–9
Sindhunata (2007) Petruk jadi Guru: Manusia & Kebatinan. Penerbit Buku, Jakarta
Sopandi S (2009) Indonesian architectural culture under guided democracy 1959–1965: Sukarno and the works of Friedrich Silaban. In: Vu T, Wongsurawat W (eds) Dynamics of the cold war in Asia: ideology, identity, and culture. Palgrave Mcmillan, London, pp 53–72
Wijanarka (2006) Sukarno dan Rencana Desain Ibukota Republik Indonesia di Palangkaraya. Ombak, Yogyakarta

Chapter 9
Patrimonialism and Marginal Urbanism: Kampung Kota in Java, Indonesia

Abstract Understanding urbanism in the Indonesian world is indispensable to include *kampung kota* as a populist live–work–play system. The reason is plain and simple because *kampung kota* does exist in most urban areas of the country that supports urbanism with labors and services; it is the indigenous phenomenon of settlement and community that provides town and city with a 24/7 live–work–play existence and experience. What is actually *kampung kota* and how are its potentialities possible to deal with global modernity? In dealing with the questions, this study is to investigate *kampung kota* as a socially constructed reality of Indonesian urbanism. The focus of the investigation is to dismantle and unfold local concepts, values, customs, and traditions that interact, collaborate, resist, and challenge modernity in shaping and sustaining Indonesian urbanity. The aim of this investigation is to explore the opportunities and constraints of *kampung kota* in carrying out sustainable development.

Keywords Kampung kota · Jati Padang · Braga · Notoprajan · Marginal urbanism

9.1 Marginal Urbanism of *Kampung Kota*

Populist urbanism in this study is understood as movements, potentialities, forces, and capacities of a live–work–play settlement system that constitutes, strives, and sustains the urbane quality of life with less state intervention. Studies concerning post-colonial urban quarters in Indonesia have been made with special focus on *kampung kota* (Archer 1986; Guinness 1981; Guinness 1986, 1993; Cybriwsky and Ford 2001; Silver 2008) that open and lead this study to the necessity for revisiting and re-address *kampung kota* as the essential component of urbanism in Indonesia. There are two main reasons for this need. First, *kampung kota* communities are the socioeconomically constructed reality of urbanism in Indonesia; there is no any urban area without the existence of *kampung kota*. Second, *kampung kota* is an urban community based on mutual self-help (Newberry 2006: 142) that uphold the socioeconomic metabolism of towns and cities in Indonesia with a specific contribution to the sense of locality; *kampung kota* has been the transformed form of traditional

B. Wiryomartono, *Traditions and Transformations of Habitation in Indonesia*,
https://doi.org/10.1007/978-981-15-3405-8_9

village in an urban context that has been surviving and developing for ages as the intermediary agency of urbanization (Santoso 2006: 82).

Guinness's studies (1981, 1986, 1993) show that *kampongs kota* in Java are more than just settlement in urban areas; they are vital urban communities with specific ecological domains and agents of state policy. This study argues that *kampung kota* is resilient in dealing with global modernity because of its inherent sense of community; the sense of communal belonging constitutes the network of cooperation for urban socioeconomic productions and services. The high population density—between 5000 and 20,000 populations/km^2—of *kampung kota*, comprises diverse professions and functions that enable its inhabitants living in a distinct ecological location. In his recent studies, Santoso (2006, 73–4) points out that *kampung kota* establishes and develops a productive local economic-oriented market based on social collaboration. The existence of *kampung kota* has been an integrated part of Indonesian urbanism since the Dutch colonial ages; it might be earlier than that under the Javanese emperor of Singasari and Majapahit from the fourteenth century.

Since the Dutch colonial era, *kampung kota* of cities such as Jakarta, Bandung, Semarang, Malang, and Surabaya has been the object of political agenda in terms of 'improvement program': health environment, sanitation, building control, and public safety. Even today in the Age of Information, *kampung kota* is still the locus of such urban improvement and development politics. The fact is simply because it is populated by the urban poor and migrants from rural areas. Since urban poverty is a crucial problem of *kampung kota*, its populations are potential voters for most political parties.

Like in many developing countries, in Indonesia, the dialectical relationship between modernity and tradition takes place in the urban area. It is more than just only economic dichotomy between the formal and informal sector. Rather, it is about everything that shapes and is shaped by domains of urban space: from ideological to practical interests, forces, and needs. Regarding its history and origin, *kampung kota* or urban village in Indonesian context is probably the most important site where modernity and tradition encounter each other in various ways of conflicts, assimilations, coalitions, and cohabitations (Fig. 9.1).

Even though its origin was mostly rooted from the pre-colonial *negara* circa before the fifteenth century, *kampung kota* as an urban phenomenon and the problem has been obviously started from Dutch colonial era, especially regarding its dialectic relationship with modernity. Global modernity for *kampung kota* is inevitably challenging because of its potentially transformative consequences on the physical environment and community values. Ostensibly, as a community, the problem of *kampung kota* remains the same; namely poverty, unhealthy environment, and economically marginalized. Until now, *kampung kota* dwellers still struggle for their existence economically, environmentally, and socioculturally. Until recently after Joko Widodo and Basuki Tjahaja Purnama have been elected as the governor and the deputy governor of Jakarta on October 15, 2012, the politics of *kampung kota* development did not change from its precedent Kampung Improvement Program—KIP—of the 1970s. The focus of the program for *Kampung Deret* is the same as the KIP, namely for the improvement of public utilities and physical housing condition. Seemingly, the nature and the capacity of *kampung kota* as an essential component of urbanism has not been explored; less attention has been given to the improvement

Fig. 9.1 Urban scene of Surabaya 2015. Photograph by author

and treatment for the empowerment of its economic network and sense of community in order to explore the capacity of *kampung kota* as the bastion of traditional locality in dealing with global modernity. The search for potential aspects and forces of *kampung kota* community become the focus of this study. Three cases will be presented for the more detailed exposition of what *kampung kota* is: Jati Padang of South Jakarta, Braga of Bandung, and Notoprajan of Yogyakarta.

9.2 *Kampung Kota* Jati Padang

Jati Padang is a *kampung kota* in the district of Pasar Minggu, South Jakarta township. The history of the *kampung kota* began in early 1960s. Prior to 1970, the location was predominantly local farming area for fruits and vegetables. When the Ministry of Agriculture was decided to be constructed in the region of Pasar Minggu, the government of the Republic of Indonesia developed housing units for the public servants. The development of housing for civil servants in early 1970s was a generative structure of the whole settlement of Jati Padang area. The housing was planned as a community of about 200 households with roads network, public utilities, and amenities. From a sociological perspective, the development of Jati Padang housing, known as *complex pertanian*, is an establishment of middle-class *priyayi*'s compound in the traditional kampung landscape. The surrounding areas of the compound were rural houses with bushes and wilderness. The influx of migrants from rural areas of Java increased to live in Jati Padang rustic areas between 1980 and 2000. Most of the migrants were job seekers for labors and services in South Jakarta.

Gradually, the area was developed as a *kampung kota* with the sub-district administrative status of *kelurahan*. Today, this settlement is an administratively integrated part of Pasar Minggu district comprising 250 ha with 6102 households, 39,981 populations and density of 16,009 persons/km^2 in 2010, and 42,655 people with the density of 17,062 in 2013 (www.jakarta.bps.go.id). In 2011, 857 families or 14% of total households fall into the category under the poverty line; most of these work in informal sectors as street vendors and temporary workers. Demographically speaking, *kampung kota* of Jati Padang is a mature and established system of live–work–play with two main roads service with the growth rate of populations 2.42 annually (www.jakselkota.bps.go.id). The intersection of the main roads becomes the referential point of orientation and movement. What is interesting to point out is the fact that the density of populations is aptly increasing with commercial activities in the region.

As *kampung kota* settlements, Jati Padang, is constituted by 101 sub-neighborhood units, *rukun tetangga*, and 10 neighborhood clusters, *rukun warga*. Based on the record of Jati Padang sub-district office in March 2014 on the occupational category, 60% of the total populations are working in urban services for public and private sectors. Less than 26% work in small businesses and trades, while the rest are street vendors, general labors, and temporary workers. According to the record in the Jati Padang sub-district office in March 2014, the number of migrants and foreigners is 54 person, mostly working in the private sector as professionals. However, in Jati Padang area, street vendors are important economic activities that support its street life and urban economy. The main roads—Jalan Jati Padang Raya and Jalan Ragunan Raya—become the commercial lines of the local and regional economy. Both are traffic through the South Jakarta's centers of urban activities: Pasar Minggu and Kemang.

Diverse shops and stalls for foods and goods as well as urban services grow along the main roads; the crossroad and the estuary of local roads are the most important locations for small business and street vendors that serve and support people living in the neighborhood. Along the main roads, the commodities and services of small businesses and street vendors vary from personal grooming, grocery, car workshops, appliances services, foods, and beverages with 170 spots in 2012. The diversity of providers along the main streets demonstrates the complexity of urban products and services (Fig. 9.2).

The economic activities of South Jakarta areas in recent decades, especially after the 1998 political and economic crises, have been growing significantly; the real-estate research by Procon Indah in 2010 pointed out the growth of South Jakarta businesses over the five years with 29%, compared to other parts of the city only 15% (https://www.thejakartapost.com/news/2010/07/29). Regarding its social milieu, geography and transit network, South Jakarta is considerably a preferable choice for office spaces and commercials. *Kampung kota* in Jati Padang is one of the best options of affordable place for office and commercial employees. According to sub-district records between 2008 and 2013, the migrants who stay in rental rooms are between 15 and 17% of the total populations. However, there are migrants who work as street vendors, *kaki lima*.

Fig. 9.2 Map of sub district of Jati Padang, South Jakarta, drawing and photographs by the author

9.3 Kampung Braga, Bandung

Kampung Braga is an indigenous settlement in Bandung which is located in the city center area. Administratively, the kampung is a sub-district of Sumur Bandung. In 2011, the populations of Kampung Braga are 6156 with a density of 11,200 populations/km^2; 70% of the total populations are at the productive category with major occupation as public servants and workers (www.bandungkota.bps.go.id). The site of Kampung Braga is 55 ha along the watershed area of Cikapundung River between Jalan Asia Africa and Jalan Suniaraja; the river runs crossing the inner-city area from the north—Lembang region, to the south—Dayeuhkolot. The history of the Kampung dates back to colonial time at circa 1810s as part of indigenous workers working for Daendels *Groote Postweg* and public servants of local Regent of Bandung. Herman Willem Daendels as the Dutch Governor-General between 1808 and 1811 urged to connect all towns of Java from Anyer on the West to Panarukan on the Eastside. The post road was the most ambitious and controversial project at the time because it was built with local resources on labor forced system. The road construction in Bandung was done in late 1810.

In order to control local regent, the Dutch Governor urged the regent of Bandung: Wiranatakusumah II to move his seat of regency from Karapyak to Bandung at the post road line. The center of Bandung was established on May 25, 1810, with *kabupaten* complex on the south, and *alun-alun*, town square, at the center as well as a grand mosque—built in 1850—on the west and grand market on the east. Meanwhile, the Dutch Rule established their checkpoint, bank, prison, and post office at the north side of the *alun-alun*. The site and location of the center remain the same until today. Kampung Braga's site is within 5 min walk from the center. Prior to the end of the nineteenth century, the site was a green and watershed area without any house on its riverbanks (Fig. 9.3).

The development of the Kampung was not significant until 1930s. It was the time when the *pedati weg* became to be the Main Street of Bandung with restaurants, cafes, theatre, clubhouse—Societeit Concordia—and an amusement center. Since 1852, Priangan region was declared by the Dutch Resident: Steinmetz for tea and cinchona, as well as coffee and rubber plantations. European settlers came to Bandung and its regions. They urbanized the town with European lifestyle with Braga Weg as their Main Street. Behind the street is the empty and watershed area along the Cikapundung River. The development of indigenous settlement was in response to the need for low-class workers of the Main Street activities on the east and market at the center area.

Regarding the significance of its European populations, Bandung was granted by the Governor-General with municipal status as Dutch town in 1906. Meanwhile, the native ruler of this traditional town remained under the Dutch resident's control. In other words, there was two governance in one town: the formal and the traditional authority. The town hall of Bandung municipality was built at the corner of Braga Weg before it had been replaced by *Societeit Concordia* building—now Asia Africa Museum, in 1940 designed by A. F. Albers. The town hall of Bandung moved further north closer to European quarters and railways station.

Fig. 9.3 The aerial view of Kampung Braga 2010. Photography by the author

The populations of European settlers at the time were no more than 1250 populations but in 1930s more than 40,000. Even though the data on native populations in Kampung Braga were not available at the time, the Main Street of Braga activities with European lifestyle from 1920 to 1941 was impossible without the support of more than hundred native workers. With well-connected link with railways built in 1884 from Bandung to Batavia, in 1941, the city attracted more and more tourists and settlers coming to the city and its farm year festivals.

The influx of native people for jobs is unavoidable. Kampung Braga has experienced a gradual change in its landscape from empty land to indigenous quarters. Most of the people living in the area today still remember that there were almost four generations staying in the same home. In 2002, the total populations of Kampung Braga are 8250 occupying 5.5 ha area. Albeit they occupy the green zone of the city, the municipality has granted Kampung Braga a *kelurahan* administration status in 1974.

Unlike traditional Javanese kampung domains, the former area of Kampung Braga has not been designated for specific use as settlement area but green zone, though almost 80% of its populations are low-class working people for city center area. Hitherto, they still live in extreme poverty with small space to stay; less than 8 m^2/household for 4–5 people. Footpaths—*gang*, connecting houses are narrow with a width less than 1.20 cm. Most building materials used for their homes are timbers, bamboo mat, terracotta, and zinc roof which are most fragile and frail for fire.

The development in the Braga area has been in decline since 1970s; *kampung* Braga is situated at the heart of the historic district of Banceuy and Braga. Most of the boutique shops at the Braga Street were closed because most customers preferred to go the modern malls, which were built in various parts of the city of Bandung. At the end of 1987, the historic prison of Banceuy was demolished and developed by a private developer for a shopping center. The prison was built by the Dutch colonial rule in 1877 and dedicated to political dissidents and statutory criminals. The first president of Indonesia—Sukarno—had been detained by the Dutch in 1929 in Banceuy Prison. Today, former Sukarno's prison unit has been preserved as a monument in the site of this former correction facility.

However, the retail business in the Braga area was still not attractive for most of Bandung's populations. The prison was an important historic building, but it did not support urban activities of the city center. The relationship of the Banceuy and Braga area with the urban poor of *kampung* Braga was not attractive. However, regarding its location and historic legacy, the Braga area is always attractive for capitalist developers for their businesses. Indeed, hotels and shopping centers have been built in the area since the 1990s. Despite the availability of commercial facilities, the livability of the area is still less vibrant. One important thing of urbanism that has not been yet worked out is the establishment of mixed uses and multifamily houses in the area so that the *kampung* Braga is socioeconomically improved with land-use intensification. In other words, *kampung* Braga needs to be upgraded as the site of multiincome housing (Fig. 9.4).

Fig. 9.4 Kampung Braga 1920s, Kampung Braga 1950s, Kampung Braga 1970s, drawings by the author

9.4 Kampung Notoprajan, Yogyakarta

Kelurahan Notoprajan is a kampung in the district of Ngampilan, city of Yogyakarta. As most kampung domains in Yogyakarta, its history is indivisible from the establishment of Yogyakarta Hadiningrat in 1775 as the capital of Javanese Kingdom Mataram under the Sultanate of Prince Mangkubumi or Hamengku Buwana I. The center of the domain has been traditionally Dalem Ageng Notoprajan. The land was granted by the Sultan to him to build his domain in circa 1813. Notoprajan is located at the northwest part of the *kraton* outside the palace wall. The seat of Notopraja and his family is built as the core of the kampung. The seat called *dalem* meaning literally innermost or self is a walled compound with complete traditional buildings with *regol*—gate, *pendapa*—reception hall, *pringgitan*—central pavilion, *dalem*—innermost, and gandok kiwa—left pavilion, and gandok tengen—right pavilion. Today, the *dalem* is not in good shape. Most of the building materials are deteriorating. The descendants of Notopraja are not able to maintain the building. Even though the *dalem* is historically significant as one important place of Prince Diponegoro's meeting spot during the Java War 1825–1830, the state and municipal government are not able to renovate the house because of the shortage of heritage fund.

In Central Javanese tradition, as *kampung kota negara*, Notoprajan is developed from *magersari*, meaning literally encircling something essential as a center, usually the landlord's residence. Accordingly, the settlement has been developed with a seat of an aristocrat at its center surrounded by the houses of his followers. The aristocrat or state officer grants his land to his followers—*pamong praja*, guards—*punggawa*, workers—*abdi*, and servants—*batur*, that enable them to build their house surrounding their landlord. The development of such a compound is usually named after their landlords such as Notodiprajan, Suryadiningratan, Kartodirjan, Ngadisuryan, and so on. The other phenomenon of Javanese *kampung* comes into being as a compound of state guild working for the King. Their compounds usually are built inside—*njeron*, or outside the *kraton* wall—*mbeteng* or *baluwerti*. Such *kampungs* are unique because their dwellers are highly skilled workers or artisans such as *pandean*—smiths/metal toll makers, *tukangan*—trade people/carpenter, *jagalan*—butchers, *gamelan rawitan*—musicians, *mantrigawen*—cooks, event organizers, etc. Some Javanese *kampungs* inside *kraton* domain are mostly the residential areas for state officers, guards, and soldiers. Their domains are named after their functional seat or groups such as *kadipaten*, *kepatihan*, *nayagan*, *palawijan*, and so forth.

The area of Notoprajan covers 3.7 ha with 8 neighborhood units—RW, and 50 household clusters—RT. The total households and populations in 2006 are 1489 and 5749. The populations of Notoprajan are relatively heterogeneous in occupations consisting of batik small businesses, craft home industries, state municipal and state workers, IT professionals, school teachers, Muhammadiyah staffs, and labors.

Historically speaking, Notoprajan has been part of Islamic Muhammadiyah movement since early 1912 founded by Achmad Dachlan. Notoprajan community has been inherent with this anti-syncretic stream of Islam since the Dutch colonial rule until today. Besides its location is close to traditional Muslim Kauman district at the State Grand Mosque of Yogyakarta, Notoprajan is the house of Muhammadiyah headquarters. Still, the organization has its school and mosque in Notoprajan's area. Albeit most of the Notoprajan inhabitants are supporters of this Islamic movement, they are relatively open to host students from outer islands with different religious affiliations living in their houses.

Besides traditional *kampungs* with a strong attachment to *negara* like Notoprajan, there are indigenous settlements that emerge and develop spontaneously in urban areas in response to job demand or as a place of refuge from rural adversity like Kampung Braga. Nevertheless, as ordinary settlement, most *kampung kota*s do not have an orderly organizing building system that symbolically represents the power of authority. Its housing form shows an irregular and impermanent structure. Narrow pathways run irregularly with meander forms that do not enable us to optimize an efficient utility network and drainage system. Poor sanitary facilities belong to the most crucial condition of a healthy environment (Fig. 9.5).

Nevertheless, the condition does not always bring urban *kampung* into a socially deteriorating situation. As traditional villages in rural areas, urban *kampungs* sustain their existence based on the spirit of *rukun*. Thus, any *kampung* dweller shares collectively the necessity for *rukun* and *kerukunan*. They develop their own norms and social codes in dealing with a highly populated density of living space. Tolerance—*tenggang rasa*, commonality—*kebersamaan*, and mutual help—*tolong menolong* become to be their traditional values system as well as the necessity of living in urban *kampung* territory. Regarding its diversity in religious beliefs, ethnicity, origins, customs, and traditions, an urban *kampung* is an Indonesian micro-laboratory of its state's motto: *Bhineka Tunggal Ika*—unity in diversity.

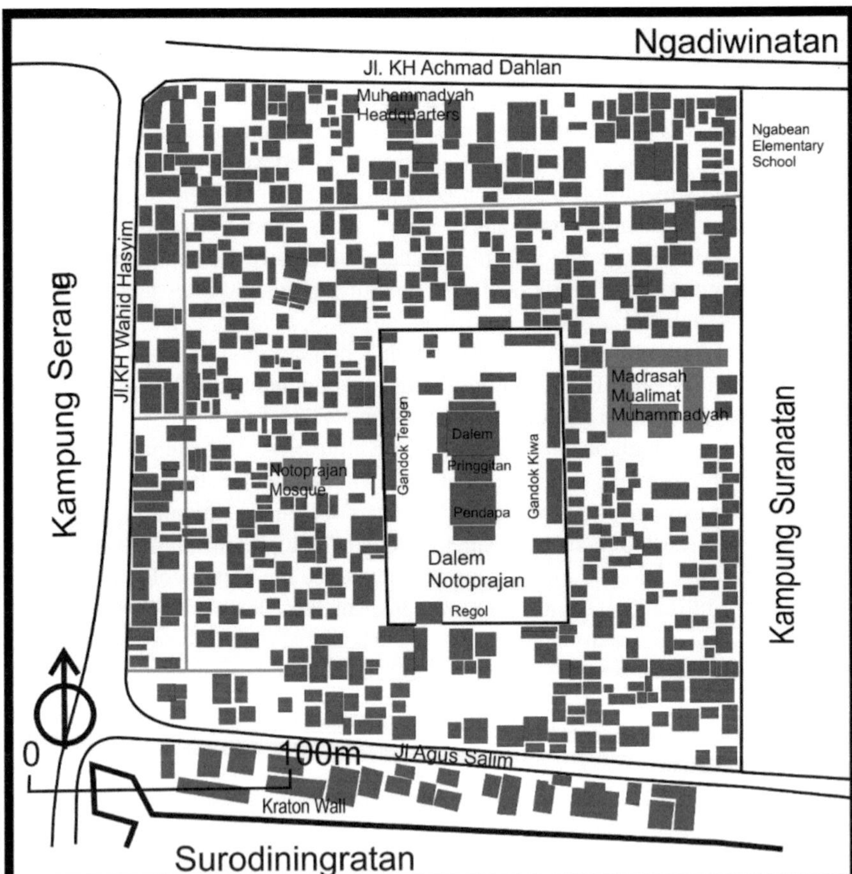

Fig. 9.5 Map of Kampung Notoprajan, city of Yogyakarta, drawing by the author

In contrast to its street-side neighborhoods and regarding their history—such as Kampung Braga and Notoprajan—urban *Kampung* does not follow the building codes and planning regulations from the local municipal authority: in regard to fire route and access for public safety. Nevertheless, law enforcement on urban *kampung* could lead to social upheavals and resentment. Since the Dutch colonial rule in the nineteenth century, urban *kampung* had been perceived as an autonomous neighborhood with its own customarily laws or *adat*.

Buildings and its infrastructure do not follow any system and procedure of safety and security. Fire is the most dangerous threat to the existence of urban *kampung*, besides flood and epidemics. In many cases, *kampungs* with a high value of land location are the object of land speculations and compulsory acquisition. Even though such violence is most experienced, the cases are mostly not brought to justice. Lack of legal status on land puts the kampung dwellers in a weak position and not too much option for bargaining position. This situation is used by some investors to bargain relocation cost and the land for new development. Such a program is often used by the municipal planning authority as an investment opportunity that enables them to restructure and redeveloping the town or city areas with new building blocks. However, it is not an easy task. In many cases, *kampung kota* has been already the world on its own. Planning intervention into the *kampung* realm is equal to changing and transforming their world (Fig. 9.6).

Fig. 9.6 Kampung Manggarai, Jakarta in 2001, photograph by the author

9.5 Kampung Kota as Home

Originally, the home in native Malay language is understood as *kampung halaman*, meaning literally village homeland. In colloquial terms, a kampung is to address village, community, and rural settlement from where the sense of home originates. *Kampung halaman* is the metaphor of birthplace or fatherland where one has own relationship to their village fellows and birthplace. The idea of home in terms of *kampung* is signified by the existence of the social network known as *kekerabatan*, the communal bond among its dwellers. The sense of home in terms of *kampung* is not absolutely a geopolitical definition, but a socially engaging relationship. *Kekerabatan* is the socially bounded relation among dwellers based on their familiarity, kinship, and engagement in collective works—*kerja bakti* and sociocultural rituals—*acara adat*. As a sociocultural entity, every *kampung* has its own customary laws and unspoken proprieties; all this is conceived as customary norms—*kebiasaan*. The highest principle of such customary norms is known as *adat istiadat*. The institution that stewards and upholds the customary orders is as *kerapatan adat* or *majelis* or *dewan kampung* consisting of *kampung*'s dignitaries and elders.

The sense of Indonesia's *kampung* is neither simply a demographic administration with a certain range of populations nor a well-planned settlement. Rather, it is a sociological notion of community with locally established customs and traditions. However, a kampung is more than a settlement with social cohesion and cooperation. Kampung is properly described as a native world with its own way to exist. This is not, by all means, to say that living in Kampung is absolutely an exclusive domain, beyond the reach of state law and order. *Kampung* is mostly created and established with the act of squatting public space or green area along the watershed zone. Spatially, a kampung is spontaneously sharing domains because of specific circumstances such as floods, urban migration for jobs, and evictions of land.

Regarding its physically built environment, *kampung kota* is akin to poor residential construction. Architecturally, there is neither a geometrically ordering system of building layout nor spatially consistent orientation. Finding a home address is barely effortless without knowing the complete name or nickname of the person sought. Sometimes, the name is not enough without being able to describe the physical and occupational identity of the person. In many cases, kampung dwellers know each other from seeing or in person. Still, finding the right person's home, we need a guide who is voluntarily ready to lead us to the door of the person we look for. In recent decades, kampung has been demographically changing in a drastic way. More and more newcomers live in kampung for seeking jobs. They are mostly migrants from rural areas who try to find their luck as workers in urban services or construction workers.

There are some traditional and historic *kampungs*, such as in the inner city of Yogyakarta, Surakarta and some Jakarta's Kampung Kramat, Kampung Condet, Kampung Kuningan, and Kampung Bali—former slave quarters from Bali—Traditional *kampung kota* does exist as the world of its own within the boundary of the urban area. The occasions that work to bring them together are mostly related to a National celebration for the Independence Day on August 17th and Islamic holidays such as Ied Fitri and Ied Adha. The event of August 17th is the time of *bersih desa* and volunteer community service; it is a newly established tradition in commemoration of Indonesian independence.

Today, the occasion becomes to be an annual community celebration like a ritual of the rebirth of their world. The celebration for Indonesian Independence Day includes communal gatherings with various contests and games. It is for many *kampung kota* a special holiday with local cultural performances. Commonly, some weeks before the day, *kampung*'s committee is set up to plan and execute several programs involving all kampung dwellers for sports contests and cultural performances. The sense of community is never utterly experienced in *kampung kota* without this occasion.

9.6 *Kampung Kota* from the Past to Present

In most cases, Kampung has been also the outcome of economically marginalized politics of the Dutch colonial rule on indigenous people—*pribumi* or *bumiputra* or *inlanders*. However, it was not a blind spot of an urban area, but considerably a backyard of the colonial residential area. Kampung was the home of low-class workers and janitors of colonial employers. Racial segregation policy for the residential area was nothing but most likely to sustain the Dutch colonial policy on European superiority over Asian and indigenous populations. The native settlement seems to be perceived as their dirty area of the backyard (Fig. 9.7).

In Semarang's Mlaten (1916–1920) as well as Bogor (1920) and Malang (1933), Thomas Karsten tries to include *kampung kota*s as an extension of urban residential units so that the native quarters have a rational block and structure with basic infrastructure (Akhihary 1990, 64). Prior to 1913, such a model has been introduced by the colonial rule with a case of Taman Sari, Batavia. Architecturally, as architect Karsten designs a prototype of an indigenous house model with local Javanese house form which is adaptive to local tropical climate and customs. In Jakarta, it was not the case during the Dutch colonial time. Niew Gondangdia or Menteng designed by P. A. J. Moojen in 1910 has excluded *kampung* in his plan. Spontaneously growing native settlements stand at the periphery of the European quarter as backyard now known as Menteng Dalam.

The Dutch policy in the 1920s for kampung kota was subtly conceptualized as an indigenous settlement improvement—*Kampung* Verbetering—in the affection of humanitarian corrective for cultivation system in 1840–1860 in Java with the Ethical Policy. The ambivalence was plausible between educating the natives on healthy sanitation and racial segregation. Due to pandemic diseases, the Dutch rule has no choice to take care of *kampung kota* because urban area as a whole is not merely a sociopolitical domain but also an indivisible environmental system. The policy for *kampung* improvement program is obviously driven by the prevention of pandemic diseases, such as cholera and pestilence and malaria, which were presumably originated from the kampung area (Doxiadis 1963, 147). Fear of pandemic spread had led

Fig. 9.7 Celebrating Indonesia's Independence Day with the bamboo climbing contest in kampung kota of Kebayoran Lama at the River Pasanggrahan, Jakarta, photograph by the author

the Dutch colonial government to improve the sanitation of the native settlements in the urban area.

Furthermore, the 'Kampong Verbetering' program was an indispensable effort to keep an eye on the natives and their nationalist movement (Kwantes 1975/1982, 418). Kampung Notoprajan and Kauman in Yogyakarta are well known as the basis of anti-colonialist movement since 1920s with Muhammadiyah faction at the upfront. Most native students and independent movement leaders live in the *kampung*. During the period between 1920 and 1942, the Dutch colonial rule was controlling the *kampung* area on a regular basis with healthy control—*operasi*—at the upfront. In Bandung area, this includes rearranging and restructuring building layout (Barker 1999, 102). Despite its rational approach to sanitation, the purpose is clear to enable *controleur* to move easily in and out of the *kampung* area.

Since the Dutch colonial era, the other crucial problem of *kampung kota* is to integrate its built environment into urban development as a whole because of its land status. Albeit the Ethical Policy was conceptualized after the speech from the throne in 1901, the Dutch regime did not take any significant action on kampung before 1920s. *Kampung* as indigenous settlement remains an economically isolated area of urban development. In 1919, the Dutch rule set up a special committee headed by Sonnenveld for repealing the Article 109 1845 Act on racial segregation. It did not give *kampung* with humanly dignified access to a well-planned urban area. As indigenous settlement, kampung remained socioeconomically marginalized from the Dutch colonial cities, such as Batavia, Surabaya, and Bandung.

The negligence of kampung as part of the urban area has been covered up with a prone humanistic concern of the Dutch on cultural preservation. In doing so, there was no hope in the future for Kampung in the East Indies as a Dutch colony free from inequality and social discrimination. Instead, the indigenous culture needed to be preserved within its own customary laws—*hukum adat*. The reason was, as suggested by van Vollenhoven from the School of Law of Leiden University, in order to protect indigenous law on land from the expansive pressure of Cultivation System 1849–1860. During the period of Ethical Policy since 1920, the Dutch did not move forward toward a unified civil law system (Lev 2000, 21–2).

The Dutch kampung improvement program is a plausibly political showcase for guilty debt in the name of public hygiene. On the other hand, prior to 1933, the Dutch body for public health affairs is part of public education. Since most native children in *kampung kota* are not mostly eligible to the Dutch schools, propaganda on the hygienic matters did not work. Indeed, the Dutch seek several options to educate native people on hygiene with the motion pictures show (Vanderbosch 1944: 278–9). The outcome is less successful in building people's awareness of health issues. The problem lies actually in the fact that *kampung kota* has not been developed as an integrated part of the municipal sanitation system. Nevertheless, *Kampung Verbetering* program was not a sustainable effort for a healthy environment because the natives were not well educated to understand the importance of public health for them.

The period after the Pacific War and the consolidation of the Republic of Indonesia until the end of Sukarno era in 1965, *kampung kota* suffered from poverty and negligence in public education as well as economic improvement program. It was, however, the time of sociocultural development of kampung community. Kampung Braga is actually the place of refugees from Bandung region who flee from the terrors of Darul Islam insurgency between 1948 and 1962. The intensive migration from a rural area to Kampung site was when fundamentalist Darul Islam movement was active in West Java insisted on Islamic state for Indonesia. Migrants from a rural area of Bandung region overflew the city and find refuge in several places along the riverbanks of Cikapundung River. Kampung Braga is one of their destinations. It was not the case for Notoprajan and most kampung beyond Bandung area.

During the beginning, the New Order era between 1965 and 1970s, mushollas and mosques played an important role as public places where kampung dwellers were able to share their burden together as a community. Fear of being accused as communist supporters in early Suharto's regime had led kampung dwellers to embrace religious faith. Unlike Jati Padang and Braga, Kampung Notoprajan and Kauman has been always the stronghold of Muslims who declared their Holy War against the communist supporters. During New Order era, these *kampungs* are also well known as the place of Suharto's opponents, then later become to be supporters of reformation movement; with Partai Amanat National—National Mandate Party—as their political vehicle and Amien Rais as their leader. Jati Padang and Braga are historically populated by moderate Muslims; in 2012, Jati Padang has 15 mosques, 24 musholas, and 2 Church, while Braga with two musholas.

Socially, as instantly growing community, kampung is indeed not free from disputes and conflicts among its populations. Struggle for power and hegemony as well as the territorial claim is part of *kampung*'s history. Contest for kampung head is always full of political tensions and intrigues. However, traditional leaders and Islamic clerics play a very important role in bringing people together. In many cases, the establishment *kampung kota* as a community has been part of historic urbanism, in terms of their contribution to the livable urban area and labor supply. Popular culture and urban public safety are the outcomes of the establishment of *kampung kota* as a community. The establishment is a work in progress since the New Order administration until now. It is obviously compelled by the influx of migration from rural areas for a job in urban services and industries.

Nevertheless, living in an established kampung is likely living in an extended family (Gannon 2003, 248). Kampung might be rightly described as a socially constructed reality of home with physically high proximity of extended family. The population of kampung in urban Indonesia varies from 2000 to 10,000 inhabitants. Average *kampung* has more or less 2000–10,000 people with population density from 90 to 200 people/ha. The density of Kampung Braga is about 150 populations/ha. Living in extreme proximity is mostly experienced in cities for more than a million populations.

Sociologically speaking, *kampung kota* is an extended family as well as an extension of the household living area in spatiotemporal context. Unlike in rural areas, *kampung kota* has its own social units or groups that sustain social cohesion and economically cooperative collaboration based on diverse productions and services. The housewives let them voluntarily join into the *arisan*, a game of money collected by rotation of winner. Men and boys become to be a voluntarily guards of their *ronda kampung*. Guarding the area of *kampung* in the night is voluntarily assigned by *kampung* head to every male in the household. It is traditionally part of community service. Boys and girls are used to be the member of a village youth group known as *karang taruna*. In the youth informal association of *karang taruna*, boys and girls are due to be active in community activities, such as spontaneous art performance, vocal group, eventual bazaar.

9.7 Kampung Kota and Global Modernity

Most of the lands of *kampung kota* are state or municipal property. Some are heritage lands—*tanah warisan* or *tanah adat*—which fall into neither state nor private property, but the traditional community. Kampung dwellers by generations are able to live and work there on a historical basis, but they have no right to transfer the land to others. Theoretically, planning and development of *kampung kota* on a state or municipal property should be not a problem. Practically, the problem of development lies in its history that *kampung kota* as a community had been in existence before the administration of land ownership was established. Indonesia' law on land and property—Undang undang no 5 Agraria—was enacted in 1960 while most *kampung kota* has been living since the end of the nineteenth century. Thus, territorial claim on land and property has been crucially problematic for most *kampung* sites. In most areas of Bandung and Jakarta as well as Surabaya and Semarang, *kampung kota*s are formerly green and watershed or riverbank areas. They are prone to flood during the rainy monsoon season, between October and April. During the New Order era (1967–1990s), most of the urban green areas along rivers and creeks have been subject to illegal settlements built by migrants from rural areas.

There was also poor filing and registering system for land ownership status and property administration at the municipal, regional and national level. Such land status came up as impediment and predicament in Jakarta area during the tenure of Ali Sadikin governorship in the mid of 1960s to early 1970s. The governor was considered serious in undertaking *kampung kota* as an integrated part of urban modernity. His policy and implementation on the Kampung Improvement Program was an enhancement of the Dutch colonial Kampung Verbetering program in the 1920s (Silver 2008, 28).

During his tenure as governor, Sadikin launched a systematic planning approach for urban restructuring program with infrastructure as its foundation. Such an approach created conflicts and tensions between the municipal government and kampung communities. His project was called with Mohammad Husni Thamrin—MHT—project, in honor the native Jakarta leader who advocated the improvement of kampung hygiene and infrastructure in the end 1930s. The MHT project was an integrated part of the First Five Year Plan of Suharto's development plan that included almost kampung in urban areas covering 1400 ha with 3.9 million of populations (Kundu and Achunine 1993, 39).

Evictions and demolitions were part of the implementation of the plan for public utilities; sanitation, water supply, garbage collection, pedestrian, road extension, building alignment, and fire route. Even though Sadikin's kampung improvement program is indisputably controversial for many because of its socioeconomic consequences, his vision on Jakarta is clear that urban development needs a master plan with strong land-use policy and urban design character. Despite sharing common ground on the healthy and manageable environment, some critics disagreed on Sadikin's Kampung Improvement Program —KIP—in 1968–9. They opposed mostly on Sadikin's top-down approach with lack of community participation and respect to local community culture and economy. The problems on evictions and demolitions are mostly unsympathetic because of its economically indifferent and culturally ignorant action.

Since 1980, some improvements to Kampung development program have been made in Jakarta, Surabaya, and Bandung with more community participatory approach. The improvements were made in terms of housing for the poor program that included community participation and technical as well as administrative coordination among governmental agencies. Kampung improvement program during the Suharto's policy has not intended for restructuring the *kampung kota* as part of the urban fabric as a whole but a nip-and-tuck surgery to address the urban poor for a political alibi and showcase for government annual report (Fig. 9.8).

The problem of poverty and poor public hygiene, as well as public safety, remains for *kampung kota* unchanged until today, especially in dealing with flood and fire risk. Instead of improving the quality of living in kampung areas, Suharto's New Order policy was focused on the relocation of urban poor from kampung to public housing projects, with site and services, walk-up apartment, affordable housing programs. It is actually another word for removal of the poor from *kampung kota* to somewhere else.

Even though post-colonial *desa* is a social institution in rural areas, its body of polity is incorporated in *kampung kota*. During the New Order, uniformity of political institution from central to local government is enforced by laws. Conceptually, it is an integrated part of the homogeneity of power at a practical level. Prior to Pre-Suharto's era, *desa* is free to practice their customs and traditions including their political body and the election of their leaders.

During the New Order regime, *kelurahan* was a strategic territorial apparatus of the central government for national development programs. In supporting the programs, various associations were established, such as family welfare association—*PKK*, village advisory council—*LKMD*, village cooperation unit—*KUD*, village small business unit—*BUUD*, and of course a military advisor for village security and territorial defense—*babinsa*. The establishment was a political will with a single vision toward a modern Indonesian society based on unified state ideology of *Pancasila*. Despite

Fig. 9.8 Kampung Mampang Prapatan, Jakarta 2016. Photograph by author

the fall of Suharto in 1998, some organizations are not in function anymore; the military advisor is still strongly represented as part of the territorial measure of national security.

Socially, *desa* or *kelurahan* in today Indonesian urban context is more or less formal in contrast to its rural similar organization. Since 1981, *kelurahan* has been a government institution in rural or urban areas. Unlike Pre-New Order village heads—*Lurah* or *kades*—since 1980s has not been more a traditionally voluntary leader but a state or local government officer paid by taxpayers' money. His or her staffs and assistants do not work voluntarily as well. Though his or her position is traditionally a community leader, *lurah* or *kepala desa*—village head—today is an extension of the modern bureaucratic institution of the Indonesian State. Politically, the head is not independent anymore as in the time before the New Order.

Family welfare association—*Pembinaan Kesejahteraan Keluarga* or *PKK*—was established in every village or *kelurahan* in 1981. Lurah's wife commonly led the association. The members of the association were mostly housewives from the communities within the *kelurahan*'s territory. The task of the association was to promote the development programs within the framework of family welfare, birth control, environmental health, modern education, and mother and child health care. The government set up an integrated service station—*pos pelayanan terpadu* or *posyandu*—in every *kelurahan* for social health care, social security, and development information. Nevertheless, *posyandu* today still exists without obviously promoting information on modernization and development.

Though the livability in urban *kampung* is deeply rooted in its origin in traditional village life, its sustainability is not visible without any gradual change and transformation of attitude toward modernity. All this is made possible by formal education for a young generation of urban *kampung* dwellers. Education is one important and strategic approach for *kampung* improvement. Besides, there are other approaches necessary as well, concerning the betterment of ecological and environmental condition. Nevertheless, the stay of many students from other rural regions and small towns in urban *kampungs* brings about a social dynamism for a gradually improved infrastructure and housing. Such a case can be found in the cities of Java such as Jakarta, Bandung, Bogor, Yogyakarta, Surakarta, Surabaya, and Malang, where well-known universities are.

Despite its social existence has a deeply rooted relation to the traditional concept of community, *kelurahan* in its practice does not comply with the traditional institution of *desa*. Instead, *kelurahan* is a modern political institution of urban community within the line of modern bureaucratic administration. Traditionally, every village had always an advisory board of elders and religious leaders who sat in the meeting hall called *Balai Desa* together with the village head. The advisory board in the urban modern context has been adopted as LKMD (the abbreviation of *Lembaga Kekerabatan dan Musyawarah Desa*). The board was traditionally a forum or a village council, which discussed any village affairs toward a single decision and opinion on any subject matter concerning the public interest. Striving toward a unity of mind called *musyawarah* is the main task of the village council (Guinness 1993,

280). All this comes out from the necessity for *rukun* as the fundamental structure of the community.

The deep structure of *rukun* is the concept and practice of syncretism that leads and drives people to the necessity for harmony, oneness, and wholeness. Accordingly, everything that comes to the *kampung* world will be dealt with, worked out, adjusted, synergized, and adopted with the necessity for being familiar with, *kampung* people accept and treat the presence and infiltration of the global economy into their world as their learning process for habitation, *kebiasaan*. Even though *kampung* people use, consume, and digest the well-packaged things of global modernity—from beverages foods, appliances, gadgets to audiovisual entertainments—they believe in the wisdom of development that everything grows safely and healthily because of the ability to manage diversity toward agreement and integration.

9.8 Kampung Kota and Informal Despotism

The most significant traditional institution of *kota* is its *pekan* or *pasar*. *Pekan* is the name for their traditional market that means literally a weekly event or the place of gathering. Traditionally, *pekan* or *pasar* takes place only once a week. Today, the role and the function of *pasar* or *pekan* in Indonesia's big cities have been replaced by the modern shopping mall. *Kampung kota* has been economically a significant contributor to the *pasar* economy. The relationship between *pasar* and *kampung kota* changes gradually because some roles and functions of this traditional economic center have been replaced by shopping malls, supermarkets, and department stores. Consequently, vendors and retailers of *kampung kota* have to move to find new places for their business. Streets and public places are their chances and opportunities.

Traditionally, a *kampung* is not confined only by its physical territory but also by its sociological domain where people are able to rely on their neighbors for a homestead. Despite its territorially identifying place, a *kampung* is another word for home in a sociological sense. In the urban context, *kampung* manifests in its informally defined territory. Most of the *kampung* dwellers are those who live in off-street neighborhoods. Economically, *Kampung* people live from petty entrepreneur waged workers, public servants, a lower rank of government officials, circular working migrants, and casual job seekers.

Despite its poor environment, the *kampung* dwellers in Indonesia are probably the socially dominating sustainers of Indonesian urban reality. Gradually, their environment as squatter housing is improved with various development programs launched by the government. *Kampung* Improvement Program in the 1970s was one of such an environment betterment scheme that provided *kampung* settlement with healthier sanitation and water supply. Living in urban *kampung* has its own ecosystem that is based on an informally sustaining economy. They mostly do pay taxes neither to the state nor to the local authority. It is also one of the reasons, why urban *kampung* has not been touched by a formal development program of the state.

A densely populated living in Javanese urban context does not always make a harmonious social life easy and manageable. Conflicts and social tensions among neighboring communities are mostly caused by economic interest. Public spaces nearby the urban *kampung* are often the areas of social conflicts because of the competition of its territorially controlled authority. Illegally controlling authority of urban public space is usually under a certain group of people in an urban *kampung* neighborhood called *preman*. They are actually a sociologically organized group for the collection of tribute from retailers and street vendors—*kaki lima*—as well as parking spots (Fig. 9.9).

Kaki Lima is mostly the object of *preman* for illegally putting off retribution. Local authority always tries to dismiss the existence of illegal retribution, but it is often without success. The reason is clear because the local authority or police are not able to control the public spaces adjacent to the *kampung* all the time. The question is why *preman* is that powerful in public space? Where are the police officers? Needless to say, that everyone in Indonesia knows well that behind at every practice of *preman* stands the security apparatus. The subsistence of the informal economy is also the existence of informally despotic authority of *preman*.

Besides its traditional origin in despotism, the apparent cause of *premanisme* is nothing but the excess of the high unemployment rate and poverty among the *kampung* dwellers. *Preman*s are mostly former criminally convicted persons who are by formal education economically marginalized. *Preman*s are possibly an outcome of *kampung* social illness if it is not a consequence of urban economy of Indonesia which is by business formal and informal. The informal economy of urban *kampung* relies on the web of small business among *kampung* dwellers and their service in petty commercials. Food, newspaper, clothes and cigarette wheels-boxes or kiosks on the sidewalk are mostly operated by the *kampung* dwellers.

Fig. 9.9 Streetscape and Kakilima of Keputran Tegalsari, Surabaya 2015. Photograph by author

Premans are those who control petty and small businesses and livelihoods. Violence and death threat is part of the daily practice of *premanism* for those who refuse to pay tribute to them. Public space in the urbanized area of Indonesia is not only ideologically nor neutral, but also economically despotic in favor of the invisible regime, the *preman*. Ideologically, public space is a conflicting domain between the formal and informal sector. Petty-cash businesses occupy the public space in response to the demand of common people for foods and goods. For state ideology, it is illegal to occupy public space for commercial activities.

On the other side, public domain for kampung dwellers is the extension of their home for economic sustainability. The street life offers foods and goods with a less expensive price because of taxation free. Municipal authority for public safety and order—*Tibum, Ketertiban Umum*—is often in confrontation with the street vendors because of their different ideological claims upon the public domain. Besides its affordability, the sustenance of petty cash is supported by the necessity for going-out-of urban dwellers in the evening. The clash between formal and informal sectors has been continued until today because the origin of this dichotomy is rooted deeply in the patrimonial culture and tradition. Security authority plays a dual role in most socioeconomically dominated territories by urban poor; formally, they act as the state and public agency and informally as the master of preman people.

Even though the existence of urban *kampung* brings about the urban streets livable with intimate social life, this livability is not about the sense and essence of urbanism because there is socioeconomic segregation between formal and informal businesses. Urbanism is not about the domination of social class either the urban poor or the rich. Rather, urbanism urges a symbiosis of diverse incomes so that urban livability is not a vibrant ghetto either of the rich or urban poor.

9.9 Kampung Kota and Income Inequality

Income inequality has been always the indicator of how laws and regulations, as well as public policies, shape our economy. In a matter of fact, income inequality is not economically self-curing process; this is not simply about choices of financial policy. Rather, it is a crucial indicator that the current economic system does not work for the need of the majority of people. The question concerning income inequality supposedly leads people to be aware of the fact that capitalist economic model does not bring about equal opportunity to all but privilege for few who are rich, which mostly by inheritance.

According to the 1945 Constitution of the country, the economy shall be organized as a collective endeavor based upon the principles of the family system. Still, in the Constitution, we read that all sectors of production which are important for the country and affect the life of the people shall be controlled by the state. The problem is not about economic growth but the distribution of income and wealth. The corporate capitalist economy of the free market is not in alignment with the spirit and goal of the Constitution because it does not provide the state with a room to play a decisive role in the distribution of income and wealth. Instead, the wealth of Indonesia's economy today is predominantly about owning and operating of big corporations (Djarum, Sinar Mas, Gudang Garam, Indofood, Indorama, Kalbe Farma, CT Corp, Mayapada, Lippo, and Mayora). Indeed, among the economic players, there are state-own corporations, such as Pertamina, Bank Rakyat Indonesia, Waskita Karya, Bank Mandiri, Bank Tabungan Pensiunan Negara, and Bank Negara Indonesia 1946.

Moreover, the crucial problem of Indonesian economics today shows that 40% of Indonesia's government bonds in 2017 was owned by foreign investors, among the highest of Asian developing markets; meanwhile, the current-account deficit was to reach 3% of the Gross National Product. In a general context, income inequality as a crucial economic issue has been put at marginal discourse in neoclassical capitalism. Is by design, capitalism engenders inequality in the result? Studies on capital in the twenty-first century by Thomas Piketty show that income inequality in a capitalist economy is soaring because the return of investment is greater than the growth of economic production (Piketty 2015); capitalism creates a widening inequality between the poor and the rich. His study on history of income and wealth distribution—for more than 18 years—confirms that despite various forces play in their dynamics, wealth inequality is always a lot higher than income inequality. Piketty points out that in the OECD countries, the rate of capital return with 4–5% is steady above the rate of growth with 2–3%. The consequence of this fact is that the middle class in most parts of the globe is losing ground, especially in Europe (Vaughan-Whitehead 2016) and in the USA (Gersemann 2005; Pew-Research-Center 2015).

Current amounts of research confirm a 40-year decline in average hourly wages for the bottom 90%, even as GDP, corporate profits, and the incomes of the top 10% have increased (Kehoe 2015). In the global context, during the past 25 years with remarkable Asian economic growths, income inequality decreased worldwide since 2000, but it has increased in the rich countries, such as the USA and Europe (Milanovic 2016). However, the country's productivity growth will be further apprehended back by the gusts of rising inequality, decaying education, an elderly population, and the rising private and public debt; we must do something about this (Gordon 2012). The Noble laureate Stiglitz (2013) points out that inequality is a choice of rules and regulations in the economic system; it is not the inexorable result of economic forces of demand and supply and globalization. Accordingly, we do not do significant actions for the rules that weaken workers' bargaining in labor markets, circumscribe monopoly powers, outsize in the financial sector for moving money from the bottom to top with the regressive tax system. How all these options are workable to restructure the country's economic system through rules and regulations?

For Indonesia, the effort to reduce poverty and income inequality has been beginning since Suharto's New Order administration at the end of 1980s. The most challenging obstacle of income inequality's reduction is not simply economic. It is cultural. Economically speaking, the main players and stakeholders of the Indonesian economy are the former cronies of Suharto; most of them Chinese Indonesian conglomerates and former high-ranking officers and ex-armed forces generals. Their economic hegemony has been well established with connection to multinational corporations. What Post-Suharto administrations have done to deal with the issue of income inequality was to provide subsidies to the poor with various economic schemes and programs. Under Joko Widodo, the administration provides social assistance programs for school children, village development, family planning, and health insurance, and small businesses; the total sum of aids in 2019 State Budget is 13.18%. All this is to elevate the poverty line of populations to a range of 8.5–9.5% (Fig. 9.10).

The poverty mitigation approach is partial because it does not go to review comprehensively the cause of poverty. The reduction of poverty is to work out the poor with certain agenda and programs that enable them to get out of the poverty category. There have been several plans and implementations that could help out the poor, such as health care, education benefit, student loan, child support benefit, unemployment assistance, social housing assistance, progressive tax, and other subsidies for energy, gasoline, and water. Most of these programs have been launched by the administrations before Joko Widodo; he reorganizes and restructures all these with focus on the poor families in a program called *Program Keluarga Harapan*—hope family program. How effective is the social program of Joko Widodo for poverty reduction? According to the ministry of social welfare, the recipients of the program has been significantly increasing from 3.5 million in 2016, to 6.00 million in 2017, and 10.00 million families in 2018 (Tribunnews 2018).

Despite Indonesian economic growth 5.2 in 2017, without well-distributed income equality, with a Gini ratio of 0.46 in 2017 and of 0.386 in 2018, productivity is a serious issue here. The key to economic growth without income inequality lies in the efficient machine of productions and services that enable to protect and serve people at large. In other words, productivity only benefits few who are on the board of directors and stakeholders; they are not directly involved in the process of productions and services? How does it happen until today? In the Indonesian context, Since the Suharto's New Order, criticism against the capitalist economy has been effectively associated with communism and national treason. Needless to say, the 1998–1999

Fig. 9.10 Seasons City and Kampung Kota Kalianyar, West Jakarta 2019

Southeast Asian economic crises had nothing to do with any social upheaval because of communism or socialism.

The main issue of the economic crises was crystal clear that the international capital movements were not regulated (Gilpin 2000, p. 162); the failures of capitalist economy will be continued until today because the root of all evil in the global economic crises lies in the contradictions within capitalist economy that are against democratization of market, organization of productions and services, and competition. Capitalist society is distinguished from others by the fact that its productive capital is accumulated in the hand of a minority of its members who enjoy the privilege to dispose of such capital in any way they see fit—in the form of rights of private property (Streeck 2016, p. 1).

The democratic economic system is necessarily established (Wolff 2012). How can we achieve this? Until recently, Indonesia has been transforming its society with political reforms but not yet doing something with the reformation of the economic production system. The organization of productions and services is still an ancient totalitarian institution with absolute authority, in which workers are not legally part of profit-sharing but simply a tool of the production system. In the Indonesian context, the obvious challenge of the democratization of the economic system is the political culture that has been dominated by the long-standing patron–client relationship between the politicians and business people since Suharto's New Order. The other obstacle of the economic reform is the increasing current account deficit because mainly of the ambitious infrastructure projects that demand imports and foreign investments.

The logic of economy should work that everybody who works in the productions and services should benefit from its profit. It is neither the government of Indonesia nor the free market economy who decides the prices in the marketplace, but the multinational or national corporations. The power of these corporations is depicted by Sheldon Wolin—emeritus prof of political sciences at UC Berkeley + 2015—in his book: 'Democracy Incorporated' as 'inverted totalitarianism' (Wolin 2008). Inverted totalitarianism is distinguished from traditional forms of totalitarianism; it does not find its appearance in an orator or charismatic leader but in the faceless concealment of the corporate state. Unlike the Hitler's Nazis, who made life uncertain for the rich and fortunate while providing social programs for the working people and poor, According to Sheldon S. Wollin (2008: 54–57) inverted totalitarianism exploits the unfortunate, reducing or weakening health plans and social services, suppressing mass education for an insecure workforce threatened by the importation of low-wage workers.

Back to the Indonesia 1945 Constitution article 33 and 34, the economic system is developed and operated on the basis of people's participation and state-intervention for economic and social justice. Yes, this is the direction of Indonesian economic system: the cooperative economic production and services system. This is a system with community-based power, in which the capital is not a dominant factor in the decision making of productions, distributions, and contributions of growth. This system diminishes duality factions between capital owners and wage labor people. All resource in the production system should be valued as significant components of contribution to potential profit or loss: capital, means/technology, skills/knowledge, and material. All people in such cooperative production system have to say as members of the community so that economic production system is not only the capitalistically profit-oriented body but also a community-owned enterprise who cares for well-being and sustainability of the world where it is. In doing so, capital is not easily moved from one to another place in the pursuit of profit. How to manage this? Restructuring the capitalization of the corporation is necessarily conducted so that by law, there is no more dominant private capital in any form of corporation.

The restructuring should enable people to mobilize public funding to buy existing private corporations that weaken the domination of private capitals and strengthen public economic power. Of course, prior to this restructuring, it is necessary to regulate the financial system against capital flight so that any attempt to move the capital from the country abroad becomes a crime against the nation-state. The bottom line of the plan of the restructuring of a private corporation is to open as wide as possible the public participation in the decision making of the production. This is considered not enough with the current opportunity of buying shares from stock exchange markets; the working people in the corporations do not have any voice on the direction of their companies is going to do with production and profit. What kind of civilization we live in when only the rich have the right to do with our economy. In a matter of fact, the rich exploits all our resources for their single interest in growing and accumulating profit. Who pays the social cost of all this profit-oriented economic system? The ordinary people pay all this cost with social diseases, street violence, alienation, etc. We need to build a strong middle-class society so that income inequality becomes fewer. The way to achieve this is to build an economically democratic society so that economic production systems are managed by the community not by the minority of people who sit in a board of directors.

The Indonesian economy is under the domination of the hereditary wealth of global multinational corporations and financial organizations. This domination is demonstrated by the dependency of the current account on the foreign investments and capital inflows. All this has led to the depreciation of foreign exchange of the rupiah against the US dollar. As the Greek experience shows us that any government confronts austerity will instantly clash with the global financial institutions that protect the 1% (Mason 2015, p. xi). The financial austerity policy from Post-Suharto administration is understandable because of dependency on foreign investments. The consequence of this policy is the stagnancy of economic growth at around 5%. In many ways, the central bank of Indonesia has been always anticipatory in dealing with the global uncertainty of free-market economy; it is not only to increase the reference interest rate but also to manage the payment and investment in terms of Domestic Non-Deliverable Forward or DNDF. By the ratification of the DNDF, transactions for export and import are financially secured from the fluctuation of the country's rupiah exchange rate (Tempo.co 2018).

Inside the country, the domination of 1% of populations is obviously shown as well. Despite the democratization in the political system since 1999, the majority of Indonesian populations is economically dominated by the 1% oligarch capitalist group. Unlike his Post-Suharto predecessors, Joko Widodo is the first Indonesian national leader who does not come from the ruling group or *priyayi*. His populist advocacy for the urban improvements has been always challenged with various cases of corruptions and bureaucratic mismanagements. As a matter of fact, the real challenge of the alleviation of the urban poor is restructuring the whole bureaucratic machinery of government from local, provincial, to the central level. All this is because urban development in the capitalist free market of economic system is not about good or bad political will and program but a set of routine negotiations and compromises in dealing with the inflation rate, retail prices of basic commodities, jobs market, and unemployment rate.

9.10 Concluding Remarks

The marginalization of kampung kota is obviously not only recognized from the physical and socioeconomic condition of the settlement but also from planning and development policy. There is hardly to find a strategic and long-term policy and program of community development from the municipal governments that concern the socioeconomic and physical improvement of kampung kota. Until today, the dichotomy of formal and informal settlement and economy has not been intentionally overcome but retained and prolonged as a subconsciously accepted reality of urbanism. Indeed, for the most elites of the country, the factual dichotomy has been perceived as a business as usual.

Despite its irregular and amorphous built form and informal land status, *kampung kota* has been an indispensable reality of Indonesian urbanism. Incorporating and integrating the reality of *kampung kota* into urban development scheme by the municipality is considered necessary. In short terms, urban livability could be improved with the amalgamation and assimilation of all incomes into a community habitation. Meanwhile, in the long terms, this incorporation and integration could prevent the socioeconomic discrepancy of urban inhabitants that could endanger public safety. The concentration of urban poor upsurges a potential source of socioeconomic unrests, radical religious movements, and terrorism. All these are mostly because of social jealousy, social injustice, and socioeconomic depression of the urban poor.

Despite its origin as the spiritual counterpart of *kota negara*, *kampung kota* is environmentally and economically be integrated within the urban structure. Architecturally speaking, the physical form and its intimate spatial character are potentially inspiring for an urban experience. However, planning and design intervention in the world of *kampung kota* should be indispensable collaborative and participatory with its existing community. Until today, this participatory approach is easier said than done. The most challenging issue of participatory development is the establishment of a democratic platform. In doing so, the municipality and the existing community are able to work together based on trust and mutual respect. Diminishing dichotomy of *kota negara* and *kampung kota* is challenging opportunity of Indonesian urbanism because of patrimonial tradition and culture that maintains the relation of municipal government and the urban poor in terms of patron–client relationship.

Whatever forces, influences, and interests work on their *kampung* lifeworld, the inhabitants have been traditionally being able to deal with their cultural strategy of syncretism. Despite the socioeconomic depression, the cultural strategy works well as long as *kampung kota* does exist as a collectively integrated system of the live-work-play community. The three *kampung kota*s: Jati Padang, Notoprajan, and Braga demonstrate their potential sense of community-based three main concepts: *gotong-royong*—voluntarily collaboration, *rukun*—harmony, and *adat-istiadat*—customary orders. All these are demonstrated by people during special occasions such as wedding day, funeral, Ied Fitri, and National independence's day. Traditional communal gatherings and rituals in *kampung kota* are essential occasions for maintaining, developing and sustaining the sense of community that constitutes the feeling of home and solidarity among them.

Economically speaking, until today, the development of *kampung kota* is more about trickle-down rather than a structural programmatic. The marginalization of informal settlements such as *kampung kota* is indivisible from the cultural resistance of the ruling group—*priyayi*—against socioeconomic equality. It is considered not exaggerated to say that the socioeconomic reality of *kampung kota* represents the income inequality of the whole nation-states of Indonesia. From daily experience, income inequality is apparently experienced from the urban landscape and people on the streets of Indonesian cities. In today Indonesia's economy, people in this country have to accept the bitter reality that one vote is equivalent to rupiahs he/she has in their income. So, nobody likes this fact. However, if you are not rich enough never go or run as a politician. This fact implies as well, if you are poor, never think or dream to go for a college education. Without a college education, one does not have a good chance to get a better income than the minimum wage.

Discourse on urban development in Indonesia is impossible without taking *kampung kota* into consideration because this settlement represents the outcome of the patrimonial practice of a socioeconomic system that is necessary to be transformed into a democratic culture. Accordingly, socioeconomic justice is programmatically implemented with a community-building approach. The right to land for the urban poor is necessary to be regarded that is not in terms of legal formal matter but based on a historic occupation in the area. In doing so, the *kampung kota* inhabitants have an economic asset from the value of the land in urban development. In addition, the problem of *kampung kota* is not simply physically oriented matter through infrastructure improvement. Rather, this is the sociopolitically imperative issue of reformation that *kampung* dwellers are necessarily regarded in equal rights with other urban citizens for economic activities.

References

Akhihary H (1990) Architectuur & Stadebouw in Indonesie 1870–1970. Warburg Pers, Zutphen

Archer RW (1986) Land consolidation for urban development in Indonesia. Human Settlements Division, Asian Institute of Technology, Bangkok

Barker J (1999) Surveillance and territoriality in Bandung. In: Rafael VL, Mrazek R (eds) Figures of criminality in Indonesia, the Philippines, and Colonial Vietnam. SEAP Publication, Ithaca, NY, pp 95–127
Cybriwsky R, Ford LR (2001) City profile: Jakarta. Cities 18(3):199–210
Doxiadis CA (1963) Review on the problem and science of human settlements, Ekistics. Center of Ekistics of the Athens Technological Organization, Athens
Gannon MJ (2003) Understanding global cultures: metaphorical journeys through 28 nations, clusters, and continents. Sage, London
Gersemann O (2005) Cowboy capitalism: european myths, american reality. CATO INstitute, Washington, DC
Gilpin R (2000) The challenge of global capitalism: the world economy of the 21st century. Princeton University Press, Princeton & New York
Gordon DL (2012) Batery park city: politics and planning on the New York. Routledge, New York
Guinness P (1981) Rukun Kampung: social relations in urban Yogyakarta. ANU Press, Canberra
Guinness P (1986) Harmony and hierarchy in a Javanese Kampung. Oxford University Press, Singapore
Guinness P (1993) Local society and culture. In: Hill H (ed) Indonesia's new order, the dynamics of socio-economic transformation. Allen & Unwin, St Leonard, pp 207–304
Kehoe M (2015) Dutch Batavia: exposing the hierarchy of the Dutch colonial city. vol 7.1, J Hist Neth Arts
Kundu A, Achunine OB (1993) National trends in housing production practices. United Nations Centre for Human Settlements UN Habitat, New York
Kwantes RC (1975/1982) De Ontwikkeling van de nationalistische beweging in Nederlandsch-Indie. Tijdschrift Bronnenpublikatie: Derde stuk 1928–1933 (Nederlands Historisch Genootschap) 19:289–293
Lev DL (2000) Legal evolution and political authority in Indonesia: selected essays. Martinus Nijhoff, Dordrecht
Mason P (2015) Postcapitalism: a guide to our future. Farrar Strauss & Giroux, New York
Milanovic B (2016) Global inequality. Harvard University Press, Cambridge, MA
Newberry JC (2006) Back door java: state formation and the domestic in working class java. Broadview Press, Peterborough, ON
Pew-Research-Center (2015) The american middle class is losing ground. http://www.pewsocialtrends.org/2015/12/09/the-american-middle-class-is-losingground/. Accessed 3 Mar 2018
Piketty T (2015) The economics of inequality. Belknap Press, Harvard University Press, Cambridge
Santoso J (2006) The fifth layer of Jakarta. Centropolis Tarumanagara University, Jakarta
Silver C (2008) Planning the megacity: Jakarta in the twentieth century. Routledge, London
Stiglitz JE (2013) The price of inequality: how today's divided society endangers our future. Norton, London & New York
Streeck W (2016) How will capitalism end?: essays on a failing system. Verso Books, London
Tempo.co (2018) Gubernur BI: Disahkan, Peraturan Domestic NDF Berlaku Hari Ini. https://bisnis.tempo.co/read/1130990/gubernur-bi-disahkan-peraturan-domesticndf-berlaku-hari-ini/full&view=ok. Accessed 28 Sep 2018
Tribunnews (2018) PKH Diharapkan Mampu Percepat Pengentasan Kemiskinan. http://www.tribunnews.com/regional/2018/04/13/pkh-diharapkan-mampupercepat-pengentasan-kemiskinan. Accessed 27 Sep 2018
Vanderbosch A (1944) The dutch east indies, its government, problems, and politics. University of California Press, Berkeley
Vaughan Whitehead D (2016) Work inequalities in the crisis: evidence from europe. Edward Elgar/Geneve: International Labor Organization, Cheltenham, UK
Wolff R (2012) Democracy at work: a cure for capitalism. Haymarket Books, Chicago
Wolin S (2008) Democracy incorporated: managed democracy and the specter of inverted totalitarianism. Princeton University Press, Princeton & New York

Chapter 10
Urbanism and Superblock Mixed-Use Development in Jakarta: Politics of Gentrification of Post-Suharto Indonesia

Abstract Despite the Post-Suharto Indonesia has been marked and characterized by the reformation for democratic governance, the New Order's liberal economic policy for urban and land development has been continuously implemented. Accordingly, the private sector has been given more power for property and land development, while the central government is only responsible for general policy but leaves its detailed implementation to municipalities in terms of regulatory partnership with the private sector. The policy embarks its direction from the vision of capitalistic accumulation and economic growth with less regard to social justice and public interest. This study examines the implementation of the liberal urban development policy with a focus on the intensification of land use. The case study is to scrutinize mixed-used developments in Jakarta; the core of the discussion and historical analysis of this study is the confrontation with the principles of good urbanism in terms of sustainable community and social justice based on Indonesian sociocultural roots.

Keywords Superblock · Gentrification · Capitalist development · Kemang Village · Podomoro City

10.1 Superblock Development and the Question of Urbanism

This chapter is a study on the issues, constraints, and opportunities of intensification of land use for urbanism in Indonesia. The focus of the study is to discuss and analyze the relationship between superblock development and political culture in the city of Jakarta. The study is to deal with the question: Why is superblock development not the data for the study was carried out from the author's personal experience working as practicing architect and planner from 1993 to 2002 and from 2010 and to 2015 in Jakarta and other parts of Indonesia.

The other sources of information have been collected from interviews with relevant engineers/planners of developers, collateral marketing materials of the related projects, and site visits of the case studies. Development policies, regulatory materials, and statistical data are available in the public domain; they are accessible

B. Wiryomartono, *Traditions and Transformations of Habitation in Indonesia*,
https://doi.org/10.1007/978-981-15-3405-8_10

Fig. 10.1 Superblock development at Tomang Area, West Jakarta 2015. Photograph by author

through the Internet and in-person from respective public offices, from national, provincial, and district level. This study argues that the intensification of land use through superblock developments in Jakarta does not fit into the framework of sustainable development; such urban transformations are only in favor of gentrification that does not lead toward social justice, safe, and healthy urbanism. The flourishing and ongoing superblock mixed-use developments are a sign that the political culture of patrimonialism still reigns the country's urban development (Fig. 10.1).

The liberal economic policy of development fits with the patrimonial practice of patron–client relationship between government and the private sector. In other words, the state does not need to take care too much for the process and quality of development but the outcome that generates jobs and economic growth. Therefore, the state intervention in spatial planning and urban development has been fending off from stewarding and guarding public interest for environmental sustainability, social justice, and the right to the city for all citizens. Gradually, the superblock land development in urban areas encroaches the public spaces and transforms them into the large-scale private realm for common uses under private management. In a matter of fact, only the have who are able to get the benefit from this management; in addition, only a few enjoy the trickle-down economy from the superblock mixed-use development in terms of employment opportunities as workers and labors. Joko Widodo's administration (2014-date) has not yet shown the change of policy for urban development; however, his bold policy on infrastructure development exemplifies his direction in opposition to the neoliberal approach to development. Urban and land development have not been yet in Joko Widodo's list of priority.

Hypothetically speaking, gentrification has been introduced as a concept of urban development to attract upper-class people from suburban back to the inner city; this is an outcome of demand and supply economics at the regional level of geography (Smith 1979). In the Indonesian context, the idea of gentrification is understood as a concept of back to the city in terms of counter-action for having a home in the countryside or suburban gated communities. Gentrification, in this sense, is conceived as the idea to attract people, especially well-paid young professionals to live close to their workplace in the city. This concept is actualized by private developers with the transformation of the formerly abandoned area but geographically strategic through land compaction, land-use adjustment, and land acquisition. Such large-scale development is impossible in urban areas without the support of development policy in favor of a higher social class than before.

Officially speaking, gentrification has a negative connotation and barely used in planning and legal documents (Kelly 2004, 150). In many cases, gentrification has been associated with urban revitalization, rejuvenation, and redevelopment. However, the sense of gentrification lies in the increasing value of land for economic growth and vibrant livability. Most issues of gentrification concern with social justice, economic marginalization, and displacement of low-income residential; such issues are conceivably unacceptable for good urbanism because they will generate social strives, distrust, and disrespect against the government and political unrests. In the Global South, gentrification is not mainly for going back from suburban to inner-city districts but for redevelopments and transformations of land uses from economically dilapidated urban spaces to a high-value real-estate commodity as an integrated part of economic liberalization (Gooptu 2011; Shin 2015; Hudalah et al. 2014). In the Indonesian context, gentrification has been confusedly understood; this concept is simply conceived as urban revitalization, rehabilitation, and restoration (Budihardjo 1997, 19). Accordingly, gentrification seems to have nothing to do with the exclusivism of wealthier social classes. To a certain extent, the concept is interchangeable with the densification of populations through the development of residential units for students (Smith 2004, 90).

10.2 Great Urbanism and Superblock

The leading figure of New Urbanism: Peter Calthorpe in his interview with Green of the Dirt ASLA (American Society of Landscape Architects) delineates three key ingredients for good urbanism: a walkable and human-scale environment, active streets, and diverse communities and uses (Green 2011). Accordingly, urbanism is a system of human habitation that works around the ecological context that preserves and enhances the quality of life. Good urbanism entails integration into societal and environmental context with human scale and outdoor walkable environment. This is not culturally confined by geographical location only for the cities in Global North, but also for urban centers everywhere. Good urbanism encourages and promotes healthy individual and community's life because its core of activities entails physical proximity and direct interaction with aural and vibrational experience.

In terms of activities, good urbanism is made possible by the diversity of its spatial land uses. Diverse uses and communities uphold urbanism as an economic household with regard to a variety of products and services. In addition, as a system of human habitation, good urbanism involves people with a variety of family types of housing, age populations, and income level groups (Ellin 2012, 2). Diversity in this sense is not an isolated system but a part of a socially connecting network that preserves and develops local traditions and historic urban fabric. In doing so, good urbanism is a development of humanly built environment with respect to humanity and social justice and with regard to the environment without any kind of seclusion from societal, economic, and ecological context and locality.

The concept of the superblock and megastructure development claims the large-scale project of densification with vertical and mixed-use buildings and massive investment in a location. Prior to such large-scale developments in 1970s American re-urbanism in New York City, the concept of superblock development has been implemented by Heinrich Schmid and Hermann Aichinger Raben for Raben Hof and Am Fuchsenweg as well as Karl Ehn for Karl-Marx Hof between 1920 and 1930; the whole complex consists of more than a thousand units of apartment, libraries, medical and dental centers, schools, theaters, laundry facilities, shops, and stores (Blau 2014, 197–200). In the American context, Jane Jacobs in her book: *The Death and Life of Great American Cities* observed such large-scale development a threat to the 'closed grained variety' and livable pedestrian that constructs the essence of urban experience (Jacobs 1961). Having learned carefully from Jacob's observation, New York City's Soho district re-urbanism was developed to transform the former district with semi-abandoned industrial lofts to be vibrant cultural quarters.

Instead of massive site clearance, high-rise towers and patrimonial power of development, the similar urban design concept of Soho district has been developed by Alexander Cooper and Stanton Eckstut with eight design principles for Battery Park City on the Hudson River 1979. The principles include the appropriation of the project as an integrated part of the lower Manhattan, integration of the project with existing streets and clocks, establishment of active waterfront amenities, representing idiosyncratic and recognizable forms, ensuring active circulation of people at ground level, reproduction and improvement of typical New York neighborhoods, putting commercial activities as focus of the project, and flexibility of land use and development (Gordon 2012, 66).

The globally accepted concept of good urbanism today lies in its actualization in urban experience at street level. Instead of generating outdoor public place, superblock development engenders artificially privatized realm with architecturally isolated megastructure; therefore, it is an antiurbanism because of its contextual ignorance and self-contained town-in-town. Superblock with its megastructure does not care much for the sense of urbanism that lies in the active outdoor public realm within the network of city's sidewalks and public transit system.

In the Post-Suharto's Jakarta, superblock development is perceived as an option of urban development by which the private sector is informally regarded as the client of the governmental patronage. This patron–client relation has been practiced since Suharto's New Order regime in power 1967–1998. The nature of this traditional relationship is for political and economic stability that keeps and sustains the actualization of urban development. In order to achieve such stability, the spatial planning and its development process and procedures are arranged and managed with strictly hierarchical top-down authority, from national to provincial and municipal level. It is argued by this study that superblock development in Jakarta is, from various perspectives, the detriment of urbanism. Why is that? This study will expound, analyze, and unveil the reasons why is gentrification in Indonesia manifests in four critical chapters concerning: the context of Indonesian urbanism, the corporate culture of developers, urban planning/design system process of the superblock development, and case studies.

10.3 The Context of Gentrification and Urbanism in Indonesia

Socioeconomically, gentrification in the Indonesian context can be seen as another word for the configuring of centripetal power. Accordingly, capital town and state are defined by its center, not by its periphery (Anderson 2007, 29). This concept has been conceived by Indonesian tradition and culture as the daily practice of patrimonialism; the concept has been incorporated as the idea of center in the socioeconomic and political system. Concentrating the gentries at the center of towns and cities has been a long-standing tradition in Java and other parts of Indonesia. The indigenous concept of town, *kuta* or *kota*, is to underscore and articulate the domain and boundary of gentries that rules the nation and state. In the patrimonial tradition in Indonesia, the presence of gentries is to define the idea of center; in Javanese context, gentries are called *priyayi* (Geertz 1960, 5–6; Simadjuntak 2013, 97); they are the elites because of hereditary privilege, charisma, power relation, and cultural refinement. In doing so, gentrification is not about social justice but simply the necessity for the manifestation of the patrimonial hegemony of the elites.

In contemporary Indonesia, the elites, *priyayi*, include high-ranking civil servants, entrepreneurs, politicians, journalists, entertainers, and academicians; they set out different lifestyle and manners from the ordinary people (Gerke 2000). Socioculturally speaking, urban dwellers in Indonesia are not aware of their rights to the city because the urbanization in this country has been developed through the surplus of jobs for the growing market of consumptive lifestyle in geographically concentrated areas. This urbanization does not transform, but maintain, reinforce, and sustain the dualistic class system of Indonesia: the ruling group, *orang kaya*, *orang besar*, or *priyayi* and the ordinary people, the *rakyat*, *orang banyak*, or *wong cilik*. In other words, the Havey's concept on the rights of the city in the Indonesian context does not apply for all people in terms of citizens but only for the ruling group (Harvey 2008). In the traditional sense, the ordinary people are not aware of their right to urban resources as citizens; their traditional awareness is under the notion of *rakyat* or *kawula* meaning the servant of the ruling group, *priyayi*. In Indonesian as well as the Javanese language, there is hard to find the correct word for society because they are socioculturally not organized as an abstract concept of society but as the indefinitely stretchable net of kinship and clientship (Anderson 2006).

Although Indonesia's Post-Suharto administrations have been the era of the political reform from authoritarian to democratic governance, the planning system of the country has not yet transformed its spiritual content from patrimonial to a democratic system. The policy of urban development remains the same that is the line of patrimonial political culture. Accordingly, most urban developments do not attract citizens for the participatory planning process in shaping and envisioning their towns and cities. This does not mean that social and political movement is not possible. In most cases, the motivation of the sociopolitical movement is generated by fear of existence rather than a nose for opportunity. Here, socioeconomic class consciousness and community awareness do not predominantly drive people toward political engagement and participation. The communal movements are mostly compelled by socioeconomic and political grievances through the corridor of local chauvinism and religious fanaticism (Bertrand 2004; Klinken 2007; Taddjoedin 2014).

Patrimonial hegemony in Indonesian society has been oppressing such political participation from grassroots; traditional authority, state ideology, and religious values system have been commonly used for such repression. Protesting and opposing the government, state officials and their policies have been culturally perceived as social diseases that serve only to destabilize unity and harmony of the world (Znoj 2007, 58; Knight and Heazle 2011, 111). Under patrimonialism, the state authority becomes the sole authority in the decision-making process of spatial use and building plan. Since the Dutch colonial city planning ordinance was introduced in 1926; the ordinance was implemented until the Republic of Indonesia's Law 26/2007. However, the existence of the planning profession and its professional body has not been officially incorporated in the ordinance so that spatial planning as a discipline is still unclear.

Long-term negligent and disinvestment areas are often the targets of architecturally gentrified development. Such areas are mostly untouched by the state or municipal program for infrastructure or affordable housing development; lack of funding is the main reason for this negligent. By the agrarian law, acquiring a large area has been regulated with a maximum of 2 ha only. The question is how the private sector is able to consolidate more than the maximum area under a single management system. One superblock development requires at least 200 ha area. From a land acquisition perspective, the flexibility of land management for large-scale development has been managed with specific local regulation that puts the restriction of area ownership according to the Agrarian Law no 20/1961 in an exceptional context. This has not been the case before 1980s. The large-scale land development for property business utilizes the shortcoming of law and regulation concerning the relationship between the restriction of area land ownership and the permit of business on the land. The consolidation of lands under single management has not been legally regulated so that private developers are able to acquire the lands in a location under various owners, but they actually belong together from one financial source. Until today, the government has not yet seriously taken the land as a strategic asset that stewards and preserves public interest through land banking and spatial land-use planning (Bisnis Indonesia 2016).

The other thing that matters for urban development is about the partnership between the public and the private sector in the process and production of urban development policies. Even though in practical reality, urban development is impossible without the active participation of business and professional association as well as communities, there is still unclear in the laws or regulatory system on spatial planning concerning how stakeholders involved and participate in the making of urban development possible. Of course, all this is in the framework of the democratic system.

The participation of citizen is granted by the law no 24/1992 and the law no 26/2007—article 65 and 66—but there is the limited direction for the alteration, adjustment, and accommodation of plans regarding the new situation, condition, demand, and aspiration. In response to actual changes, the laws provide a limited option for establishing an agency of stakeholders that is authorized by law as a committee of planning adjustment. The unclearly formulated the power of urban citizen and community in the law No 26/2007, in terms of their active role and function in the process and production as well as the operation of urban development policy, reflect, and underscore the patrimonial centralized authority of the state, from national to the local level. At any level of textual analysis, the law No 26/2007 signifies more about the role and function of government as patrimonial patron for its citizen on spatial planning, than as the public servant of good governance that regards potentials in society as partners and associates for common interests in urban development. Substantially speaking and regarding the output of spatial planning, the laws 24/1992 and 26/2007 have given a limited consideration for acknowledging and recognizing the function and role of planning professional expertise behind the production of planning documents.

The significance of citizen participatory includes the possibility of shaping and developing the vision of development. In such capacity, the community is able to work together with other stakeholders on the adjustment and restructuring plans and programs according to the vision for the public good. Such collaboration is necessary, for example, for the cases of large-scale development in the urban area. Until recently, large-scale developments have gone through the planning process and system with less community participation so that superblock developments have come into the reality of the city. Complaints and criticisms on superblock development have been publicly presented by experts in various media. However, the 'antiurbanism' development is going on because legally speaking its developers have fulfilled all requirements of municipal planning and development system. Deficiency of vision on good and healthy urbanism is probably one of the main reasons why large-scale and megastructure development finds its niche in Jakarta.

The annually increasing influx of job seekers and other migrants from all parts of the archipelago entails Jakarta's urbanism to develop a compact city development. In 2010, 49.8% of Indonesia's population lives in urban areas, while the projection for 2015 and 2020 is 53.3% and 56.7% (Bappenas_BPS_UN Population Fund 2013, 48). Regarding the scarcity of land in urban areas, high-density development is an option by which the intensity of land use is made possible. High-density and mixed-use development is an option that concentrates populations for the more walkable built environment. The compact mixed-use development comprises commercial, residential and recreational components. The development policy for superblock system in Jakarta has been considered the private sector's participation in the urban development of Jakarta. This line of thinking is incorporated in the Governor's Decree no 38/2012 and Municipal regulation no 6/1999 article 43 paragraph 4. Such a public–private partnership is theoretically beneficial to each other. The reality is mostly different that such a partnership is in the sense of patron–client relation. The officials and politicians control the system of information and procedures for the development permits and business licenses. Access to the system is formally time-consuming due to several reasons.

Most property developers do simultaneously two approaches: formally and informally. Informally, the business people used to contact the officials or politicians prior to their formal process of application. The informal agreement between the patron and the client of development permits is the key to the successful development process. In return, the business people are traditionally indebted to the officials/politicians for their informal and patrimonial assistance; their gratitude to the patron has various forms, from money, property, vehicle, etc. Lawfully, such forms and actions of gratitude fall into the category of bribery and graft.

The patron–client relationship mentioned earlier has been practiced since Suharto's era; such practice is well known as *kolusi* or collusion. After 1998, there is a reformative movement against the practice of corruption–collusion–nepotism. Since the informal and patrimonial relationship between politicians and business people has been developed for decades and generations, the practice of corruption, collusion, and nepotism is not easy to disappear from business in Indonesia because it has penetrated deeply into all government system of Indonesia, from legislative, judicative, and executive institutions (Winarta 2009, 272).

10.4 Corporate Patrimonialism and Superblock

From 1990 to date, most superblocks in Jakarta have been built by powerful property developers, such as Pakuwon Jati, Lippo, Ciputra, Podomoro, Summarecon Agung, and inland group. Despite the complexity of land acquisition in inner-city areas, they have eventually acquired the lands for their superblock projects. Some of them are located at strategic areas such as Podomoro City in Tanjung Duren–Grogol area, Senayan City in Senayan area, and Ciputra World Kuningan City in Kuningan CBD area. The companies' leaders of superblock development are mostly well-experienced entrepreneurs since Suharto's New Order times. Among them are Ciputra—born in 1931—Mochtar Riady of Lippo—born in 1929—Alexander Tedja of Pakuwon—born in 1945—and Trihatma Kusuma Haliman of Agung Podomoro—born in 1954.

The developers have been always supportive of the programs of ruling administrations after the fall of Suharto's regime in 1998 until present at the national, provincial, and municipal level. The patron–client relationship between state officials and business people has been part of Indonesian tradition. The root of such tradition has been developed since 1619 when Jan Pieterszoon Coen of the Dutch V.O.C. and his Chinese friends: Ben Kon or Su Ming Kang and Jan Kon worked together built Batavia (Parmer 2012, 195). Prior to the arrival of the Europeans, the patrimonial partnership between indigenous rulers and the Chinese merchants have been commonly practiced at most ports of Java, Sumatra, and other islands; the *shahbandar* or port authority position was commonly held by Chinese chief (Cribb 2000/2013, 246).

The patrimonial relationship between state officials and property businessmen is common practice in Indonesia or in Southeast Asia; they build symbiotic mutuality based on trust and loyalty. In many ways, the relationship is informal and private; they bypass official ranks and bureaucratic hierarchies (Bertrand 2013, 19). The property companies in Indonesia are mostly family businesses with strict patrimonial organization and tradition. The patron–client ties between the state officers and business persons are mutually beneficial. The officers get concrete support from their clients on job creation and investment for their city. In return, the property businessmen have their advantage from their patrimonial relation to the officers concerning smooth access and the process of development permits and building approvals.

The subtlety of patrimonial relationship has been developed throughout years and decades in such a way that publicly this patron–client tie is seemingly not something against law in terms of bribery and graft as well as collusion and nepotism. The core of patrimonialism is about the privilege of access to resources including information. The patron–client relationship between high-ranking officers and business persons relies on this privilege and seeks the breaches and opportunities between regulations and real conditions.

Corruption, collusion, and nepotism after Suharto's fall are likely only shifted of power relation from executive to the legislative agency as well as from old to new regimes. The issues for policy-makers are fundamentally not about capacity of program design but politics (Robinson 2006); the reformation agenda for democratization and good governance have not yet successfully dissolved the predatory and illiberal practice of patrimonialism. Castle Asia's team (Castle and Manuwoto 2011) documents and comments scandals and political affairs in Indonesia 2010 delineating the complexity of the patrimonial relationship between politicians and business people in Post-Suharto era; the development of superblocks in Jakarta is by no means impartial from such political complexity. Despite the reformation process in Post-Suharto's Indonesia, some vestiges and attitudes of traditional patrimonialism still persist and are seemingly not to fade away in the near future (Mackic 2010, 83).

10.5 Planning/Design and Superblock Development

The development of superblock in Jakarta is regulated by the DKI Jakarta Regional Zoning Plan Number 6/1999 Article 43 paragraph (4). Accordingly, the development of superblock is conducted through the utilization of land use using a mixture of housing and other facilities such as a shopping mall and/or workplace. The residential uses 35–65% of the total area of available space. The program requirements when applying the mixed-use development within the framework of superblock should include common utilities, common social facilities, common parking, and open green space (Fig. 10.2).

The first step for superblock development that the developer should do is to acquire the principle permit for superblock development; this permit is required to qualify the capacity and competence of the company to carry out the project from legal, technical, and financial requirements. The second step is to get the assignment permit for the utilization of land—*Surat Ijin Penunjukkan Penggunaan Tanah* or SIPPT—from the local authority or municipality. Prior to the establishment of the proposal for location assignment permit, the superblock developer should submit the letter of recommendation from the National Land Agency—*Badan Pertanahan Nasional*—concerning the status of the land is ready for the development. The next step after having attained the location permit is to get the development permit. In order to attain this permit, the prospective developer should have already acquired sufficient land for the whole project.

For the development permit, the developer of superblock needs to submit a block plan or master plan. The most important issues include planning/design criteria for

Fig. 10.2 Podomoro residential superblock, Jakarta 2016. Photograph by author

the location concerning the issues of building volume, intensity, height, the proportion of uses, and green open space. The plan should be supported by various studies concerning the quality and performance of architecture, structure, utilities, and infrastructure. The block plan permit should include the general requirement of building permit; the plans and programs should be supported with technical drawings and outline specifications of structure and materials. In order to get the approval for the block plan, the developer and it's planning/design team should go to the meetings and consultations at least with three boards of planning/design review; the boards are for architecture/urban design, structure/construction, and utilities/infrastructure.

The most important thing is to ensure the public that the development of the project would be beneficial for local and regional growth. All this is described in the outline of the project comprising studies on the environmental and traffic impacts of the development; the project should be examined by municipality and representatives of professionals from various perspectives and aspects, such as the regional economy, urban planning, environment, society, and culture. Within a certain period of time, usually within six months, the local authority could terminate the permit in case there are no activities for the project. After the block plan, permit is the step of the development process that leads to the permits for the physical development of the whole project; these include building, structural, and utilities permit. Prior to the attainment of the permits, the developer should do the first pile test in the location. Complaints and protests against the development of the project are usually deliberated and negotiated during the meetings of environmental impact assessment (Fig. 10.3).

From the perspective of spatial planning, superblock development does not integrate its plans and buildings into the whole structure and fabric of the city. The massive and gigantic building system does not connect and incorporate the existing public realm and street system within the superblock system. Instead of creating a continuous network of walkable street pattern and block system with the open public realm, superblock development isolates such system in its area under private control. Superblock establishes a ghetto of luxury mixed-use development under privatization of land and air of the city which are supposedly open under a municipal management system.

Privatization of 5–10 ha area of development includes the self management of utilities and security system. Why all this happens in Jakarta and other cities in Indonesia is for the simple reason because the municipality is not able to provide appropriate infrastructure and security for large-scale development. In the sense of the patrimonial relationship between local government and developers, such privatization is regarded as part of the vision of gentrification and investment for more jobs in the city. Agung Podomoro Land underscores the vision of gentrification with its slogan: 'back to the city' (Podomoro City 2012). With such a motto, the developer intends to attract middle income employees and their families to stay in the city instead of living in suburban areas and commuting to the city. The privatization of large-scale area of mixed-use development is, at any rate, an establishment of an inclusive community. The inclusivity of living in superblock complex is even publicly highlighted as a marketing gimmick, such as 'white glove service' at Kemang Village

Fig. 10.3 Center Business Center Sudirman Superblock 2012. Photograph by author

(Lippo Homes Group Kemang Village 2015) and 'city' for Podomoro's mixed-use superblocks (Agung Podomoro Group 2012).

Even though the residential component within the mixed-use system includes low-income units either for rental or sale, the alienation of the complex from the rest of the urban environment is obvious. The development of middle-income units is actually not based on socially responsible planning approach but simply for the reason of the captive market in the area, such as Agung Podomoro City. What is interesting is that superblock development is also to integrate schools and university in its complex. Kemang Village and Podomoro City have an educational component—*Sekolah Pelita Harapan International*—in their superblock system. Such integration is actually in response to Jakarta's traffic jam for commuting.

Maximizing mixed-use in superblock development is not only commercial and residential components but also all possible urban activities that include educational, sociocultural activities, and public services. In order to attract customers, each superblock development characterizes its products and services with a certain theme that connotes exclusiveness or populism. The notions of 'city,' 'village,' 'world' have been used by developers to signify the exclusivity of its community; they likely want to highlight that their superblock complex as an entity or a community. The message behind this exclusivity is to attract and convince customers and prospective buyers that the superblock is a self-contained system of live–work–play.

10.6 Environmental Sustainability and the Superblock

Architecturally speaking, compact high-density and mixed-use development is a planning/design approach to establish private urban realm in a building system. In doing so, maintenance and operation in the mixed-use complex will be under control of the private agency. Superblock development unveils the concept of urbanism inside architecture with its environmental artificiality. Even though the Jakarta Governor's Decree 38/2012 regulates that the engineering performance of high-rise buildings and superblocks should be in compliance and provide with the smart building management system, energy consumption for superblock complex is still enormous.

The load for the power supply for most superblocks does not rely on non-fossil-based energy. The artificial lighting and air conditioning system of superblock complex consumes intensive use of fossil based energy for the electricity; such a complex is constructed with bulky and gigantic structure with less open green space. Even though daylight comes through the skylight, the superblock building system is mostly well-isolated and covered from direct air circulation from the atmosphere. In short, the superblock building system does not reduce the carbon emission of energy used in its artificially constructed environment.

Environmentally speaking, superblock developments burden the underground water reserve of Jakarta. Such a massive development pumps up their water supply during the construction from a deep well on site. The gigantic structure gives the site an enormous load that causes land subsidence in the area. Complaints from family houses around the superblock confirm the scarcity of water supply from their ground well. Based on the research results of the Department of Mines Jakarta and the Hydrogeological team of Institut Teknologi Bandung, the land in Jakarta fell 1.4261 cm annually; this is due to groundwater extraction of 17.5 % and because of the heavy load of the building for 82.5 % (Republika Online 2009).

From a sociological perspective, sustainable community entails diversity in ages, income groups, and employment fields. Such development is not preferable for sustainability because it is more or less a process of exteriority of the exclusive community from the public realm and the whole urbanism. The construction of superblocks segregates urban populations through the architecturally privatized realm of social space. The casual interactions in such containment mostly take place in the mall which is socially determined by patron–client communication and services.

Superblock developments in Jakarta take place in the areas where the demand for places of gathering and urban recreation is relatively high. Most of the mixed-use superblocks are located at the traffic junctions or at strategic business districts, such as Kemang Village, Gandaria City, Mega Kuningan, Ciputra World, and Kuningan City in South Jakarta, Podomoro City, Taman Anggrek, and St Moritz in West Jakarta. The captive market of a condominium in South Jakarta is mostly for the expatriate populations and their families, while in West Jakarta, most residential units are sold to employees and students of universities in the area. However, the Lippo group developed St Moritz and Kemang Village with a specific target for their high-class mixed-use components. In the interest of economic progress and growth, such development is a generator of new jobs and investments. Socioeconomically, compact and high-density mixed-use development widens and amplifies the gap between income classes. Economically, the sustainability of the activities in the superblocks lies in the job creation and transactions based on the capitalistic consumerism of urban populations.

Since various malls in Jakarta have been built everywhere in the last three decades, an oversupply of commercial retails overshadows commercial activities in the superblock shopping areas. Sustainability of economic activities in the superblock malls relies crucially on the visitors for the eatery and casual shopping. In order to attract customers, some superblocks routinely hold art performances and cultural events in their main atrium. Between 2000 and 2011, the demand of average retail space in Jakarta showed an annual growth of 12.2% that is equivalent with average growth above 200,000 m^2/year, while the total supply of retail space was 4.04 million m^2 (Colliers International 2012).

The superblock development is mostly constructed on the land with the area between 5 and 12 ha. The captive market for such developments is the people of middle-class income category with approximately $4450 monthly (Harinowo 2008). The problem of an apartment in the superblock is not to build it but how to sell it. The perplexing condition of selling apartment units is about the status of ownership. The challenging aspect concerns the cultural lifestyle of living. Most Indonesians value landed houses as the most preferred choice for their home. Living in a high-rise and multifamilies setting needs timely adjustment and adaptation. Instead of buying units at high-rise apartment, most Indonesian families choose rental landed houses. Indonesian developers adopt Australian strata title system for their apartment units. Such status does not offer the sense of security of ownership because collective possession of common elements is culturally still ungraspable as property. A common profile of the owner is a young urban professional person (Deviani 2015) or expatriate individual. Law No. 16 of 1985 and Government Regulation No. 44 of 1988 regulate the strata title matters. Accordingly, foreigners are only eligible to own units of condominium building as long as the land status is the right to use, not the right to own.

10.7 Case Studies

Among more than 12 superblock developments in Jakarta, there are two cases that would be discussed in this study: Podomoro City in the district of Grogol Petamburan, West Jakarta and Kemang Village in the district of North Cipete of South Jakarta. Based on the 2010 population statistics (BPS DKI Jakarta Selatan 2014), Podomoro development is situated in the west part of the city with 2,328,936 population and density of 17,749 people/km^2, while Kemang Village is located in the south part of Jakarta with 2,104,092 inhabitants and a density of 14,620 people/km^2; the total inhabitants of the city in the same year are 9,786,690, while the average Jakarta population density in 2010 is 14,750 population/km^2.

Agung Podomoro Group Company is the developer of the superblock of Podomoro City in West Jakarta. The company was ranked top of developers by the Building and Construction Interchange in 2012. The Company was founded and developed by Hilman family that has 24 subsidiaries with projects spread across Jakarta, Karawang, Bandung, Bali, Balikpapan, and Makassar. In the last ten years, the company even noted the completion of more than 50 property projects. The captive market of the company are the middle-class consumers. However, its scale of operation includes a range of projects ranging from low cost to high-end apartment South of Jakarta, the high end, and the neighborhood mall, shop houses, hotels, and office towers.

The superblock of Podomoro City is constructed in a traffic junction of West Jakarta within the land of 21 ha. The superblock is situated in the district of Grogol Petamburan of West Jakarta administration. The built-up area for the megastructure covers 16 ha land for 42 stories office tower with a gross area of 80,000 m^2, 45 floors shopping mall with 167,000 m^2 of gross floor area, 12 floors four stars hotel with a total area of 31,000 m^2 containing 420 rooms, and three condominium towers. The mixed-use superblock is extended with three residential blocks consisting of slab tower apartments, shopping arcade building, and office towers.

The superblock of Podomoro City highlights the concept of 'park' as its thematic development; it is in the hope that the mixed-use complex is imagined as the oasis of the less greenery urban landscape of west of Jakarta. Water and plants play an important role in featuring the architecture of the mixed-use complex. Design features have been utilized as a marketing gimmick for the superblock development. Since such features are not from public funding, but from privately patrimonial business, their cost must be paid off by the customers of the superblock; the allure of architectural consumerism has been explored by the developer as the patron of the planners, architects, and designers. In other words, creativity and artistic ingenuity have been commercially exploited for the return of investment. The lavish and glamorous displays in the mall and its other parts of the superblock are not free, but all these are subject to be paid back in various commercial forms, from foods, services, and goods.

As a whole, the superblock is located at the Tomang junction in the central area of the subdistrict South Tanjung Duren. The whole area of the district is 9900 ha consisting of seven subdistricts, 73 neighborhoods, and 862 compounds. According to the local authority office of Grogol Petamburan, the total population of the subdistrict in 2009 is 183,369 population, 46,120 households, with a density of 183.5 people/ha. Economically speaking, the gentrification through superblock development should be more than 400 people/ ha. Podomoro City has more than 1000 residential units in various types from moderate to luxury units as well as from bachelor to three bedrooms apartments. The main target of their market is young professionals and students. Three universities: Tarumanegara, Krida Wacana, and Trisakti are within walking distance of the superblock of Podomoro City; each of them has 20,000, 3000, and 30,000 students. Since 1993, the Tomang junction area has been gradually growing as the generator business activities of the West Jakarta region; prior to that year, the 176 ha area of subdistrict South Tanjung Duren has been mostly quiet low-rise residential area and west land. Today, there are three commercial superblocks: Citraland, Anggrek, and Podomoro City in the location.

Theoretically speaking, such generators are more than enough to perform livable and vibrant urbanism with a well-connected pedestrian network and outdoor landscape amenities. In reality, the three megablocks are private islands of mixed-use activities about and for themselves. Each development enclave has its own main access and parking management system. Where are the role and the function of the municipality for the public good in the Tomang junction area? The Jakarta development policy only entails the superblock developments to connect and facilitate the interface structure between the public and private realm with public transit structure, shelters, and pedestrian easements.

The other case of superblock development is Kemang Village. This megastructure is developed within the area of 15.5 ha adjacent to the eastern side of Krukut riverbank; it is situated in the district of Mampang Prapatan of 7730 ha. Architecturally speaking, the superblock stands out as a massive building complex with high-rise towers standing visually contrast to its surrounding indigenous kampung structure with the low-density settlement. According to local authority office of Mampang Prapatan district in 2013 (BPS DKI Jakarta Selatan 2014), the residential consists of 38 neighborhoods—*rukun warga*—and 411 compounds—*rukun tetangga*—or 3899 households, which predominantly occupies almost 88.41% of the land use of the whole area of this district. The exclusive character of the superblock is highlighted with private access roads of approximately 300 m from public transit route from Jalan Kemang Raya and Jalan Antasari Raya.

The development comprises medium size shopping mall of 80,000 m^2 areas, five-star hotel with 275 suites, and International Bachelorette School for 1000 students, 250 beds private hospital, 13,000 m^2 of area for exotic spa and amenities, three slab apartment towers—each of 29 floors—and four condominium towers consisting of two buildings of 41 floors and two others of 39 floors. The exclusivity of the Kemang Village is established by the social image of its location and setting; the area has been known as the niche of the international community and expatriates of Jakarta. Several high-middle class families of the city find their comfortable home in this area. Grocery stores, specialty shops, and other commercial services have been provided with imported goods and foods for the international community living in the district since the 1970s. The unique livability of the district has been traditionally established along the main road with human-scale development but not the well-supported pedestrian network.

Diversification of land use has been growing time to time without a clear and definitive planning/design framework and development policy. The local demands and market forces in the Kemang area prescribe and transform the previously residential land use to the mixed-use ribbon development along the main road. The main appealing aspect of the location is its greenery and relatively low-density indigenous settlements. The main challenging for new development in the period of 1970–1990 was obviously its accessibility and integration in the main road and utility system of the whole city. Gradually, the municipality improves the infrastructure though water supply and centralized sewage system are still in progress until recently. Nevertheless, the establishment of South Jakarta ring highway in 2014 supports the development in Kemang area with convenient access to the airport of Sukarno–Hatta.

The growth of the area has been going on as the spill-over of 1960s established Kebayoran Baru New Town and Central Jakarta. The needs of expatriate and high-middle-class residential area entailed local developers to build several compounds in the kampung structures. This opportunity is growing in response to the increasingly corporate and commercial activities along the South Jakarta ring highways; since 2000s several multinational companies and their associates establish their headquarters along those arteries. The integration of new infill developments is mostly in alignment with the existing topographical terrain and its greenery.

The superblock development in Kemang area has begun in January 2009; the investment for the project is eleven trillion rupiahs or one billion US dollars. The shopping mall has been operating since 2010, while in the same year, the three condominiums: the Ritz, the Empire, and the Cosmopolitan have started its marketing for their units. Regarding its construction scale and investment in 2009, Kemang Village superblock is the largest development in the district of Mampang Prapatan.

The developer of the superblock is Lippo group. Since its early establishment in 1990, the company has been always under the patrimonial leadership of Riyadi's family. Like the Podomoro, the Lippo group belongs to the conglomerates of business in Indonesia that enjoyed Suharto's policy of liberal economic development; they were able to get a loan from state-owned banks for their business. Moreover, the Domestic Investment Law 1968 provided the domestic investors with the same incentives and guarantees as the Foreign Investment Law 1967 (Wie 2006, 10); this Law contained various attractive incentives, including generous tax concessions, such as tax holidays, duty-free imports of capital goods, and guarantees, including the free transfer of dividends and profits and a guarantee against arbitrary nationalization of foreign enterprises (Sadli 1972, 204).

Under the New Order's liberal economic policy of 1967–1998, several domestic business groups began to come into play under the political patronage of senior government officials and senior military officers; they involved and operated in various sectors, including textiles, electronics, transport equipment, pharmaceuticals, real-estate developments, often as joint ventures with foreign investors (Robinson 1986, 144). According to Building and Construction Interchange of Asia 2012 (BCI Asia 2012), the Lippo group company has become one of the sixth largest in Jakarta. Many mega projects the company are currently in the progress of completion, ranging from the superblock of St. Moritz to Kemang Village. The Company recorded a market capitalization of Rp. 16.4 trillion in the Indonesia Stock Exchange. Lippo has a variety of real-estate business; from the division of Hospitals contributed revenue of Rp. 402 billion in the first quarter of 2012, while the residential division of township Rp. 547 billion. Lippo also still has a business large-scale integrated developments, commercial, as well as non-core asset management.

In most cases, the continuity of superblock developments in various places of Jakarta shows that the public and private partnership for gentrification in Jakarta pay more attention to the interest of business activities and less consideration for place-making in the public realm. Instead of intensifying the existing land use with well-connected public connectivity and mid-rise building system, building all-in-one system of urban activities in superblock system becomes the preference of urban development policy in Jakarta between 1990 and present.

From the demographic figure of Jakarta 2013, it is doubtful to consider the superblock development by private sectors as an option of the housing of middle-class populations. Rising family in the superblock complex has not yet been an option for Indonesian middle-class populations. Living in a high-rise apartment is an option for young urban professionals with no kid as well as students on a rental basis. According to the market survey of apartments, residential units in superblocks are mostly for investment (The Jakarta Post 2007). Most young families in Indonesia prefer to live in landed houses rather than in high-rise units. The buyers are mostly people at the managerial level; they put their units for rent as their business.

Regarding the continuity of megastructural developments, there is likely no room for a municipality to intervene the planning and design process of the superblocks that work in the interest of great urbanism at policy and technical level (Krismantari 2010). In other words, the municipality does not challenge, negotiate, and work together with the private sectors at pro-active way concerning the realm between the private and the public realm for its safety, convenience, health, societal and environmental sustainability, and esthetics.

10.8 Concluding Remarks

The patrimonial planning system in Indonesia gives an opportunity for the patron–client relationship between officials and property business people; both have something in common for providing urban inhabitants with great places for work–live–play. Patron–client ties play an important role in the politics of urban development. Good urbanism is not the main objective of such a patrimonial relationship but creating jobs, providing a place to live, and economic growth for the elites. Superblock development is one of the outcomes form such patrimonial relationship. Accordingly, superblock development is a massive response to the growing population that consists of single-person households, young professionals, and students. An integrated system of live–work–play becomes a demand of many young people in urban areas against longer commutes, debilitating traffic congestion, and smog/air pollution. The superblock developers read and observe this market demand to persuade a high-density and mixed-use development for a self-contained community.

Architecturally speaking, superblock and megastructure developments show that the planning system has not been effectively working to manage urban development for good urbanism based on social justice. Favoritism of development policy for the wealthier social class—*priyayi*—is obvious. All this is subconsciously or intentionally cultural and historical that city and state in Indonesian context are defined by the presence of center—either as a prime location, extravagant architecture, and elites of society. The flourishing of superblock developments in Jakarta and other Indonesian cities affirms and confirms that gentrification in Indonesia is to represent the patrimonial hegemony of the elites. The democratization of the public election system and the liberalization of the economy have not yet worked effectively to transform the society and state toward social justice and equal opportunity; urban development policy has been still working in favor of the wealthier social class—*priyayi*. The root of all evils for this antiurban development lies in the predominantly patrimonial culture and tradition of Indonesian society.

The patrimonial practice in the politics of urban development has been demonstrated with a strong line of command from the central government to local municipal administration. During the Suharto's New Order, the practice was obvious. Despite the reformation of the political system after 1998, the politics of development has not yet been transformed with a democratic process. What happened with the superblock developments shows the patrimonial practice is to be continued with a different actor: the local government. Participatory planning approach to urban development is not fully incorporated within the decision-making process of planning and building. As a matter of fact, the governorship still plays a significant role in the urban development policy. The legislative body remains at a passive function with fewer options in fighting for the public interests of social justice and environmental stewardess. Conflicts of policies between the central and local government come into play without a clear objective that matters for good urbanism so that there is the right to live, work, and play for all inhabitants. With superblock development, the public domains have been marginalized; most urban amenities have been spatially privatized as the commodity of the capitalist economy. Meanwhile, the urban poor becomes the servant player within the trickle-down economy in the food courts and retails businesses.

The democratization of planning system should be an option for better urbanism that makes development plans and processes publicly accountable and open for adjustments and negotiations. Nevertheless, democratization in the planning system is likely insufficient because the patrimonial culture and tradition have not been radically removed from the decision making the process of state and corporate bureaucracy; it has been working in Indonesian society as part of the lifestyle and cultural values system. The political transformations do not penetrate deeply into the whole system of state bureaucracy and private sectors. The President of Indonesia 2014-2019: Joko Widodo during his campaign trails, suggested the Indonesian nation to do 'mental revolution' for transforming patrimonial society to the democratic nation-state.

References

Agung Podomoro Group (2012) Agungpodomoro.com. Accessed 24 Apr 2015. https://www.agungpodomoro.com/index.php?i=3&s=16&cid=9&lang=ind

Anderson B (2006) Imagined communities, reflection on the origin and spread of nationalism. Verso, London and New York

Anderson BRO'G (2007) The idea of power in Javanese culture. In: Holt C (ed) Culture and politics in Indonesia. Equinox, London, pp 1–70
Bappenas_BPS_UN Population Fund (2013) Indonesia population projection. BAPPENAS RI, Jakarta
BCI Asia (2012) Big ten of construction developers in Indonesia. Building and Construction Interchange of Asia, Hongkong. https://www.bciasia.com/
Bertrand J (2004) Nationalism and ethnic conflict in Indonesia. Cambridge University Press, Cambridge
Bertrand J (2013) Political change in Southeast Asia. Cambridge University Press, Cambridge
Bisnis Indonesia (2016) Indonesia Darurat Bank Tanah, Aug 30. Accessed 30 Aug 2016. https://industri.bisnis.com/read/20160513/45/547112/indonesia-darurat-bank-tanah
Blau E (2014) A capital without nation: red architecture in Vienna and spatial politics between world wars. In: Milkenberg M (ed) Power and architecture: the construction of capitals and the politics of space. Berghahn, New York, pp 178–207
BPS DKI Jakarta Selatan (2014) Mampang Prapatanin figures. Badan Pusat Statistik Kota Administrasi Jakarta Selatan, Jakarta
Budihardjo E (1997) Preservation and conservation of cultural heritage in Indonesia. Gadjah Mada University Press, Yogyakarta
Castle J, Manuwoto A (2011) Indonesia: political pulse 2010. Equinox, Jakarta
Colliers International (2012) Research and forecast report: Jakarta retail market. Accessed 24 Apr 2015. https://www.colliers.com/~/media/files/apac/indonesia/pdf/jakarta_retailmarket_report_2q_2012.pdf
Cribb R (2000/2013) Historical atlas of Indonesia. Curzon Press, Surrey
Deviani I (2015) Apartemen Jadi Gaya Hidup Kaum Muda, Mar 06. Accessed 27 Apr 2015. https://www.mediaindonesia.com/mipagi/read/8984/Apartemen-Jadi-Gaya-Hidup-Kaum-Muda/2015/03/06
Ellin N (2012) Good urbanism: six steps to creating prosperous places. Island Press, Washington, DC
Geertz C (1960) The religion of Java. University of Chicago Press, Chicago
Gerke S (2000) Global lifestyles under local conditions: the new Indonesian middle class. In: Chua BH (ed) Consumption in Asia: lifestyles and identities. Routledge, London and New York, pp 135–158
Gooptu N (2011) Economic liberalization, urban politics and the poor. In: Ruparelia S, Reddy S, Hariss J (eds) Understanding India's new political economy: a great transformation? Taylor and Francis, London, pp 35–48
Gordon DLA (2012) Battery Park City: politics and planning on the New York 2012. Routledge, New York
Green J (2011) The Dirt_American Society of Landscape Architects, Aug 02. Accessed 25 Apr 2015. https://dirt.asla.org/2011/02/08/interview-with-peter-calthorpe-author-of-urbanism-in-the-age-of-climate-change/
Harinowo C (2008) The Jakarta Post, Sept 16. Accessed 22 Apr 2015. https://www.thejakartapost.com/news/2008/09/16/economic-growth-the-rise-indonesian-middle-class.html
Harvey D (2008) The right to the city. New Left Rev 53:23–40
Hudalah D, Winarso H, Woltjer J (2014) Gentrifying the peri-urban: land use conflicts and institutional dynamics at the Indonesian Mentropolis. Urban Stud 1–16
Jacobs J (1961) The death and life of the Great American cities. Random House, New York
Kelly SB (2004) Community planning: how to solve urban and environmental problems. Rowman & Littlefield, Lanham
Knight N, Heazle M (2011) Understanding Australia's neighbours: an introduction to East and Southeast Asia. Cambridge University Press, Cambridge
Krismantari I (2010) Superblock: dream or nightmare? Accessed 07 May 2015. https://www.thejakartapost.com/news/2010/10/23/superblock-dream-or-nightmare.html

Lippo Homes Group Kemang Village (2015) Kemang Village.com. Accessed 24 Apr 2015. https://www.kemangvillage.com/white_glove_services.php
Mackie J (2010) Patrimonialism: the new order and beyond. In: Aspinall E, Fealy G (eds) Soeharto's new order and its legacy: essays in honour of Harold Crouch. ANU E Press, Canberra, pp 81–98
Parmer NJ (2012) Historical perspectives of Chinese populations in Southeast Asia. In: Amstrong MJ, Amstrong RW, Mulliner K (eds) Chinese populations in contemporary Southeast Asian Societies: identities, interdependence and international influence. Routledge, London, pp 18–54
Podomoro City (2012) Podomoro City. Podomorocity.com, Jakarta. Accessed 5 May 2015. https://www.podomorocity.com/html/site_plan.php
Republika Online (2009) Republika Online, Feb 12. Accessed 24 Apr 2015. https://www.republika.co.id/berita/shortlink/30913
Robinson R (1986) Indonesia, the rise of capital. Allen & Unwin, Sydney
Robinson R (2006) Corruption collusion and nepotism after Suharto: Indonesia past or future? Spring. Accessed 22 Apr 2015. https://www.iias.nl/sites/default/files/IIAS_NL40_13.pdf
Sadli M (1972) Indonesia. In: Drysdale P (ed) Direct foreign investment in Asia and the Pacific. Australian National University Press, Canberra, pp 201–205
Shin HB (2015) Economic transition and speculative urbanisation in China: gentrification versus dispossession. Urban Stud. Accessed 04 Aug 2015. https://eprints.lse.ac.uk/62608/
Simadjuntak D (2013) Beyond wealth and pleasant posture: exploring elite competition in the patronage democracy of Indonesia. In: Abbink J, Salverda T (eds) The anthropology of elites: power, culture, and the complexities of distinction. Palgrave Macmillan, New York, pp 95–112
Smith N (1979) Toward a theory of gentrification: a back to the city by capital not people. J Am Plann Assoc 538–592
Smith DP (2004) "Studentification Ication": gentrification factory? In: Atkinson R, Bridge G (eds) Gentrification in a global context. Routledge, London, pp 73–90
Taddjoedin MZ (2014) Explaining collective violence in contemporary Indonesia: from conflict to cooperation. Palgrave Macmillan, New York
The Jakarta Post (2007) City planners factor in Superblock solution. Accessed 07 May 2015. https://www.thejakartapost.com/news/2007/09/01/city-planners-factor-superblock-solution.html-0
van Klinken G (2007) Communal violence and democratization in Indonesia: small town wars. Routledge, New York
Wie TK (2006) Policies for private sector development in Indonesia. ADB Institute discussion paper 46. Asian Development Bank, Tokyo
Winarta FH (2009) Suara Rakyat Hukum Tertinggi. Penerbit Buku Kompas, Jakarta
Znoj H (2007) Deep corruption in Indonesia: discourse, practices and histories. In: Anders G, Monique N (eds) Corruption and the secret of law: a legal anthropological perspective. Ashgate, Surrey, pp 53–76

Chapter 11
Capitalist Agenda Behind the Seawall Development in Jakarta Bay, Indonesia: The Marginalization of Urban Poor

Abstract The availability of land for a planned community has been always a crucial issue for urban developers in Jakarta. Creating land through reclamation is considered a less expensive way to acquire that is free from local resistance and land price speculation. Despite environmental issues and resistance from local fishermen, the reclamation project proceeds within the framework of the National Capital Integrated Coastal Development that is disputable in response to climate change. This study attempts to divulge and examine what is actually at stake in the project. The research question is: to what extent is the National Coastal project worth doing, and why the coalition between private and public sector so strong is through this project is most likely toward the marginalization of urban poor. Despite the urge for reformation, this study argues that the public policy on urbanism in North Jakarta has been historically trapped by the long-term planning of neoliberal Suharto's New Order legacy from 1990s.

Keywords Seawall · Jakarta bay · Reclamation development · Capitalist development

11.1 The Study on Integrated Coastal Management

This paper is a study on the project of integrated coastal management for the national capital of Jakarta since 1995; the project is one of the neoliberal development projects of Suharto's legacy (Botz 2001, 33–4; Wie 2006). Although the Post-Suharto's administration under Joko Widodo sets out its political agenda to strengthen the presence of the state for public good, the continuity this project as planned is considered as the reactivation of the New Order's development policy; this indication is confirmed by the installment of Sri Mulyani Indrawati as key person in Joko Widodo's administration (The Jakarta Post 2016).

The study begins with a historical exposition of the project and its contextual situations concerning what and why the project is regarded as strategic for the future of Jakarta as Indonesia's national capital. The study argues that whatever to take in the public policy for anticipation and management of climate change there is a necessity for public transparency and accountability. Since the project has been regarded as

B. Wiryomartono, *Traditions and Transformations of Habitation in Indonesia*,
https://doi.org/10.1007/978-981-15-3405-8_11

less accessible for the freedom of information, this study, furthermore, argues that there is a hidden agenda behind the execution of this project on behalf and in the name of public interest.

Studies on the relationship between spatial land-use planning and climate change for coastal areas have been characterized by wide and open interests with divergent objectives and agenda. In UK, despite climate change becomes a material consideration of spatial planning, the discourse for more prescriptive policy statements had been unlikely provided by the central government, but there were the networks of professionals, academics, citizen activists, and local governments that established the coalition of political processes in raising the profile of climate change within regional spatial plans (Wilson and Piper 2010, 77–9); the topics of spatial land-use planning for climate change comprise infrastructure, flooding, coasts, water resources, biodiversity of land and water, economic development, tourism, transportation, waste management, energy system, and built environmental design. In contrast to the UK, the Netherlands has been well known for its innovative series of national environmental policy, programs, and plans since 1989 (Straaten 1992); the Dutch government had unveiled a systematic integration of environmental matters and resolutions.

Regarding 60% of the landscape below sea level, the Netherlands is one of the countries around the globe with the pervasive risk of climate change. Rotterdam's Delta Project is one important example of how spatial planning on water and land has been well integrated to manage environmental impacts of fluvial flood risk on agriculture, water resources, biodiversity, and urban environmental quality (Wilson 2009, 230). The discourse on climate change adaptation and mitigation in the context of spatial planning is likely seamless through water management community in their familiar concern of safe, livable, and healthy community (Wilson and Piper 2010, 88); all this is made possible because, in this country, the water management community and network has been well established as a long-standing tradition.

11.2 Reclamation and Seawall Construction

The transformation of the landscape through land reclamation and seawall construction has been internationally known as part of the interdisciplinary integrated coastal management system. Under this system, the coastal landscape and water will be regarded as a functional habitat with the myriad of resources and activities, which are necessarily maintained, conserved, developed, and sustained to stewards its ecological balance between economic growth and livable biodiversity for sustainability. The main concern of the system is the functional integrity of economic development and ecological conservation so that human intervention into landscape and water ecosystem does not endanger the sustainability of the coastal region; despite its relatively small coverage in percentage of global land area, this zone is highly productive and sustains a high biodiversity (Mokness et al. 2012, preface). The concern on the coastal region has been triggered by the fact that currently, 60% or 3.6

billions of global world populations live and work in the coastal area—the area within 60–150 km depth along the coastline (Green 2009, 2).

The public policy of development and conservation in the coastal region is considered as strategic direction for sustainability and anticipation of climate change. The formal movement for integrated coastal management was initiated with the United Nations Conference for Environment and Development in Rio de Janeiro, Brazil 1992. One of the important Agenda 21 of this international movement is to accelerate global cooperation for sustainable development in the Global South and related domestic policies. Since the degradation of the environmental quality of the coastal areas is getting worse in terms of entropic states, global cooperation becomes necessary to constitute the circle of responsibility for sustainable environment and society through public awareness, the improvement of governance, adaptation/mitigation to change in the coastal zone, and linking science and management.

The issues concerning land developments—in the context of seawall construction and land reclamation—for coastal management include the conflicts between the environmental and social sustainability and economic developments. Integrated coastal management is necessarily implemented as a public policy that is to prevent, minimize, and resolve the potential conflicts between environmental conservation and economic growth because of the increasing activities in the coastal landscape for residential, commercial, industrial, and touristic developments. Regarding its environmentally and economically strategic geography of the coastal zone, the integrated coastal development is necessarily governed by a government agency that is able to work in interdisciplinary, intersectoral, and interdepartmental manner; most importantly, such a governing body for coastal management needs the support of interdisciplinary expertise and well-equipped resources, especially in response to the predictable climate change (Cicin-Sain et al. 1998, 57–66; McInnes et al. 2003, 1–15).

The construction of the seawall and land reclamation in Jakarta Bay is developed with Rotterdam's Delta Project in vision. Rotterdam's Delta Projects stands out as one of the exemplary works for the integration of land reclamation and seawall construction into spatial land-use planning of water city; this project has been considered a successful adaptation and mitigation of climate change based on the synergy of research and public policy. The Dutch government did not deal with the project simply as pragmatic problem solving of fluvial flood risk. Rather, the Netherlands worked on the synergy of hydrological management and spatial land-use planning with a visionary approach to build and develop a safe, healthy, and livable built environment. In order to actualize this vision and prior to the execution of the project, the Dutch government delivered research grants to build the public awareness and knowledge on the relationship between water and spatial land use in terms of living, working, and recreation, all this is envisioned and experienced within the concept of living with water (Hendriks and Buntsma 2012, 154–8).

The water-related disasters and its impacts on people and environment have been always the national issue for the Netherlands. This has brought about collective awareness and cooperation among citizens and politicians to establish the water boards in every region. The board plays an important role in water management and other natural disasters in its region. Members of the board are elected from community members of the region; they are responsible to oversee the dikes, flumes, and ditches in their area.

The works and effectiveness of the water boards are considered as the oldest democracy in action in the Netherlands. From a historical perspective, the flood risk and impacts have brought politicians and parties together to establish and develop national policies on flood protection and water management. Rotterdam's Delta Project that took several decades of its completion since 1996, is not only a Dutch reputation for hydraulic engineering and science but also an exemplary work of collaboration among stakeholders based on public awareness and participation. The experience of the Delta's Project is the success story of living and working with water throughout the landscape; this experience has been expanded internationally to construct similar works in South Korea, Bahrain, Bangladesh, in Jakarta Indonesia.

11.3 Historical Setting and Background of the Project

The national capital of Indonesia, Jakarta, is described by Rosenau and Wildsmith as a site of tensions between global mercantilism of megacities and local efforts at urban planning; these tensions demonstrate the fragmenting and integrating challenges they incessantly have to confront (Rosenau and Wildsmith 2003, 190). The urban development in North Jakarta is a vivid case study that illustrates the tensions with obvious marks of indigenous legacy, colonial heritage, modern national vestige, post-colonial trace, and global awareness. All this is experienced in the daily lifeworld of North Jakarta when one crosses the landscape along the highway from the International Airport of Soekarno–Hatta through the wasteland of Muara Kamal Angke, the outskirts of Pantai Indah Kapuk planned community to Kampung Rawa Bebek, Ancol, Warakas Pasar Ular, and Tanjung Priok. The inherent tensions of the urban experience in North Jakarta are likely impossible to be merely resolved with spatial planning but in an interdisciplinary project within the framework of integrated coastal management. From this point, the National Planning Agency of Indonesia conceptualized, designed, and scheduled the National Capital Integrated Coastal Development Plan as long-range strategic planning in 1994–1995. The plan is intended to manage and anticipate the environmental and socioeconomic impacts of climate change in the coastal areas of Jakarta.

As a matter of fact, the complexity of issues, constraints, and opportunities of North Jakarta is not only demonstrated and characterized by its fragmentary urban forms, structures, functions, materials, and communities, but also by its extreme gap between the poor and the rich. Despite this widespread discrepancy, all communities share something in common for the necessity of a strong community on a territorial basis for public safety. The whole municipality of North Jakarta covers an area of 146.66 km^2 with populations of 1,645,312 in 2010 and 1,711,036 in 2013 (BPS Jakarta Utara 2016); the total populations include all people regardless their status either Jakarta resident or non-resident. More than 30% of low-income populations in

Table 11.1 Populations under poverty in Jakarta (BPS DKI Jakarta 2016)

	2009	2010	2011	2012	2013
North Jakarta	76,200 or 5.34%	92,600 or 5.62%	84,730 or 5.07%	87,200 or 5.14%	90,900 or 5.30%
DKI Jakarta	339,600 or 3.80%	388,200 or 4.04%	355,200 or 3.64%	366,300 or 3.70%	371,700 or 3.72%

Source Jakarta Statistic Agency BPS Jakarta https://jakutkota.bps.go.id/linkTabelStatis/view/id/8

the North Jakarta area are migrant workers and fishers who mostly are not registered as Jakarta citizens (Table 11.1).

In dealing with climate change, land-use policy and planning in Jakarta attempt to build 31 km gigantic seawall in the north part of the city. This is an opportunity for public and private partnership for building the adaptive capacity of the city for climate change. However, from a business perspective, such a development is considered an opportunity for creating new lands and job opportunities. Since the nature of the land development of the project is on the empty reclamation areas, the planning process does not involve public participation as a basic requirement. In most cases in Jakarta, land-use development deals with the contentious processes of negotiations between the land property rights and the need of community at large. This involves a long process of land acquisitions and legal settlements.

The reclamation of 5100 ha land becomes an option to prevent new developments from potential litigation of land ownership and social conflicts in north capital city. Despite potential environmental crises and public debates, Jakarta's project of coastal defense strategy and development has been going on since 2007 until the Indonesian Committee for Corruption Eradication—KPK—has found the bribery case for the project on March 31, 2016. All this is considered as the top of the iceberg of the patrimonial planning system and practice in Indonesia for the marginalization of urban poor. Accordingly, public participation and transparent implementation have been not crucially important in the project planning process. This study argues that land-use policy and planning in Indonesia has been more associated with a patron–client relationship for the circulation and accumulation of the capital rather than for wider public interest for good urbanism. In order to avert corruption, collision, and nepotism in the future, this study suggests that the planning system in public participation and transparency of the planning process becomes a basic requirement of land development. Consequently, the planning system is necessarily reviewed and amended to accommodate public participation as an integrated part of urban development.

For several decades, the modern Jakarta metropolitan area has experienced severe floods. The other acute problem of drinking water in this city includes the salt-water intrusion in its shallow and deep aquifers that have reached inland up to 10–15 km from the coastline (Asian Development Bank 2009, 49); penetration in deeper aquifers (40–140 m) has been recorded up to 5–13 km inland in the areas of Cengkareng, Kelapa Gading, and Grogol. Meanwhile, the annual combination of rising sea level of 0.25–0.57 cm and the subsidence of land due to the extra exploitation of groundwater of 1 cm reaches the high risk of flooding covering 90 km^2 area of the North Jakarta (Pusat Pengembangan Kawasan Pesisir Laut, ITB September 2007).

The necessity for building the seawall is to anticipate the climate change for the city of Jakarta. Originally, the conception of the wall was to provide the estuary of Jakarta with a newly reclaimed lake. Since the seawall structure would be more useful if it could be inhabited, the outcome of the master plan is to create a newly reclaimed land of urban development; the master plan was initially developed by a Dutch consortium led by Witteveen+Bos as the main contractor and Grontmij, with subconsultants KuiperCompagnons, Deltares, Ecorys and Triple-A (Waterfronts NL 2015). The form of land is to emulate the Indonesian state emblem: *Garuda* (Indonesia Investments 2016). This emulation will be observable from the airplane when people come to land on the Soekarno–Hatta International Airport from North Jakarta sky.

The origin of this symbolic request is more important than the underlying necessity for such representation as a collective compulsion of Indonesia as a nation-state. The master plan for the 2700 ha, as delineated for development in the 1985–2005 Jakarta spatial plan was to use nearly 50% of the land for residential development, 15% for commercial and industrial usages, and roughly 5% to expand the port facilities, 14% for public facilities, and the remaining 16% for green open spaces, including the making of a conservation area for mangrove forest to protect the coastline against erosion (The Coordinating Ministry for Economic Affairs, Indonesia 2014). This project began on the North Jakarta land reclamation in 1996.

The conception for the reclaimed land development in North Jakarta has been developed during Suharto's New Order administration in 1980s. The formal presentation of the project of 2700 ha was exposed by the Jakarta Governor, Wiyogo Atmodarminto to President Suharto in March 1995; the project was formally endorsed by the Indonesian government with the Presidential Decree No. 52/1995 and Jakarta Local Regulation No. 8/1995 (Kompas Megapolitan: Putri Rosalina and Heru Margianto 2016). From this point, the development was intended to provide the North Jakarta area with sufficient land for new large-scale development for upper-middle-class residential demand. Previously, such developments have been done in former mangrove areas of Ancol for recreation areas of 552 ha and Kapuk for subdivision development of 400 ha; both were well known as successful projects that are able to transform the image of the coastal areas of Jakarta from gloomy and sketchy locations to vibrant and livable vicinities.

The reclamation for Kapuk development was challenged by several non-government organizations for environmental issues, but the project proceeded with the support of the local and national governments. From the onset, the formalities of the reclamation project in North Jakarta have been full of contradictions, overlapping, and disharmonies to each other (Lembaga Bantuan Hukum Jakarta 2016). The execution of the reclamation project was not indicated in the land-use plan of Jakarta, *Rencana Umum Tata Ruang*, 1985–2005. Based on this, the project was considered as an ad hoc program without integration and coordination in a long-range planning strategy at the municipal and regional levels. The Ministry of Environment issued the letter No. 14/2003 that did not endorse the project because of the unseemliness of environmental impact assessment. However, the reclamation project proceeded with the support of Jakarta's governor under the legal shield of presidential decree mentioned earlier.

The group of six developers for the reclamation project litigated the Ministry of Environment for its decision letter and won the case in the Jakarta court of justice in 2007; the ministry did not accept the court decision and appealed the case to the Supreme Court. The final decision was won by the ministry on July 28, 2009, that the reclamation project violated the principles of environmental impact assessment and the law of environmental conservation. Despite this legal outcome, the reclamation project continued as business as usual and far away from public attention.

Surprisingly, the case of impropriety for environmental impact assessment was lifted up again by the Supreme Court with new decision No. 12/PK/TUN/2011; the court stated that the reclamation activities in North Jakarta were legally acceptable. This means that the Jakarta government and the group of developers should work out their plan to be in compliance with the requirements of environmental impact assessment. More precisely, the project violated Article 36 paragraph 1 of the law No. 32/2009 concerning environmental conservation.

The coalition of local government and private sector for the reclamation project continued their work with the support of President Yudhoyono's administration. Based on the Presidential Decree No. 122/2012, the national government approved the reclamation project to develop subdivisions on each island. Especially for the island G, the Jakarta governor issued the Permit No. 2238/2013 that allowed the development on the islands for further details. What is interesting to note is the fact that the reclamation project is more attractive than the construction of Jakarta giant seawall as such. As a national project, the media coverage for the construction of the giant seawall and reclamation has been weakened by other national issues since 2014.

The transition of administration for Yudhoyono to Joko Widodo in 2014 took momentum for reviewing the project from an environmental and social perspective. All this was considered by the new administration as a necessary condition in the interest of public accountability and well-synchronized system of laws and regulations concerning the seawall development project and reclamation of 17 islands in the North Jakarta area. At the end of September 2015, the coordinating ministry for maritime, fishery, and environment issued a moratorium that canceled the reclamation project for commercial and residential development. The ministry endorsed only the project for infrastructure and power plant in the reclamation area. Meanwhile, the commission for the eradication of corruption continues to reveal more and more evidence for bribery indictment and abuse of power among the public officials in the process and execution of the project.

How effective is the 48.27 km giant seawall construction project for anticipating the climate change? Is there any other option to manage the rising sea level and the subsidence of land in the coastal area of Jakarta? The US$280 million seawall project has been studied and reviewed by the Ocean Engineering team of *Institut Teknologi*

Bandung (ITB) with an outcome and recommendation that the construction is not a worth doing solution; instead of building this structure, the ITB team suggests to cancel the reclamation and do not build the giant seawall (News Institut Teknologi Bandung/Najma 2013).

The reason for the cancelation is their concern about the environmental impacts of reclamation; the effects include the decreasing river streams and increasing sedimentation in the estuary areas. According to the ITB Ocean Engineering team, the problems of monsoon or tidal flooding and land subsidence will not be resolved with the heavy construction of seawall and reclamation, but simply by dredging the rivers on regular basis. Moreover, the team proposes another alternative, namely River Dike; this is to make levees along with the coastal areas experiencing land subsidence or subsidence and enhance the locations with river embankment. Levees are necessarily designed with masts stuck into the ground first, so the construction is sturdy, despite embankment subsidence but will still be structurally standing. This design is less expensive and does not need to shut down current activities of fishers and existing facilities, such as Muara Karang power plant.

Nevertheless, the success story of Delta Project in Rotterdam is likely, not impossible to be implemented in the Jakarta Bay; the Zuyder Zee project was not only able to construct large area from flooding but also provided Rotterdam with more than 160.000 ha new land for various uses from agriculture, industrial, residential to cultural. Moreover, the project brought about technological innovations in hydraulic engineering approach from deterministic to probabilistic method; the innovations include: large-scale application of geotextile and asphalt malt for scour protection, application of geo-containers, the use of discharge caissons, high capacity of dredging sand, and technology of granular filters (Pilarczyk 2012, 13–14).

It is worth noticing that the Delta water work is a national project of the Netherlands in terms of investment and coordination. In this case, the transformation of water to land has been managed and administered by the government as the patron. This is not the case for the land reclamation and seawall construction in Jakarta Bay; several island developments have been transferred from the government to private sectors so that the status of the new land is under the ownership of the developers. In addition, the spatial land-use planning of the new land has not been prepared, developed, and established by the government through a democratic planning process based on public interests and citizen participation, but in the interest of property business of private developers with the formal legislative process of Jakarta municipal council. In other words, the Jakarta Bay project's land reclamation and seawall construction are not to transfer the Dutch experience in Delta Project from its philosophical foundation of development, but only to utilize the Dutch hydraulic engineering for the interest of private sector's property development. In this case, the local government works as the regulator and inspector or the project. A potential collision between the private sector and key public officials becomes vulnerable because of the dominant role and function of the private sector in the land reclamation. This potential has been discovered by the Commission for the Eradication of Corruption recently.

11.4 The Right to the Land

One important challenging aspect of community planning in Jakarta is the scarcity of available land and the status of land ownership. The land registration in urban areas has been most problematic. One most obvious reason is the fact that land registration has been overly complex because it lacks a well-organized system of administrations and land surveys. The uncertainty of tenure in urban areas has been impaired by an unwieldy and complex land registration process. There is a lack of clear rights and procedures for registering common rights and non-transparent expropriation measures. In order to deal with these shortcomings, Suharto's New Order established the National Land Agency, *Badan Pertanahan Nasional*, in 1988. The agency has been beginning to register the parcels of land since 1990. Until recently, the progress of land registration has been achieved with less than 50% and challenged by the available fund for systematically mapping, surveying, and registering rights to land.

The complexity of land tenure and its status has been beginning under the Dutch colonial rule. The V.O.C did not care about land tenure and controlled the productions of land and populations by force at the coastal areas of the archipelago. At the end of the eighteenth century, the Dutch colonial control over territorial began in Java from 1808 to 1811, under Herman Willem Daendels as governor-general. Land tenure in Java was introduced by Stamford Raffles who was British Lieutenant Governor (1811–1815); he replaced the Dutch compulsory system for agricultural production with a cash-based land tenure system (Kanō 2008, 312).

The Raffles' land tenure was adopted by the Dutch with several adjustments based on agreements with village chiefs; individuals pay the land tax to the village heads, then the chiefs deliver the money to the nearest tax office. Since 1872, land tax in Java was widely practiced by the Dutch colonial rule on the basis of village and district territoriality on an annual basis. This land tenure system is still practiced in various rural areas until today; in West Java, such a system is known as *girik*. The land registration is actually to transfer the traditional and colonial land tenure system to modern administration with an accurate map and legal documentation. The process of this registration does not always run well because of disputes over land status and ownership. In most cases, there are lands that are not traditionally cultivated by individuals, but by the traditional community; such lands were claimed as communal property in terms of *tanah adat*. In this kind of status, the use and control of the land are not transferable through transaction and business deal, but through local customary laws. To a certain extent, such lands are exempted from tax.

The process of land registration and certification has been always problematic because the administrative process is mostly not transparent; this opens the opportunity for the abuse of power and corruption among the public officials. Under the New Order, the role and function of the National Land Agency had been often identified with the extension of executive power for the legal justification of land use and ownership for private sectors which were the clients of the people in power (Rosser et al. 2004, 3–5).

Reclamation of land in the North Jakarta Bay is considered a shortcut of land development with less social and administrative issues with land registration and acquisition. Private and public sectors in Indonesia share something in common for this project with a single main issue: the environmental impacts and responsibilities. There are environmental and social issues concerning the project of reclamation in the North Jakarta. The issues include disputes with local fishers, overlapping works with power plant construction, and indictments for conspiring to bribe public officials. The issues blew up when the Indonesian Commission for the Eradication of Corruption caught at hand a prominent public official and a leading contractor for bribery indictment on April 1, 2016 (BBC Indonesia Isyana Artharini 2016). The bottom line of the blunder of the land reclamation lies in the transfer of ownership of new land from the government to the private sector because of the investment of construction and development. In other words, the state does not present as the guardian of public interest because she lacks the money to develop the reclamation land.

11.5 Reclamation Project and Planning System

The planning system in Indonesia has been a historical evolution of the Dutch planning administration, *Staadvorming Ordonatie* (SVO) and Stadsvorming Verordening (SVV) in 1949; SVO works as principles, while SVV is for the implementation of the ordinance. However, the integration of these planning legal frameworks in urban development has been not effectively enacted until 1976 when the New Order administration dealt with massive rapid urban development and transformation of the landscape, especially in Jakarta, and other provincial capital cities in Java and Sumatra between 1976 and 1998. During this period, there were several regulations and government decrees concerning various aspects of spatial planning and land-use policy. All these were not considered as a well-integrated system of the legal framework for urban and regional development in the country.

The legal frameworks at the level of the ordinance include the Presidential Decree No. 41/1973 on Batam Island Special Region, the Presidential Decree No. 13/1976 on Planning and Development of Greater Jakarta Area, the Presidential Decree No. 8/1985 and No. 53/1989 on Housing and Industrial Estate. At the implementation of urban and regional development, the ministries issued several government regulations, such as Ministry of Interior Affairs' Regulation No. 2/1987 on city planning and urban development, and Ministry of Public Works' Regulation No. 60/1986 with similar theme and areas of concern.

Prior to the issuance of the Law No. 24/1992, a spatial planning system for urban and regional development did not have a clear and firm legal framework; confusions, misunderstandings, and disorientation have led physical transformations of the landscape in urban and rural areas toward environmental and social crises. Suharto's patrimonial governance used to manage the sociopolitical conflicts, crises, and resistances concerning urban and regional development with the authoritarian rule on behalf of public security, national stability, and sustainable development.

To some extent, the reclamation project in Jakarta Bay did not violate any principle in the Law No. 24/1992 on spatial planning. However, the project was necessarily adjusted to implement the Laws of Decentralization No. 32/2004, No. 23/2014, and No. 9/2015 concerning local governance and decentralization. The amendment of the legal framework for spatial planning was made with the Law No. 26/2007. Based on this current legal framework, most decision-making processes and executions take place in Jakarta provincial governance. Under Post-Suharto's administration, the national government does not have direct control over the realization of the Jakarta seawall and reclamation project.

Democratic governance entails involving the court of justice for any litigation against the project. Since the power of public interest has been formally incorporated by the members of Jakarta municipal council, then, local communities, local business associations, and non-government organizations do not have a chance to work in the planning processes and decisions. The problem is that the relationship between citizens and their city councilor is challenged by a cultural gap of patrimonial tradition. Accordingly, the city councilor is not to act as a public representative in terms of a public servant, but that of the patron to his/her community. The office of city councilor is mostly not in the location where his/her community is. In the law, a public meeting for the project of land development is mentioned as a formerly required process, but its significant outcome is not considered as a legally binding step for making a decision either the project is sustained or canceled.

In a matter of fact, the Law No. 26/2007 does not include reclamation in the spatial planning context; it is considered as a new subdivision or a new town development on the empty land. By nature, the spirit of the law on spatial planning is still to incorporate the patrimonial power of government as the sole patronizing agency for its citizens. The law no. 26/2007 does not care for community development, but simply about the legal framework of development for the blueprint of negotiations, accommodations, adjustments of the private sector's business plan. In other words, citizen and community participation in the planning process has not been regarded as active involvement.

Moreover, the reclamation is free from issues of land acquisition from people, technical coordination with existing infrastructure and public utilities, and social and environmental complaints of communities which live and work in the nearby location. The reclamation project is away from most people of North Jakarta. The only community who is affected by the execution of the project is the group of traditional fishers. This group has been managed by the local government that they will get a fair compensation package with housing, job, and school for their children. The most crucial problem for the reclamation project is an environmental issue. This is considered more technical, rather than political. This implies, as long as the developer of the land of reclamation is able to follow the necessary requirements of the environmental impact assessment, the project is legally worth carrying on. Even though the law has been designed to work at the local governmental level, this does not mean to incorporate local communities' vision and concern because the planning system has not well understood and sociopolitically practiced as their development process and framework toward better urbanism, built environment, and civil society.

11.6 The Spatial Planning of the Reclamation Islands

Planning can be understood as the collective management of development toward urbanity that is sociopolitically constructed and based on shared values and public interests. The reclamation project in Jakarta Bay is far away from this understanding. From the onset, the project has been exclusively in the domain of political elites. Information about the project is inaccessible without a special connection to the elites at the national and provincial levels.

Participation and access to information concerning the reclamation project are not provided within the legal framework either as a board member or planning committee. All this is considered as an exclusive, private, and confidential process of planning and development because the laws and other government regulations do not regard and include citizens and communities as active components of planning and development. To a certain extent, the patron–client relationship between local government and private sector becomes interchangeable; in most cases, the developers of land are the patrons and put key public officials under their pay role.

The most recent bribery case that is under the investigation of the Commission for the Eradication of Corruption—KPK—shows that members of Jakarta legislation have been under pressure from the reclamation developers; the developer group insists the legislative members arrange their contribution to public utilities less than 15% of the total project cost (Tempo.Co 2016; DetikNews 2016).

Behind all land and housing developments by most private sectors, the marginalization of urban poor has been typically the main objective of land developments. The gated and self-contained community becomes an important gimmick of property marketing for new residential development; in this country, social exclusivity is traditionally regarded as a cultural value that is not simply for the practical reason of security. All this is because the social gap between the haves and the have-not is extremely wide. The marginalization of urban poor has been considered as the heart of property business in Indonesia since 1970s under Suharto's New Order regime, especially in North Jakarta. This includes the residential developments of Pantai Indah Kapuk and Pantai Mutiara, Despite the Law No. 1/2011 and Ministry for Housing Development No. 7/2013 entail the harmonious composition between high-, middle-, and low-income housing, in reality, this requirement has been modified and adjusted with various ways for profit. The common practice to achieve this requirement is to compensate with penalty's payment to the local government or negotiations with the key public officials behind the close door.

After the establishment of the Commission for the Eradication of Corruption on December 16, 2003, the bribery issues and other misconducts by public officials have been under the loop of a public inquiry. The case of bribery reveals the cause of the gridlock of legislative reviews and deliberations for the spatial planning of the islands; the developers of the ongoing construction tried to negotiate with the members of provincial legislation concerning the amount of their contribution to the government for the development of infrastructure, public housing, and public utilities in North Jakarta. In this matter, the governor of Jakarta was firm with its regulatory position for 15% contribution of the taxable value of the property; this is derived from the Presidential Decree No. 52/1995 mentioned earlier. The executive body of Jakarta turned down the amendment of legislation regarding the amount of contribution; the internal dispute between them was not impartial from the alleged patrimonial relationship between legislative officials and the developers of the reclamation project.

In the past and under Suharto's New Order administration, this kind of patron–client ties has been silently practiced without any counter from media and public inquiry. However, access to information is still an issue for media concerning the public accountability of the reclamation project. Informal meetings between the patrons and clients of reclamation project have been often revealed after the suspects were busted by the Commission for the Eradication of Corruption.

Until recently, the dispute on the reclamation project is more about their business deal. The content of the spatial plan is barely discussed in the meetings of the legislative body. Commonly practiced, the spatial planning document is prepared by the planning office of the provincial government, Bappeda DKI Jakarta (Pusat Informasi Perencanaan Pembangunan BAPPEDA Provinsi DKI Jakarta 2013); this document is called Rencana Umum Tata Ruang Wilayah or General Spatial Plan of Subdivision for 20 years. The most important things in this spatial planning document include land uses, spatial structure, main infrastructure, environmental conservation, and public utilities.

The legislation deliberates, ratifies, and legislate the spatial document as the legal framework of the physical development of specific subdivisions. The intervention of private sectors in the process of legislation happens mostly behind the formal consultation. Bylaw entails every development proposal should be in alignment with the General Plan of the city, especially the subdivision spatial plan. Indeed, there is room for negotiations and deliberations for adjustment of the proposed development plan with the municipal spatial plan. Such modifications and amendments need the formal ratification from the legislative and executive body. In most cases, public meeting and media accessibility become a ritual and formality that mostly do not affect directly to the content of the development plan; the representatives of media, business associations, and local communities do not have the legal right to sit in the board of spatial planning committee. In doing so the spatial plan is not to incorporate public interest through citizen participation for land development. In other words, the new land development for subdivision is a kind of coalition of business and political interest without public accountability through the act of freedom for information and public inquiry by mass media.

As the outcome of the private and public partnership, the reclamation in Jakarta Bay lacks intensive public reviews. This can be seen in the master plan, that each island has been designated for specific land use and structure. The Spatial Plan for Strategic Coastal Area of North Jakarta or RTRW 2030, the islands should have specific main land use and urban form that create and develop diverse activities and built environments with waterfront characteristics. From a business perspective, the program and development on reclaimed land are a massive investment for its infrastructure and land compaction. Consequently, the developer has to make the best value of land and building out of this project.

The marginalization of the urban poor is likely unavoidable. Only the have had the access to get the benefit of this reclamation project. Indeed, the 15% contribution of the saleable and taxable property products of the reclamation goes to social and affordable housing projects, and other public utilities in the Jakarta area, but not in the location. This implies that the reclamation project is a development of gated communities on behalf of the National Capital Project of Giant Seawall Development.

Even though the Spatial Plan 2030 of North Jakarta includes the plan and program for public transit infrastructure, the physical plan of each island has not yet clearly delineated the urban design framework for the pedestrian-friendly environment and diverse residential units that the city and its public space are for all citizens. The law and regulations of spatial planning do not cover the quality assurance of good urbanism for the execution of the project, from the construction to its operation and maintenance; the regulatory frameworks still lack on criteria for site plan application, urban design system, and evacuation scheme for natural disasters or other dangers in the context of climate change and pedestrian-friendly environment. In addition, the local and provincial governments are not provided with a sufficient system for public accountability of public officials that consists of permanent public auditors, integrity commissioner, ombudsman, and lobbyist registrar.

11.7 Landscape Transformation and Conservation

Landscape transformation in the coastal region of North Jakarta has been recommended by the Ministry of Environment and Forestry as necessary to be integrated with the existing character of this zone. Conservation of the coastal ecosystem and native flora fauna, natural water bodies, distinctive geological features, and other locally occurring activities such as fishery and traditional maritime transportation is particularly significant in preserving the landscape character and identity of the region (Green 2009, 6). This conservation is by no means an anti-development and change. Rather, the necessity for conservation is necessarily understood as the process of adaptation and accommodation of new activities within the framework of the similar ecological system in which the biodiversity and ecological processes in the newly established landscape environment are kept with low entropic states.

How to achieve this environmental benefit is apparently and possibly not economical for the logic of capitalism. Until recently, the developers are transforming the landscape from its natural ecosystem free of charge and transfer of negative impacts onto the taxpayers who are not only losing the ecological services but also have to pay for the post-nature restoration measures (Cudlin et al. 2011, 27). Economically speaking and based on experience in the Pantai Mutiara Development, the price of land in the reclamation project will be sold at a much better price than that in the real estate in Jakarta areas. The environmental impacts of land reclamation development are mostly experienced after more than ten years after its completion. The concept of ecological services is necessarily implemented and integrated into the spatial land-use planning system and statutory that ensure and conserve natural ecosystems against harmful cosmic radiation, maintain biodiversity, produce fertile soil and biomass, clear air, and water, mitigate climate extremes, and decompose organic waste (deGroot et al. 2002).

The ongoing land reclamation of the island G has been monitored by the Ministry of Environment and Forestry concerning several environmental issues; the ministry issued a moratorium for the project on April 18, 2016. The issues include sedimentation and silting, shortage of clean water supply, the use of non-environment-friendly materials from the island of Tunda in Banten Province, conflict with local traditional fishers, and other issues concerning the existing maritime traffic, conflict of underwater infrastructure of the electricity plant and communication, administratively unsynchronized and overlapping regulatory matters. According to the ministry of environment and forestry, only 10 out of 17 islands of the land reclamation project have been qualified to pass the Assessment of Environmental Impacts (Setiawan et al. 2016).

Based on Oslo-Paris Commission for the Protection of Marine Environment of the North-East Atlantic 2008, the environmental impacts of land reclamation, with special reference to the Netherlands' Port of Rotterdam and Maasvlakte, comprise the everlasting loss of endangered habitats, substantial negative effects on the common tern and black duck, fluctuations in coastal currents, amplified noise and reduced air quality (by the increase of NO2, SO2, and PM10 concentrations) during the construction phase as well as adversative effects on benthos organisms and habitats from sand extraction, elevated fine silt concentrations (OSPAR Commision 2008).

11.8 The Developers of Reclamation Project

After Suharto came to power in 1967, the private sector played a bigger role in economic development than before. Unlike mining, petrochemical, pharmaceutical, forestry, and agricultural industry, construction and real-estate businesses did not grow significantly before 1980. As any domestic business groups, real estate and building contractors enjoyed to utilize access to foreign investments through a patron–client relationship between high-rank public officials and local business persons; Suharto overturned Sukarno's foreign investment policy with a liberal foreign investment law in 1967 (Wie 2006, 9–12). The law comprised various attractive incentives, including substantial tax concessions—such as tax holidays, duty-free imports of capital goods—and guarantees, including the free transfer of dividends and profits and a guarantee against arbitrary nationalization of foreign enterprises (Sadli 1972, 204).

Under Suharto's New Order, big and national scale business groups were mostly to involve the regime's officials through their family members or cronies. Such groups have mostly established and consolidated their expertise, field, and networks until today. Most developers of the private sector for the reclamation project were likely not impartial from their relation to Suharto's regime. Under the authoritarian New Order, collusive, corruptive, and nepotistic practice between business people and key public bureaucrats were common. The rapid growth of this practice was possible because of the patronage of the influential political elites, especially President Suharto; under this patronage, the conglomerates enjoyed the privileged treatment, including largely subsidized credits from the state banks, protection, and monopoly positions, and assured government procurement (Wie 2006, 14). The stock exchange boom in 1989 showed the real capacity of these conglomerates after some of their subsidiaries had gone public.

The National Capital Integrated Coastal Development of Jakarta Bay is a development project within the bilateral framework of governmental cooperation between the Republic of Indonesia and the Netherlands (NL Agency, Ministry of Infrastructure and Environment 2012); this project was to follow up the Suharto's legacy from 1995 mentioned earlier. As a whole, this large-scale new development plan was a plan for new waterfront urbanism located along with the coastal areas of North Jakarta, comprising a total area of around 2700 ha reclaimed from the Jakarta Bay. The project involved the creation of a new area of high-rise buildings, including the revitalization of the historic districts in North Jakarta and expansion of the recreational vicinities along the northern seaside. From the onset, this land development was intended to be developed by private developers applying state of the art infrastructure technology. Even though the seawall project has been less regarded as the national project that concerns and deals with the climate change and other environmental issues in Jakarta metropolitan area, public attention and concerns are more focused on the reclamation of islands at the coastal line of North Jakarta.

As a matter of fact, the reclamation project was a capitalistic approach to develop a land of development with less social cost with a profitable estate. The former Jakarta governors: Joko Widodo (2012–2014) and Basuki Tjahaja Purnama (2014–2017), did not directly against the reclamation project but they tried their best in honoring the legal commitment from the previous administrations. What Basuki did was to negotiate with the developers concerning their contribution to the local government with a 15% retribution fee from the property tax imposition basis. The purpose of the fee was to enable the local government to support the poor in the city with the financial aids for their housing, public health, and child education support.

What the former governor Jakarta (2014–2016)—Ahok Basuki Tjahaja Purnama—did, was to find the win–win solution between the capitalist developers' interest and the public gain concerning the reclamation projects. Ahok was fully aware that the projects were not socially and environmentally good for Jakarta. On the other side, he—as a government officer—was legally bound with the previous development policies and commitments to the developers. What he tried to do in the best interest of public good was to set up a possible cross-subsidy from an economically profit-oriented development to socially and environmentally supportive programs. Despite a non-inclusive solution for the whole reclamation projects, the legislatively drafted ordinance of fifteen percent development contribution of the developers for reclamation projects was considered a thoughtful option. The draft of Ahok's development contribution was diminished when Anies Baswedan came to power as a newly elected Jakarta governor in 2017. The legal draft for the fee was canceled by the Jakarta Parliamentary House. The case of the cancelation was lately known by the public because of corruption that involved some members of the local legislature. The building development in the reclamation areas was temporarily canceled because the newly established lands did not have any land-use planning scheme as the buildable legal basis or reference.

Unfortunately, Ahok was not able to pursue the negotiation concerning the building permit for the reclamation areas because he lost the election for the Jakarta governorship for the period between 2017 and 2022 to Anies Baswedan. The newly elected governor requested to cancel all building developments in the reclamation areas in June 2018. The National Land Agency refused to cancel the building permits in the reclaimed islands C, D, and G because the constructions on the man-made areas were not against the law. Most reclamation developers agreed to pay the fine concerning the building permits in the newly reclaimed islets. As a matter of fact, it is not the main issue of the reclamation project. The public request for the project is clear that is to stop the development as a whole program because it is not beneficial for the most public interest. Basswedan's administration did not have any scheme to make the reclamation project supportive for the poor in the capital city. Instead of pursuing Ahok's development draft for a cross-subsidy with 15% fee of development, Baswedan issued 932 building permits in the island C and D with a local governor decree; he defended in issuance on the reason that the reclamation of the Jakarta coast was proper in terms of a national government project as according to Presidential Decree No. 52/1995 on reclamation as well as Jakarta Bylaw No. 8/1995 on reclamation. Indeed, Jakarta local government had an agreement between the private developers and the Jakarta administration in 1997. According to the deal, the private corporations developing the reclaimed isles will receive 35% of land-use rights.

Map of Jakarta Bay and the giant seawall project. *Source* Bappeda DKI Jakarta 2010–2030 and Bappeda Tangerang Regency 2010–2030 and Pelindo II 2011

The developers of the series of islands in Jakarta Bay are private corporations (PT Muara Wisesa Samudra, PT Jakarta Propertindo, PT Taman Harapan Indah, PT Manggala Krida Yudha, and PT Jaladri Kartika Pakci) and state-owned companies. Most of these companies have been the ole players for large-scale land development since Suharto's era. They have been granted with a principal permit to develop the land-based on reclamation in 2012. Only the islands of C, D, and G are under construction. The rest islands are still on paper. So far, there is no progress report available on the construction of the seawall and the mythic bird of *garuda* form land development (Table 11.2).

11.9 Concluding Remarks

The execution of Suharto's National Capital Integrated Coastal Development is a massive effort. The initial purpose is to manage the environmental impacts of land subsidence, seawater intrusion, raising sea level, and tidal flooding in North Jakarta. The response of this Suharto's New Order administration worked out the potential impacts with a grand plan for building a giant seawall. How the construction is able to manage the diversity of existing and potential marine ecosystem still an open question. The most crucial issues concern on the impacts of the reclamation that will slow down the stream of 13 rivers and the stagnant water in the large lagoon with the

Table 11.2 Developers, areas, and permit status of reclamation project

Name of Island	Builder and developer	Total area (ha)	Planning permit status
A	PT Kapuk Naga Indah	79	Principal permit 2012
B	PT Kapuk Naga Indah	380	Principal permit 2012
C	PT Kapuk Naga Indah	276	Construction permit
D	PT Kapuk Naga Indah	312	Construction permit
E	PT Kapuk Naga Indah	284	Principal permit 2012
F	PT Jakarta Propertindo	190	Construction permit
G	PT Muara Wisesa Samudra	161	Construction proses
H	PT Taman Harapan Indah	63	Construction permit
I	PT Jaladri Kartika Eka Paksi	405	Construction permit
J	PT Pembangunan Jaya Ancol	316	Principal permit 2012
K	PT Pembangunan Jaya Ancol	32	Construction permit
L	PT Manggala Krida Yudha	481	Principal permit 2012
M	PT Manggala Krida Yudha	587	Principal permit 2012
N	PT Pelindo II	411	Construction process
O	PT Kawasan Ekonomi Khusus (KEK) Marunda	344	Principal permit 2012
P	PT Kawasan Ekonomi Khusus (KEK) Marunda	483	Principal permit 2012
Q	PT Kawasan Ekonomi Khusus (KEK) Marunda	369	Principal permit 2012

Source DKI Jakarta Provincial Government Office April 2015 (Jakarta Business 2015)

seawall area. Why should the integrated coastal management include reclamation as an option of the problem-solving scheme?

The already built-up environment of North Jakarta has been overloaded with exclusively planned communities of Pantai Indah Kapuk by Salim Group, Pantai Mutiara by Intiland, Ancol recreational and residential development by Pembangunan Jaya Group. By its exclusive inhabitants and access, all these developments are mostly considered a gentrified zone with less care for integrating local and existing communities in the whole regional planning system.

Architecturally speaking, the planned communities which were established during the Suharto's era, have been developed as socially gated subdivisions. The relationship between the existing planned communities and the National Capital Integrated Coastal Development project becomes clear when the scheme of the whole project exemplifies the reclamation plan for 17 islands at the coastal line of North Jakarta, from Muara Angke to Marunda. The reclamation project will include the participation of the private sector in urban development. This sounds valid from job creation and economic growth in the North Jakarta area. However, since the participation is apparently only to involve business tycoons with their established groups, there is something in this scheme. The reality is that the giant seawall and reclamation are arguable as the best solution for tidal flooding and land subsidence in North Jakarta.

Most criticisms of the National Capital Integrated Coastal Development of Jakarta Bay are about the reclamation project. Technically speaking, the discussions and debates on the necessity for giant seawall are less regarded by public and media coverages in the country. Nevertheless, the question either giant seawall is necessary or not, is actually relevant in the context of socioeconomic and environmental sustainability for the whole development in the coastal areas of Jakarta. This region has been well known as the socioeconomically and environmentally acute zone because of its shortcomings of drinking water supply, public sanitation, highly polluted air, water, and soil. The poverty in these regions is recorded as the highest in Jakarta.

Why does the spatial planning for the integrated coastal development 2010–2030 integrate the development of affordable housing in dispersing areas within the reclaimed islands? Instead of this integration, the Jakarta provincial government utilizes and allocates the contribution of the reclaimed land developers to build infrastructure, public utilities, and affordable housings somewhere outside the project areas (Viva.co.id bisnisnews 2016). There is likely no room to question the policy of the Jakarta provincial government to proceed with Suharto's legacy project. Despite there are several socioeconomic and environmental issues and internal legal contradictions concerning the sustainability of this seawall and reclamation project (Jakarta Legal Aid Institute LBH 2016), the executive and legislative body of Jakarta government still hold the integrated coastal management plan as a realistic and good for the future of capital city, especially in response to climate change.

The reason why the project is considered worth doing and continuing lies in the facts and evidence of abuse of political powers by public officials, as revealed by the Committee for the Eradication of Corruption since March 28, 2016. This is considered as the top of the iceberg of the patrimonial practice of corruption, collusion, and nepotistic collaboration between the private sector and key public officials through urban development projects. So far, the committee has not gone deeper into the cases of bribery, abuse of political power, and other illegal activities since Suharto's New Order era.

The environmental impacts of land reclamation and seawall construction in Jakarta Bay are obviously predictable that human activities will dominate the use of landscape. This will endanger the quality of the littoral ecosystem based on coastal biodiversity. Since the land reclamation is now in progress, the most important thing is how to ensure the future functional integrity of biodiversity on the newly established landscape. This is considered as the essence of landscape development that transforms the water ecological system into the land environment through reclamation. In other words, to develop a new place for work, live, and play is not simply to build the built environment. Rather, from the landscape ecological principle, this is to provide the open green space that allows natural components, functions, and processes to grow with less human intervention.

The necessity of area for the land of conservation on the land reclamation is possibly not economic for property business, but it is a must for the sustainable future of the built environment where we work, live, and play. It is argued that new development is necessary to integrate the less privileged communities into a grand development of the National Capital Seawall Development in Jakarta Bay so that the transformation of the landscape in the coastal area of the city happens for the benefit of all citizens in terms of sustainability and social justice.

The current practice of land development in Indonesia does not reinstate the government from national to provincial and municipal level the power to guard public interest as mandated by the constitution because they are not able to acquire, manage, and own the land as an asset for large-scale developments. Even though the land reclamation in Jakarta Bay is potential for the dominant role and function of the government to acquire new land, this opportunity has been transferred to the private sector. In doing so, the government is considered fending off the call for protecting and guarding public interest for sustainable development. The deadlock for the litigation of spatial land use on the land reclamation shows that the public interest is at stake and under the benevolence of private developer's interests for a profitable business. In a matter of fact, the current planning and its regulatory system do not give a room for citizen participation in a transparent way; the system is obviously relied on the personal integrity of the key public officials instead of on the public accountability based on freedom of information and democratic planning process.

References

Asian Development Bank (2009) The Economics of climate change in Southeast Asia: a regional review. ADB Publications, Manila

BBC Indonesia Isyana Artharini (2016) KPK turun tangan, proyek reklamasi Teluk Jakarta terancam? April 4. Accessed 12 July 2016. https://www.bbc.com/indonesia/berita_indonesia/2016/04/160403_indonesia_kpk_reklamasi

BPS DKI Jakarta (2016) Beranda » Kemiskinan » Persentase Penduduk Miskin Menurut Kabupaten/Kota Administrasi, 2009–2013. Accessed 22 July 2016. https://jakarta.bps.go.id/linkTabelStatis/view/id/54

BPS Jakarta Utara (2016) Beranda » Kependudukan » Jumlah Penduduk Menurut Kecamatan dan Jenis Kelamin, 2013. Accessed 22 July 2016. https://jakutkota.bps.go.id/linkTabelStatis/view/id/11

Cicin-Sain B, Knecht RW, Jang D, Fisk GW (1998) Integrated coastal and ocean management: concepts and practices. Island Press, Washington D.C.

Cudlin P, Sejak J, Pokorny J (2011) Valuation of ecosystem services towards ecological citizenship. In: Westra L, Bosselman K, Soskolne C (eds) Globalisation and ecological integrity in science and international law. Cambridge Scholar, Newcastle Upon Tyne, UK, pp 26–38

de Groot RS, Wilson MA, Boumans RMJ (2002) A typology for the classification, description, and valuation of ecosystem: goods and services. Ecol Econ 41:393–408

DetikNews (2016) Sidang Suap Raperda Reklamasi, July 18. Accessed 19 July 2016. https://news.detik.com/berita/3256033/m-sanusi-akui-anak-buah-aguan-minta-pengaturan-paripurna-agar-kuorum

Green RJ (2009) Coastal towns in transition: local perceptions of landscape change. Springer Business Media & CSIRO, Dordrecht, NL; Collingwood, Victoria, Australia

Hendriks MJA, Buntsma JJ (2012) Water and spatial planning in the Netherands: living with water in the context of climate change. In: Ludwig F, Kabat P, van Schaik H, van der Valk M (eds) Climate change adaptation in the water sector. Earthscan, London & Sterling, VA, pp 143–158

Indonesia Investments (2016) Giant Sea Wall Jakarta; National Capital Integrated Coastal Development (NCICD). Accessed 24 July 2016. https://www.indonesia-investments.com/projects/public-private-partnerships/giant-sea-wall-jakarta-national-capital-integrated-coastal-development-ncicd/item2307

Jakarta Business (2015) Daftar Pengembang Pulau Reklamasi Di Teluk Jakarta, December 10. Accessed 20 July 2016. https://jakarta.bisnis.com/read/20151210/384/500479/ini-daftar-pengembang-pulau-reklamasi-di-teluk-jakarta

Jakarta Legal Aid Institute LBH (2016) Reklamasi. Accessed 24 July 2016. https://www.bantuanhukum.or.id/web/tag/reklamasi/

Kanō H (2008) Indonesian exports, peasant agriculture and the world economy, 1850–2000: economic structures in a Southeast Asian State. NUS Press, Singapore

Kompas Megapolitan: Putri Rosalina and Heru Margianto (2016) Jalan Panjang Reklamasi di Teluk Jakarta, dari era Soeharto sampai Ahok, April 4. Accessed 12 July 2016. https://megapolitan.kompas.com/read/2016/04/04/10050401/Jalan.Panjang.Reklamasi.di.Teluk.Jakarta.dari.era.Soeharto.sampai.Ahok

La Botz D (2001) Made in Indonesia: Indonesian workers since Suharto. South End Press, Cambridge MA

Lembaga Bantuan Hukum Jakarta (2016) Reklamasi Teluk Jakarta, Proyek Ambisius Penuh Pelanggaran, April 1. Accessed 12 July 2016. https://www.bantuanhukum.or.id/web/reklamasi-teluk-jakarta-proyek-ambisius-penuh-pelanggaran/

McInnes RG, Jakeways J, Claire M, Street C, Houghton H (2003) Making coastal management work-experiences from the implementation process on the isle of wright, UK. In: McInnes RG (ed) International conference on coastal management 2003: proceedings of the international conference on coastal management. Thomas Telford, Reston, VA, pp 1–15

Mokness E, Dahl E, Støttrup J (2012) Global challenges in integrated coastal zone management. Willey, Chicchester Sussex

News Institut Teknologi Bandung/Najma (2013) Muslim Muin Ph.D.: 'No need to build Jakarta Giant Sea wall', May 31. Accessed 22 July 2016. https://www.itb.ac.id/en/news/3918.xhtml

NL Agency, Ministry of Infrastructure and Environment (2012) Jakarta coastal development strategy. Terms of Reference Final Report. Programme Water Mondiaal & Programme Partners voor Water, The Hague/The Netherlands

OSPAR Commision (2008) Assesment of the environmental impact of land reclamation, biodiversity series. Vers. Publication Number: 368/2008. Accessed 25 August 2016. https://qsr2010.ospar.org/media/assessments/p00368_Land_Reclamation.pdf

Pilarczyk KW (2012) The impacts of delta works on the recent developments in coastal engineering. In: Kim YC (ed) Coastal and ocean engineering practice. World Scientific, Danvers MA, pp 1–38

Pusat Informasi Perencanaan Pembangunan BAPPEDA Provinsi DKI Jakarta (2013) Rencana Tata Ruang Wilayah 2030. Accessed 21 July 2016. https://bappedajakarta.go.id/?page_id=1270

Pusat Pengembangan Kawasan Pesisir Laut, ITB (2007) Survey GPS land subsidence Jakarta. Geodetic Engineering, Institut Teknologi Bandung, Study, Bandung

Rosenau JN, Wildsmith D (2003) Jakarta as a site of fragmegrative tensions. In: Bishop R, Philips J, Yeo WW (eds) Postcolonial urbanism: Southeast Asian cities and global processes. Routledge, London, New York, pp 187–204

Rosser A, Roesad K, Edwin D (2004) Indonesia: the politics of inclusion. Working Paper, Brighton, Sussex BN1 9RE: Institute of Development Studies

Sadli M (1972) Indonesia. In: Drysdale P (ed) Direct foreign investment in Asia and the Pacific. Australian National Unicersity Press, Canberra, pp 201–205

Setiawan D, Sulistiowati T, Triyono A (2016) Reklamasi Teluk Jakarta tersandung masalah, August 25. Accessed 25 Aug 2016. https://fokus.kontan.co.id/news/reklamasi-teluk-jakarta-masih-merana

Tempo.Co. (2016) TERKUAK: Aguan Diduga Dalang Suap Reklamasi, Ini Buktinya, April 14. Accessed 19 July 2016. https://m.tempo.co/read/news/2016/04/14/063762580/terkuak-aguan-diduga-dalang-suap-reklamasi-ini-buktinya

The Jakarta Post (2016) Budget cut spillover looms, August 5. Accessed 22 Sept 2016. https://www.thejakartapost.com/news/2016/08/05/budget-cut-spillover-looms.html

van der Straaten J (1992) The Dutch national environmental policy plan: to choose or to lose. Environ Polit 1 (1):45–71

Viva.co.id bisnisnews (2016) BISNIS: Ini 'Upeti' Agung Podomoro ke DKI Demi Garap Pulau G, July 18. Accessed 24 July 2016. https://bisnis.news.viva.co.id/news/read/798030-ini-upeti-agung-podomoro-ke-dki-demi-garap-pulau-g

Waterfronts NL (2015) Jakarta coastal development strategy. Accessed 7 July 2016. https://www.waterfrontsnl.com/project-jakarta/

Wie TK (2006) Policies for private sector development in Indonesia. ADB Institute Discussion Paper 46. Asian Development Bank, Tokyo

Wilson E (2009) Use of scenarios for climate change adaptation in spatial planning. In: Davoudi S, Crawford J, Mehmood A (eds) Planning for climate change: strategies for mitigation and adaptation for spatial planners. Earthscan, London & Sterling VA, pp 223–235

Wilson E, Piper J (2010) Spatial planning and climate change. Routledge, London & New York

Chapter 12
Violence Against Architecture: Demolition of Cultural Heritage of Otherness in Indonesia 1600–Present

Abstract This paper is to examine the violence against the architecture from the perspective of historical development and cultural anthropology in the Developing World. This study argues that this violence is the outcome of the process of contestations, negotiations, and constructions for cultural integration and acculturation that have failed to overcome the concept of *otherness*. The cases are mainly carried out from Indonesia between 1600 and present. The aim of this study is to describe, examine, and dismantle the backgrounds and events of violence against the work of architecture of otherness by patrimonial regimes; in the argument that the ferocity happens because they do not want to see themselves in the other. In order to see otherness in ourselves and vice versa, this study argues that the concept of the patrimonial nation based on monolith culture should be deconstructed toward democracy with an emphasis on multiculturalism, social justice, through the sovereignty of law.

Keywords Violence · Architecture · Otherness · Dutch colonial · Sukarno · Suharto · Post-Suharto

12.1 Violence and Architecture

In the 'Developing World,' violence against architectural work is considered as a politically marginalized issue. There are some issues, such as massacre, rape, famine, and poverty, which are likely more urgent and important matter than the destruction of cultural heritage. In a matter of fact, political instability has been mostly the cause of various destructions in the 'Developing World' in terms of colonialization, civil war, and the struggle for independence. One of the roots of all evil for these devastations is the exercise of domination by a patrimonial regime or group over others for the control of territory, populations, and other resources (Leftwich 2014). Architectural and cultural heritage has been mostly a part of collateral damage for such domination.

What is the architecture for the patrimonial authority or regime in the 'Developing World'? In order to deal with the question, this study attempts to frame this issue within the concept and practice of patrimonial hegemony throughout history in

B. Wiryomartono, *Traditions and Transformations of Habitation in Indonesia*, https://doi.org/10.1007/978-981-15-3405-8_12

Indonesia. Based on this framework, the relationship of patrimonialism and its violence against the cultural heritage of others will be examined through philosophical, historical, and sociological perspectives covering from global historical evidence to specific events in Indonesia.

This study presents an explorative query on the relationship between patrimonial regimes and their aggressiveness against the built environment of other culture. With a particular focus on architecture in Indonesia from 1600 to present, the context of this investigation seeks to unfold and dismantle the aspects and reasons of violence against the architecture of otherness based on its purpose. The study departs from the premise that architecture is never impartial from the ideological preference of its ruling regime and society because it is representative of their sense of locality, values system, and cultural identity. Therefore, violence against architecture is the case of criminal justice that needs legal consequences. Until today in Indonesia, the justice system for the architectural and cultural crime has been considered fail to serve.

12.2 Violence Against Architecture in History

The Bibliotheca Alexandrina was one of the ancient institutions targeted by violence against architecture and urbanism in history. The library was well known in its function in the dissemination of knowledge and in the preservation of ancient papyrus rolls and inscriptions. There was no available information on the interior and exterior layout of the building complex with respect to Ptolemaic Palace (Lulat 2005, 50–1). The complex had consisted of a museum, the main library building, a daughter building, and warehouses. Julius Caesar had indirectly caused the first destruction of the Ptolemaic Alexandria library in ancient Egypt in 48 BC, wherein the attempt to set fire to the Egyptian fleet, the fire had spread out into the city and burned down the library (Green and Karolides 2009, 325). The next fire was intentionally by Emperor Aurelian (270–275 AD) and followed by the Emperor Theodosius in 391 AD. Greek–Roman historian Plutarch (46–120 AD) reported the inflammation of the library a century later after the event (Cherf 2008, 55–8).

Even though the library building was made of stone wall system with arch colonnades, its roof system was partly flammable as described in Lucan's *Pharsalia* poetic account (McKenzie 2007, 77). What is the most tragic was not only the demolition of its building structure in 48 AD but also the precious collections of 700,000 papyrus rolls (Tatarkiewicz 2006, 232), where texts written in papyrus were gone forever in the dust and rubbles. The collections included more than 200,000 papyrus rolls that were bestowed from Anthony on Cleopatra from the library of Pergamum.

The burning down of the library of Alexandria was impartially the impact of Caesar's expansion of his patrimonial power over Egypt; the library was rebuilt and served as a learning center until 389 AD when the Emperor Theodosius ordered his provincial governor to destroy it in the campaign to unite empire under one religion (Tomkins 2012: 56). When the Muslim army under the order of the Caliph Omar Ibn al-Khattab arrived in Egypt in 642 AD, they did come across the library of Alexandria (Shalaby 2011, 234). Theodosius then ordered to burn down places of pagan worship. Despite the confusion concerning the cause and reason of the devastation of the Alexandria library complex, it is clear that the violence against architecture and its institutional function had been unavoidable in the struggle of political and cultural hegemony.

Another incident similar to the burning down of Alexandria library occurred when Spaniard patrimonial forces led by Hernan Cortez conquered the Aztec city of Tenochtitlan in Mexico 1519. Yet another similar case was also when Napoleon Bonaparte conquered Spain in 1811; he destroyed the library of the Benedictine monastery at Montserrat of Catalonia (Green and Karolides 2009, 325). Further demolition of libraries in history also happened when the German invaded Belgium on August 25, 1914. The library of the Catholic University of Louvain was burnt down with 70,000 volumes, 300 manuscripts, and other precious documents such as early printed Bibles, rare Jesuitical materials, and Low Country political pamphlets (Summerfield et al. 1998, 531).

The library has been ideologically associated with knowledge while functioning as a learning center. The presence of the library in society represents the house of ideas and historical lessons. The historical cases in the burning down of library as mentioned above are symbolically significant in cutting off the connection of society from their past and destroying the cultivation of knowledge and practices. The demolition of the library, intentionally or not, designates the violence against its collection. Since the collections consist of information and historical materials, destroying the library is nothing but the eradication of media for the cultural sustainability of certain nation, ethnic group, race, and ideological system. An interesting inquiry is in understanding the motive behind the action of demolition.

The historical relationship of patrimonial authority to the destruction of the city was invoked by the burning of Rome on July 18, 64 AD. Emperor Nero was accused by rumors that he had set the fire while watching from his Palantine and playing lyre as the city burned. However, the source of the rumors, Cassius Dio, had actually misinterpreted Tacitus' and Suetonius' account of events that had happened some 165 years after the great fire (Dando-Collin 2010, 1–5). The rumors are not confirmed. The story of patrimonial regime's violence against architecture was confirmed to be done by Genghis Khan and his troops (1167–1227); most villages, towns, and cities he conquered were destroyed and burnt down, such as in Volga Bulgaria, Persia, Russia, Armenia, Georgia, Crimea, Kievan Rus, and North China (Hartog 2006). What Genghis Khan had done was not only to burn down cultural institutions but flatten the whole society, their heritage, and their culture. For humanity, while murder, looting, pillaging, and raping is a crime, destroying the architecture of otherness is a tragedy.

Violence is not always in the form of burning and destruction but also in the closure of institutions, so that the occupants of the building and the people are forcefully scattered or separated. Institutions for the design like the soul or spirit of the activity and presence of the place. The closure of the Bauhaus School in 1933 during the Nazi reign (Droste and Bauhaus-Archiv 2002, 228) was not only in the locking of old buildings in the former factory at Berlin Steglitz, more so than that it was actually the assassination of the ideas for modern industrial society. This incident was not the first time the Bauhaus had been overruled in Berlin. The school had suffered from a series of persecution since the school began operation in Weimar (1919–1925), following the move to Dessau (1925–1932), and finally to the city that became the center of power of *Dritten Reich* in 1933.

12.3 The Dutch Colonialism and Violence Against Architecture

Even though since 1602 the Dutch sea power *Vereenigde Oostindische Compagnie* (V.O.C) had controlled the spice trade in most port cities of Indonesia, prior to 1806 they did not establish Dutch colonial settlement in the form of town and city, but citadels or strongholds at coastal areas with shops, barracks, workshops, and warehouses. The Dutch patrimonial trade control was made easier with the patron–client relations; the Dutch had similar patronage to the Chinese merchants and settlers in the archipelago. In order to attain their monopoly of the spice trade, the Dutch V.O.C never hesitated to exercise military forces in order to burn down indigenous settlements and their spice plantations. J. P. Coen and his V.O.C forces consisting of Japanese mercenaries demolished the village and killed indigenous people of Banda Neira in 1621 and Ambon in 1623 (Parmer 2012; Bown 2009, 49); it was the most memorable incident in the history of colonialism in Indonesia.

Batavia, the first Dutch colonial settlement in Indonesia, was established at the estuary of the River Ciliwung after the V.O.C under Jan Pieterszoon Coen attacked, captured, and destroyed the indigenous settlement of Jayakarta May 30, 1619, including the palace, mosque, and 'kampung' (Heuken 2007, 36). The Dutch V.O.C then expelled the local rulers and other dignitaries from the location. Then, the V.O.C forces burnt down the Portuguese stronghold and several other European office representatives in town (Great Britain Public Record Office 1870, 553). The entire indigenous settlements within the town of Jayakarta were flattened, and their inhabitants fled to hinterlands or other places.

The only populations left in the new town were Chinese settlers, slaves from various places of the archipelago, and former Portuguese militia *mardijkers*. Jan Pieterszoon (J.P) Coen built the castle and *bastide* of Batavia with the help of Chinese populations. Coen's patrimonial patronage mobilized Chinese clients and associates, led by Su Bin Kong or Ben Kong and Jan Kon to construct Batavia as the first Dutch V.O.C company town and headquarters in East Indies (Parmer 2012, 28–30). Coen then renamed the town from Jayakarta to Batavia; the concept of the Dutch settlement was given from the directors of the V.O.C in the Netherlands, Herren XVII as a 'place of rendezvous ' or a hub of the Dutch Company trading network (Blussé 2013, 356). In other words, Batavia was originally not intended as the Dutch State colonial town, but the Company's personnel in the Castle of Batavia, not in Batavia Town Hall ran and managed most administrative and financial businesses and public affairs of the town.

The most expansive Dutch colonial urban settlement came to the reality in Dutch East Indies in the island of Java between 1808 and 1811 under the governor Herman Willem Daendels. The most significant project of Daendels as Governor of French–Dutch East Indies was the construction of 1000 km long post road from Anyer in the West to Panarukan in East Java. Despite the sacrifice of forced labors, the urbanized settlements of Java grew and developed along the Daendels' post road. Unlike his predecessor: J. P. Coen, Daendels did not build Batavia as a walled city; he demolished the castle and parts of the *bastide* of the V.O.C's town in 1808. Daendels utilized the materials of Old Batavia for the new development of the town of Batavia in the hinterlands of Jakarta, now Gambir, and Gondangdia and Menteng area.

Since the establishment of the Dutch East Indies colonial rule in the 1800s, the architecture of archipelago had been diversified and enriched with various European building styles and types. Prior to the Pacific war in 1942, the Indonesian archipelago was probably the only living architectural museum of East and West in the tropical region. Indeed, during the Dutch colonial reign between 1800 and 1942 in the Southeast Asian archipelago, casualties, and violence against indigenous architecture occurred in various places including Gowa's Sultanate palace in 1667 and Bantam's Surosowan Palace in 1680. Previously in 1511, the Portuguese sea power burnt down the Malacca's Sultanate. Struggle for economic hegemony and trade control over spices in Indonesia was the reason behind the destruction of indigenous palace architecture and capital towns during colonial times.

Nevertheless, the European colonialism in the archipelago was also the beginning of an appreciation of indigenous culture and tradition. It began with British Lieutenant-Governor Sir Stamford Raffles when he was working in Java (1811–1814) and Sumatra (1820–1822). Raffles wrote *History of Java* in two volumes published in London 1817. The Raffles' book comprised not a strictly historical account of Java, but also about land, society, culture, and tradition. On the certain case, the accuracy of the information, such as land tenure, that he wrote was doubtful probably because of mistranslations of the Javanese inscriptions and texts; most issues on his book have been superseded and resolved by the Dutch scholars in their publications after the nineteenth century.

Despite his sympathy and fascination for the landscape, society, culture, and architecture in Java, he traveled through the countryside of Central Java with a specific interest in archeological traces and relics. Raffles was well known among the Javanese court circle of Yogyakarta as the first Western commander who was able to seize the Sultanate Palace of Yogyakarta on June 21, 1812. According to local court sources, Raffles did not only destroy the architecture of the palace but also plundered several collections from court archive (Cribb 2000: 111). After the fall of the Javanese center of power in Yogyakarta, Raffles controlled Java with the principle of 'divide and rule' (Ricklefs 2008, 111).

During the Japanese occupation in Indonesia from 1942 to 1945, several Dutch colonial buildings were burned down because of the war. However, some iconic buildings and structures were burned down intentionally because they were associated with the Dutch as an opposing authority for the Japanese. The bronze statue of Jan Pieterszoon Coen in the Waterloo Square, presently known as Lapangan Banteng, was an important landmark of Dutch colonial military headquarters complex. The bronze statue was torn down not because it was a symbol of the Dutch conqueror over the East Indies because it serves as a valuable asset for the war. During the Japanese occupation, most of gold, silver, steel, and other precious metals were compulsorily collected by Japanese officials from people in Java, Sumatra, and other parts of the archipelago.

12.4 Sukarno and Violence Against His Own Historical Site

Under Sukarno's Guided Democracy, several remarkable buildings from colonial times were demolished under his rule. Sukarno ordered the demolition of the most important witness of the proclamation for Indonesian independence at Jalan Pegangsaan Timur 56. Shortly after the proclamation of Indonesia's Independence, members of the national movements for independence: Johanna Masdani, Mien Wiranatakusumah, Ratulangi, Zubaidah, and Gerung, agreed to establish a memorial statue in front of the house wherein 1946 they had sponsored and built the monument for the commemoration of one year Indonesian Independence (Kompas.com_Megapolitan 2009). One year later, Prime Minister Syahrir inaugurated the monument on August 17, 1946.

Instead of supporting the initiative of the establishment of the Proclamation Monument, Sukarno ordered the major of Jakarta to do nothing. In its place, he also instructed to change the name of the edifice from the House of Proclamation, *rumah proklamasi*- to the House of Independence' pioneers, *rumah perintis kemerdekaan*. Ten years later, on January 1, 1956, Sukarno ordered the Jakarta mayor for the demolition of the House, which was previously the residence of the Dutch East Indies officer: Baron van Asbeck. The flattening included Jo Masdani's historical monument in front of the House (Kompas 2009, 62) and Sukarno ordered to demolish them in 1946. It was canceled when the Dutch colonial state tried to come back to Indonesia and re-control Jakarta during 1946 and 1949.

Even though Sukarno's residence had been declared by the law as a historical site and cultural heritage in 1948, Sukarno insisted on his order to the major of Jakarta, Dr. Soemarno, for the demolition of the house on August 15, 1960. Several months before the flattening of the house and its monument, Jo Masdani, Mien Wiranatakusumah, Zua Ratuangi, Zubaidah, Gerung, Maria Ulfah, and Setyati sent their protest letter to Sukarno (Kompas 2009, 64) which received no response even though the proclamation had been in perfect standing in front of the historical house since 1946.

Until now, people did not know as to why Sukarno tore down his own house (1942–1946). Sukarno mentioned in his speech in front of the National Planning Board, *Dewan Perancang Nasional*, on August 13, 1960, concerning the demolition that the action was necessary for a new beginning. Furthermore, he mentioned to the board that he needed more space for the garden surrounding the monument. Sukarno did not have an opportunity to build his monument for the historic proclamation until his fall in 1967. In remembering Sukarno, the Jakarta Governor Ali Sadikin rebuilt the replica of the Jo Masdani's obelisk structure on the site in 1970; the replica of the monument, *tugu proklamasi*, was inaugurated by him on August 17, 1972. Since then, the idea of the monument for the proclamation never died until 1980. Ali Sadikin was Sukarno's choice for the Jakarta governor; he was assigned to replace Dr. Soemarno Sosroatmodjo. In consolidating his New Order's power, Suharto needed to gain public support, especially from the generation of the 1945 revolution.

Sustaining and realizing Sukarno's projects were important for Suharto; he had ordered the projects of the Mosque of Istiqlal by Silaban 1968 and the Conefo Building for People Assembly Hall and House of Representatives by Soejoedi in 1969. Suharto was in an uncomfortable position concerning the case of the Proclamation Memorial House and Site, and under public pressure, he had ordered the State Secretariat to launch an open design competition. The committee of the competition was confirmed with Presidential order no 30 1979. The most important requirements of Suharto's order were to accommodate Sukarno's idea of that the Proclamation Monument should be built as a columnar structure 17 m in height. The winning result of the design competition was the chosen work of Boediono Soerasno, Sidharta, Sugiyo, Sumartono, and Nyoman, where the construction of the whole complex began in 1979 until 1980. On August 16, 1980, President Suharto inaugurated the monument.

The reason for the demolition of the original artifacts of the Proclamation Memorial House and Site remains unclear. In fact, Sukarno and his Prime Minister, Syahrir, did not always come into terms concerning their political attitudes toward the colonial heritage. Syahrir's administration attempted to save the House and its site at its original form in the line of thinking that architectural work is impartial. On the other side, Sukarno's attitude was ambivalent. Some colonial buildings were still in good operation and maintenance under Sukarno's administration (1959–1965) and most of them were used for a similar function such as the former Dutch East Indies governor residences for presidential palaces in Jakarta and Bogor. It is probable that Sukarno has something personal to do with the Proclamation Memorial House at Jalan Pegangsaan Timur No 56 as he used to live in that house with his wife Fatmawati and their children from 1942 to 1946. In that house, he had also written the draft for the proclamation before he became Indonesia's first president. In addition, it was in that same house that Fatmawati knitted the first Indonesian flag for the first Independence Day of the new Republic of Indonesia. Most of the meetings and other preparations for Indonesia's Independence took place in the living room of Sukarno's house.

In order to visualize his vision of a new Indonesia, Sukarno commissioned the architect Friedrich Silaban in 1961 to design the building adjacent the historical site of the proclamation. The construction of the building came shortly within one year later. In 1963, Sukarno inaugurated the building and named it *Gedung Pola Pembangunan Semesta Berencana*. The building was planned by Sukarno as the exhibition house of arts and media. The architecture of the building is a modern concrete structure with strong tropical climate adaptation. In contrast to modern Dutch East Indies tropical buildings, Silaban applies a strict rectangular pattern of concrete sun shading and simple joint details.

Unlike the neoclassical architecture of Concordia and Harmonie Building, the architecture of the four-story building, Gedung Pola, is a clear-cut monumental edifice without any attachment to historical styles from Europe or elsewhere. Sopandi argues Silaban's Gedung Pola, in its transparent monumentality shares something in common to Guiseppe Terragni's Casa del Fascio at the Lake Como built between 1932 and 1936 (Soepandi 2006, 71). To a certain extent, Terragni's Casa is strongly composed of asymmetrical form consisting of dualistic integration of solid and rectangular frames. On Gedung Pola, Silaban treats the façade in a symmetrical arrangement. In addition, Terragni's design for the Casa is spatially playful and creates a variety of solid-void scenes between the frames. On Silaban's design, the geometrical order is strictly integrated with the building structure.

Today, the historical sites of the proclamation for Indonesia's Independence are revitalized with an artificial site with iconic and symbolic elements. Original and authentic artifacts are absent from the scene of the historical site; this is probably the actualization of Sukarno's imagined Indonesia as a free nation-state. However, the visualization of Sukarno's patrimonialism is demonstrated by his apt perception of the monumentality of modern buildings. In the name of 'nation-building,' Sukarno imposed his vision on his architects, such as Friedrich Silaban, for Indonesia's architecture that unveils the spirit of freedom and grandeur modern expression.

The mosque of Istiqlal and the Senayan Stadium, *Gelora Bung Karno*, are the examples of Sukarno's heritage for grandeur architecture. Unlike the stadium complex that was built on empty land, the mosque of Istiqlal is situated in the former Wilhelmina Park with statues and monuments. Despite its negligence during the Pacific war 1942–1945, the park was featured with its English garden design in the former Dutch East Indies archipelago. The Ciliwung River flows through the site featuring potential of the urban recreational park for Jakarta, where 1930–1900 photographs collection of the Tropenmuseum in Amsterdam shows the features of the park.

Under the name of 'nation-building,' the site of the park and some remnants of the citadel of Prince Frederick Hendrik was seized to ground. Sukarno selected the site of the park of Wilhelmina for the mosque of Istiqlal because of its adjacent location to the Catholic Jakarta cathedral. He wanted to set up a place of religious tolerance where a mosque, the Christian church, and Hindu–Buddhist temples stand close together within an area. The former site of Wilhelmina Park was considerably perfect, but the idea was not realized during Sukarno's time in power.

12.5 Suharto's New Order and Violence Against Architecture

In post-colonial Indonesia, architectural heritage had experienced functional and financial hardship at the beginning of the Republican administration from 1950 to 1959. One of the colonial heritages under historical hardship was the *Societat de Harmonie* Building. The clubhouse was designed by the Dutch military engineer Schultze. Prior to 1985, visitors to Jakarta might still enjoy the view to the classic imperial building that stood tall at the end of Jalan Veteran, formerly *Harmonie*, at Rijswijk straat, *Societet de Harmonie* Building serves as the clubhouse for socialites of the Dutch East Indies. The members of the society included prominent European business people and government officers in the Southeast Asia region. The establishment of the society's clubhouse began in early 1814.

On January 18, 1815, the club was inaugurated by Lieutenant Governor-General of Britain, Sir Stamford Raffles. Among the socialites of Batavia between 1811 and 1815, Raffles was well known as Founder and President of Batavia society for arts and sciences. Raffles pioneered the admiration and appreciation for indigenous arts and architecture, especially from Javanese society and culture. The impact of the publication of *The History of Java* by Raffles was apparent in the form of self-criticism among the Dutch society concerning the attitude of the Dutch colonialism to native people in the Dutch East Indies. In fact, some ethnographers and art critics utilized the art forms from Indonesia as the source of their strength and inspiration to the Dutch colonialism (Bloembergen 2006, 88).

The origin of the clubhouse building was the initiative of Dutch Governor-General Rainer de Klerk in 1776 and his intentions were to move the club from Old to New Batavia (Susetio 2012). Originally, the club was located at the Jalan Pintu Besar, nearby the Old Batavia Town Hall. Regarding its unsafe location and vulnerability of site during monsoonal floods, The Dutch Governor Daendels agreed to relocate the clubhouse in the New Batavia area, adjacent to the Rijswijk: a military post-control from Old Batavia to New Batavia or Weltevreden. In 1808, the construction of the clubhouse began. For the construction, Daendels mobilized his government staffs and Dutch military officers to sign up for their membership in the club. In doing so, Daendels was able to collect the fund for the establishment of the club.

The construction did not come to completion. Due to the political development in Europe in 1811, Daendels had to leave Indonesia. Raffles took over the completion of the construction for the club in 1812 where he sustained the Daendels' idea of membership for the club with a broader scope comprising people in the circle of arts and sciences. For military and high-ranking government officers, Raffles also sustained the establishment for their club called *Societat de Concordia*. The building was inaugurated by the Governor-General Jean Chrétien, Baron Baud on September 21, 1833, and eighteen years after Raffles left Batavia.

The location of Concordia Building is now in present-day Banteng square. Both clubs performed weekly concerts and live-music every two or three days in a week

(McMillan 1914, 281), but the Harmonie and Concordia Buildings were closed during the time of Japanese occupation. Some contents and decorations were plundered and taken down; The Japanese authority had used the buildings for their military office until the end of the Pacific War in 1945. The Concordia Building was then demolished under Sukarno's order in 1960. One hundred and seventy years later on April 1985, the *Harmonie* Building had similarly face the fate for demolition.

The plan for the flattening of the *Harmonie* Building had been made by the request of the Presidential Secretariat office to the Jakarta municipal authority since the early 1980s. The Land Use and Spatial Planning Committee for the location did not support the request and need for adjustment. Of course, the plan of demolition was unexpected by many, including Indonesian societies and associations for the arts and architecture. They signed the petition to the Jakarta Governor in an attempt to cancel the demolition. Under the New Order administration, there is no room given for debates or arguing concerning the plan for culling down the building. Any critical opinion is subject to suspicion against national security and public harmony. During that time, the architecture community felt powerless to make a petition to the authorities both at the national and provincial levels.

The violence against the architecture of otherness, including the Concordia Building in 1960 and other colonial heritage buildings, was possibly driven by the resentment against colonialism and imperialism under Sukarno's Non-Alignment movement. A similar case of violence also occurs to buildings built by the governor-general of Japan in Keijo or Seoul at the time he had occupied Korea in 1926–1946, that received the fate to be razed to the ground between 1995 and 1996. The reason for the demolition from the people and government of the Republic of Korea was simply that they did not want to see the symbol of Japanese colonialism of the past, and as such the magnificent neoclassical building complex fallen to the ground. The former Governor office in Keijo was demolished by the government of South Korea on behalf of the Korean victims of Japanese occupation in Korea 1910–1945.

Arguably, the *Harmonie* Building was not destroyed by the hatred of colonialism, and instead by the cold hand of pragmatism—the city needed re-alignment of streets and the widening of the existing road. From the site, the State Secretariat of Indonesia needed more parking space for their daily business. Unlike the juxtaposition of the *Harmonie* Building, the Capitol Seoul aforementioned was intentionally built to break and obstruct the strong visual axis of the Gyeongbokgung Palace from its main entrance to urban centers. Positions and intentions of the construction as such are architecturally in order to break the power center and harassing architectural Korea. The palace was destroyed in Japanese invasion and occupation in 1595 and in 1911 (Glendinning 2013, 448).

During Suharto's New Order from 1967 to 1998, the story of violence against architectural representations of 'otherness' had been marked out by its opening and its end. The anti-communist demonstrations on streets between 1964 and 1967 victimized buildings and structures with signs and symbols of the Communist Party of Indonesia, its supporters and its organic associations. The Embassy Building of the People's Republic of China in Jakarta Kota was the main target of mass hostility and violence against the architecture of otherness in 1965 (Vic 2005, 285). The transition of power was accompanied by riots and mass violence against humanity and architecture. The incidents occurred on May 4–8 1998 in Medan, on May 13–15 in Jakarta metropolitan area and in Surakarta, Central Java. The main target of the killing, raping, and architectural destruction was associated with the minority of the Chinese population in urban areas of Indonesia.

The riots and hostility against the Chinese were actually triggered with economic envy and disparity between the populations and the political oppression of Suharto's regime. The Chinese citizens and their properties became the scapegoat of mass frustration against the corruption, nepotism, and collusion of the New Order's business because of their weakest link to the actual target: Suharto and his military regime. Scapegoating has been used by Indonesian mainstream on behalf of indigenous sociocultural identity for their economic and political frustration against the militaristic patrimonialism of Suharto's New Order. During this period several domestic business groups, most of them Chinese Indonesians, began to arise under the political patronage of senior government officials and senior military officers; they operated in various sectors, including textiles, electronics, transport equipment, and pharmaceuticals, often as joint ventures with foreign investors (Robinson 1986, 144).

The policies of Suharto's administration toward Chinese populations were a double standard. While the Chinese tycoons were given extensive opportunity in the strategic businesses, in other parts, such as politics and culture, their activities were harshly restricted and some even banned (Wie 2006, 12). These included the prohibition of celebrating Chinese New Year in public areas, the Chinese schools, and Chinese language newspapers. This policy reflected the need to minimize social tensions, which in the past had often erupted in anti-Chinese riots, and strive for social harmony (Elson 2001, 161). However, the New Order's double-standard policies toward Chinese brought about the exteriority of Chinese minority from the Indonesian society that had led the concept and reality of Indonesia's nation to the brink of collapse. All this was actually against its state ideology of *Pancasila*, for humanism and social justice, and its state motto: unity in diversity.

Joint Fact-Finding Team, *Team Pencari Fakta Gabungan*, noted that there were 1217 people were killed on fire, 31 people missing, and 52 ethnic Chinese women were raped (Perempuan 1999). Even in the city, from where Madame Suharto was residing in Surakarta, the rage against the New Order's signs and symbols was uncontestable in history. The residence of former Suharto's Minister, Harmoko, in Solo Baru area, was burnt down by people as the climax of burning and destruction of shophouses along the main streets (Aritonang 1999).

As predicted, the Bank of Central Asia (BCA) buildings in Jakarta, Surabaya, Medan, Surakarta, and other towns including Salatiga, Padang, and Palembang, were the target of the mass violence against architecture as BCA was popularly and presumably associated with the Suharto's New Order. The hatred and rage against the rich and the powerful were seemingly conditioned by the political and military elites for their vested interest in the transition of power. The other source, non-governmental volunteers for humanity, reported that the May 1998 incident had caused 1,109 deaths from the burning building, 27 deaths by gunshot, 91 wounded, and an additional 31 missing (Tim Gabungan Pencari Fakta 1998_TGPF 1998). The report from Jakarta's

Metropolitan Police counted 463 dead and 69 wounded. The sum of property loss and damage in Jakarta's Metropolitan Area was estimated at US$238 million.

Concerning the violence against architecture, the municipality of Jakarta reported at least 5723 buildings and 1948 vehicles burnt down and destroyed, while Jakarta Metropolitan police reports counted 3862 buildings and 2693 vehicles.

Shockingly, the May 1998 riot against the Indonesian-Chinese populations was not its first incident. There were several occurrences involving the killings and destructions of architectural properties of Chinese populations in Indonesia in the periods between 1946 and 1965; however, only a few have been reported and investigated (Cribb 2005, 44–5). After the Pacific War, hostility and violence against the Chinese happened in several towns and cities of Java. During the Japanese occupation of 1942–1945, the Chinese populations were treated as potential Japanese nationals (Touwen-Bouwsma 2013, 55–6). The violence against Chinese minorities and their architectural heritage continued between 1966 and 1967 (Heidhues 2006, 84); but most of the incidences have neither been officially investigated nor filed for justice, such as what happened in Bandung on May 13, 1963.

The incidents began with a dispute between Indonesian Chinese and native students in the Campus of Institute Teknologi Bandung (ITB). Racial tensions build up into mass violence against the minority of Chinese populations and their properties in Dago, Braga, Otista, and Asia Afrika areas of Bandung; the riots with demolition of architecture spread out to other towns and cities in West Java, such as Sumedang, Tasikmalaya, Cirebon, and Sukabumi (Darmayana 2012). During the Sukarno's Guided Democracy period (1959–1965), the root of racial tensions was not only because of cultural differences between ethnic groups but also as a result of multi-ideological streams and socioeconomic jealousy.

The Chinese population, emerging as the capitalist class, was commonly the target of political resentment and economic hardships of the majority of populations under Sukarno's patrimonial regime. During the New Order reign, the racial tensions became worse as Suharto's regime was ambivalent in dealing with Chinese populations. On one side, the ruling elites of the government and businesses worked for hand in hand in terms of Ali-Baba partnership (Fox 1980, 730); in most cases, Ali-Baba was a business partnership between Suharto's generals and Chinese tycoons or *cukong*.

On the other side, the New Order's bureaucratic regimes unfairly and discriminatively treated the Chinese populations in terms of civil rights and opportunities; most government offices and military positions were limited and restricted for the Chinese minority. To a certain extent, the Chinese minority had been often the target of extortion, harassment, and humiliation when they would apply for various legal permits and licenses, and was also the target of compulsory adoption of the appellation from Chinese to Indonesian name. In Bahasa Indonesia, the common address of calling Chinese person '*cina*' is actually a derogatory term; the respectable word for Chinese individual or ethnic group is *Tionghoa*.

Suharto and his New Order regime restricted traditional and cultural practices of Chinese populations. Since the compulsory transition of power from Sukarno to Suharto in 1967, all Chinese elements and traces in language, graffiti, and ritual were legally banned and prohibited; the ban and prohibition included Chinese architecture, books, pamphlets, broadcasts, and newspaper (Heidhues 2006, 87). The dichotomy and prejudice between the Chinese minority and the majority populations have begun since the Dutch East Indies colonial rule, especially during the Javanese war during 1825 and 1830 (Setiono 2008, 87). However, the scapegoating has become more noticeable during the post-colonial Indonesia (Tan 2008, 239). The resentment accumulated, widened, and enlarged until its anticlimax in the tragedy of May 1998.

Until today, the justice of the May 1998 tragedy for humanity had not found any light. The destruction of remarkable buildings and massacre against Chinese populations in various cities of Indonesia are the dark sides of Indonesian history. The May 1998 tragedy marked a historical disgrace for Indonesia as a multicultural nation-state. Concerning the May 1998 tragedy of humanity and social justice, the Indonesian government has not ever issued an official statement of apology to the Chinese community in Indonesia.

In the Reformation era, President Abdurrahman Wahid (2000–2001) recalled the legislation banning the practice of Chinese tradition and culture in Indonesia. During the Reformation era (2000–present), assimilation of Chinese and Indonesian culture has been going on in various fields; democratization in various levels and areas has significantly weakened the influence of patrimonialism in Indonesian politics and society. All this effort is in the hope that democracy will enable Indonesia, as a nation-state and society, to respect civil rights based on regard and empathy to its historical assets and heritage.

Even though the patrimonial regime was abolished, the Post-Suharto administration were not able to prevent violence and hostility against minorities and their properties, such as in the case of the destruction of the Ahmadiyya mosque of Al Furqon in the district of Parakan Salak in the regency of Sukabumi, West Java on April 27, 2008 (Novel 2008). The similar case happened for the Catholic Church of St. Albertus in Bekasi on December 18, 2009 (Hutabarat 2009, 1). Other incidents of hostility against the architecture of otherness are also the demolition of the Islamic Syiáh boarding school in Sampang Madura on December 29, 2012 (Hamid 2011).

The violence and hostility against the architecture of minorities, including Ahmadyah, Syiah Muslim, and Christian populations, show that the patrimonialism is not always about political regime as it is about the mainstream and majority of society. The democratization of the government system was not fully able to able to manage patrimonialism in the majority of populations because the supremacy of law and human rights do not work as the superstructure of justice for all; patrimonialism utilizes the tyranny of the majority in society for its own sake of self-righteousness. Violence and hostility against minorities and their architectural representation have not yet found justice until today. Not limited to Indonesia, we can see similar cases of violence, such as the bombardment and burning of the architecture of Bagdad in 2003 and the demolition of buildings around the globe with explosives. The reason why is violence and hostility against the architecture of otherness is likely because the attackers are ontologically not able to see themselves in others and vice versa.

The Post-Suharto's Indonesia has not been free from violence against architecture. Horizontal conflicts based on religious differences become stronger than the times of Suharto's New Order. The religious intolerance is one important cause for the most recent attacks on several Buddhist Chinese Vihara in Tanjung Balai, Sumatra on July 30, 2016, the Churches of Oikumene in Samarinda on November 13, 2016.

12.6 Concluding Remarks

The presence of architecture is indivisible from the perception of its society either as their self-image or as the otherness. Accordingly, annihilation and extermination of others and their representations are ontologically equivalent with the self-destruction of humanity that does not count on us or others. Overcoming the dichotomy of self and otherness is likely impossible without the deconstruction of perceptions concerning self-identity in and through architecture. Patrimonialism is one of the predicaments for preservation and conservation of cultural heritage because of its monolith cultural ideology that leads toward its narrow-minded perception concerning otherness. Democratization, multiculturalism, social justice, and the supremacy of law would prevent the idea and practice of a patrimonial agency or regime from the direction, movement, and action of destruction against otherness either as humankind or their representations such as architecture and culture. The prevention of such devastation is effectively made possible if societies work collectively and transculturally against self-destruction of humanity.

Until recently, the criminal justice system in Indonesia does not serve society as a whole for violence against cultural heritage and representation.

Historically, the relationship between patrimonialism and architecture has been established within the necessity for the muscular monumentality, strict order, and masculine identity reflective of fatherhood. Since patrimonialism depends crucially on the personality of the father at the apex of its power hierarchy, the violent nature, and behavior of his regime are inseparable from his personal fear and distress.

Violence against architecture might be seen as the necessity for diminishing the presence of otherness. Why does violence happen? The question leads this study to scrutinize the dynamics of the interpersonal and intercultural relations. The nature of violence is likely indispensable for the need for supremacy and hegemony in the dichotomy of self-identity and otherness. Martin Heidegger (2002) puts the Pre-Socratic doctrine of Parmenides (515–460 BC) that *being*, thinking, and identity belong together (Heidegger 2002); Heidegger concludes that the principle of identity presumes the meaning of identity itself. In other words, identity is about perception in terms of thought. Heidegger interrogates the relationship between thought and *being* that is actually in question because the presence of man is neither subjective nor objective, but an inclusively ontological totality of beings.

Instead of the subject–object way of thinking, Heidegger points out the necessity for metaphysics to think the ontological difference between *Sein* (being), and *Seinde* (beings). Accordingly, the most crucial matter of identity and difference is to distinguish reality from illusion, observation from perception, and description from opinion. Heidegger speaks of the identity of all beings in terms of the transcendence of the world where all beings are one in the present state. Accordingly, the problem of otherness does not exist in the state of presence because the structure of being present is care (Heidegger 1927/2010, 181–200). Since presence does not endure for the permanent basis that is impossible to bring 'the other' into 'the same' in the reality of the lifeworld. Levinas recognizes this problem and suggests to pass the other into infinity in the absolute distance so that it will be respected (Levinas 1991, 289–91). The architectural heritage of the others is likely not acceptable as something present in infinity because of its graspable condition; it is not the presence of infinity like God. The problem of otherness, in the sense of strangers, gods, and monsters, lies deep in the human psyche; however, by sounding out the certain borderlines between 'Us' and 'Others,' we may become more ready to acknowledge the strangers in ourselves and ourselves in strangers (Kearney 2003, 1–20).

Nevertheless, the relationship between patrimonialism and violence is unclear without considering the relationship between its narcissistic demand and aggressivity. Jacques Lacan elaborates on this relationship; he sees that between patrimonial subject and narcissism there is self-image (Lacan 1999, 90). This image is an emotionally unstable identity. Aggressivity is considered as the act to stabilize this identity. Lacan finds that the degree of aggressive tendency is correlated to the intensity of narcissistic identification, where the more demand for self-image is consumed, the more aggressivity for domination is exercised. The necessity for this self-image identification comes from the need for self-image building to overcome patrimonial self-insecurity and fear of being ordinary. Accordingly, this exteriority egoistic image sets aside others and establishes otherness either as subordinate and follower or as opponent and competitor. For patrimonialism, it is barely to maintain partnership and alliance because its self is emotionally insecure so that it is unable to include others in the extended self. Frustrations in the inability to attain the idealization of self-image could drive the patrimonial ego to do either the violence or hostility against others' representation or self-destruction of its own image. Being unable to manage or transcend the dichotomy of 'us versus them' leads the patrimonial regime to exercise violence for domination and the annihilation of others, in reference to humankind and their properties.

Architecturally speaking, patrimonialism entails the idealization of its self-image through the accumulation of power. The imaginary, patrimonial ego encourages existentially security, represented with grandeur monumentality or muscular aesthetics, and patrimonial demands for impoverished affirmation and increasing attention. For egoistic patrimonialism, monumentality and masculinity boost the sense of self-worth for public attention. Spectacle and splendor appearance and performance express this need for attention. At the same time, patrimonialism necessitates order and identity. In doing so, its system is recognizable from its architecturally exaggerated scale and dimension for its actual function, structure, and element.

Architectural monumentality and order expressive of chauvinistic identity draw a clear cultural line between 'us and them.' Without this ontological demarcation, patrimonialism does not gain its existential strength and self-confidence. The otherness of architecture is possibly associated with its architectural presence and mission that is ideologically not in compliance with the self-image of the ruling regime. The demolition of other architecture asserts a form of cultural superiority. Organizationally, patrimonial regimes utilize a pyramidal hierarchical structure with the father at its apex. Such a structure enables them to control its organization efficiently, while employing a visually recognizable image that is indicative of a patrimonial organization. To a certain extent, identity in patrimonialism is usually strengthened with strict affiliation and collective agreement, or with pledged membership. Loyalty, consensus, and conformity are the first priority conditions of such membership. The necessity for this conformity is nothing but the internal sense of insecurity of the patron about its patronage. The emotional insecurity of the self is the source of fear, jealousy, hatred, and anger.

References

Aritonang D (1999) Runtuhnya Rezim Daripada Soeharto: Rekaman Perjuangan Mahasiswa Indonesia 1998. Pustaka Hidayah, Malang

Bloembergen M (2006) Colonial spectacles: The Netherlands and the Dutch East Indies at the world exhibitions, 1880–1931. National University of Singapore Press, Singapore

Blussé L (2013) Port cities of Southeast Asia 1400–1800. In: Clark P (ed) The Oxford Handbook of the cities in the world history. Oxford University Press, Oxford, pp 356–362

Bown SR (2009) Merchant kings: when companies ruled the world, 1600–1900. D & M Publishing, Vancouver

Cherf W (2008) Earth wind and fire, Alexandrian fire storm in 48 AD. In: el-Abbadi M, Fathallah OM (eds) (2008) What happened to the ancient library of Alexandria. Martinus Nijhoff, Leiden

Cribb R (2000/2013) Historical atlas of Indonesia. Curzon Press, Surrey

Cribb R (2005) Legal pluralism, legislation and the root of violence in Indonesia in violent internal conflicts in Asia Pacific: histories, political economies, and policies. In: Adam AW, Anwar DF (eds) Violent internal conflicts in Asia Pacific: histories, political economies, and policies. Yayasan Obor Indonesia, Jakarta, pp 41–57

Dando-Collin S (2010) The great fire of Rome: the fall of the Emperor Nero and his city. Da Capo Press, Philadelphia PA

Darmayana H (2012) Peristiwa Rasialis "10 Mei 1963" Dan Provokasi Imperialis. Accessed 14 Apr 2015. https://www.berdikarionline.com/gotong-royong/20120813/peristiwa-rasialis-10-mei-1963-dan-proyek-imperialis.html

de Hartog L (2006) Genghis Khan: conqueror of the world. Tauris Parke, London-New York

Droste M, Bauhaus-Archiv (2002) Bauhaus 1919–1933. Taschen, Berlin

Elson RE (2001) Soeharto—a political biography. Cambridge University Press, Cambridge

Fox J (1980) Indonesia-Australia perspectives, vol 1–3. Australian NationalUniversity, Research School of Pacific Studies, Canberra

Glendinning M (2013) The conservation movement: a history of architectural preservation: antiquity to modernity. Routledge, London

Great Britain Public Record Office (1870) Calendar of state papers of colonial series, East Indies, China and Japan 1617–1621. Longman, London

Green J, Karolides NJ (2009) Encyclopedia of censorship. Facts on File, New York

Hamid U (2011) Tribunnews, December 30. https://m.tribunnews.com/nasional/2011/12/30/usman-hamid-usut-kekerasan-terhadap-warga-syiah

Heidegger M (1927/2010) Being and time. In: Schmidt DJ (ed), Stambaugh J (trans). SUNY Press, Albany, NY

Heidegger M (2002) Identity and difference. In: Stambauch J (trans). University of Chicago Press, Chicago, IL

Heidhues MS (2006) Chinese voluntary and involuntary associations in Indonesia. In: Khun Pearce KE, Hu-Dehart E (eds) Voluntary organizations in Chinese diaspora. University of Hongkong Press, Hongkong, pp 77–120

Heuken A (2007) Historical sites of Jakarta. Cipta Loka Caraka, Jakarta

Hutabarat BA (2009) Pembakaran Gereja dan Tanggung Jawab Pemerintah. Suara Pembaruan 24:1

Kearney R (2003) Strangers, gods and monsters: interpreting otherness. Routledge, London

Perempuan K (1999) Seri Dokumen Kunci 2; Temuan Tim Gabungan Pencari Fakta Peristiwa Kerusuhan Mei 1998. Publikasi Komnas Perempuan, Jakarta

Kompas PB (2009) Bung Karno diantara Saksi dan Peristiwa. Penerbit Buku Kompas, Jakarta

Kompas.com_Megapolitan (2009) Monumen Tugu Proklamasi, Riwayatmu Kini…, August 16. Accessed 14 Apr 2015. https://megapolitan.kompas.com/read/2009/08/16/21303348/Monumen.Tugu.Proklamasi..Riwayatmu.Kini

Lacan J (1999) The seminar of Jack Lacan on feminine sexuality, the limit of love and knowledge, Book XX, Encore 1972–1973. Norton, New York

Leftwich A (2014) Theorizing the state. In: Burnell P, Randall V, Rakner L (eds) Politics in the developing world 4e. Oxford University Press, Oxford, pp 181–194

Levinas E (1991) Totality and infinity: an essay on exteriority. Kluwer, Dordrecht

Lulat YG-M (2005) A history of African higher education from antiquity to the present: a critical synthesis. Greenwood, Westport CT

McKenzie J (2007) The architecture of Alexandria and Egypt, C. 300 B.C. to A.D. 700, vol 63. Yale University Press, London & New Haven

McMillan M (1914) A journey to java. Holden and Hardingham Aldephi, London

Novel (2008) EraMuslim, April 28. https://www.eramuslim.com/berita/nasional/ketum-pbnu-pembakaran-masjid-ahmadiyah-tindakan-kriminal.htm#.VSssiZP-nYk

Parmer JN (2012) Historical perspectives of Chinese populations in Southeast Asia. In: Amstrong MJ, Amstrong RW, Mulliner K (eds) Chinese populations in contemporary Southeast Asian Societies: identities, interdependence and international influence, pp 18–54. Routledge, London

Ricklefs MC (2008) A history of modern Indonesia since c. 1200. Palgrave Macmillan, New York

Robinson R (1986) Indonesia, the rise of capital. Allen & Unwin, Sydney

Setiono B (2008) Tionghoa dalam Pusaran Politik. TransMedia Pustaka, Jakarta

Shalaby A (2011) Islam in Egypt, Nubia and Sudan. In: El Hareir I (ed) The spread of Islam throughout the world. UNESCO, Paris

Soepandi S (2006) Indonesian architectural culture during the guided democracy (1959–1965). In: Vu T, Wongsurawat W (eds) Dynamics of the Cold War in Asia: ideology, identity and culture. Palgrave Macmillan, London, pp 53–72

Summerfield CJ, Devine ME, Levi A (1998) International dictionary of University Histories. Taylor and Francis, London

Susetio D (2012) Kompas.com_Lansir_Pendirian dan Pembongkaran Gedung Harmonie. Accessed 14 Apr 2015. https://nasional.kompas.com/read/2012/02/21/1320540/Pendirian.dan.Pembongkaran.Gedung.Harmonie

Tan MG (2008) Etnis Tionghoa di Indonesia: Kumpulan Tulisan. Yayasan Obor, Jakarta

Tatarkiewicz W (2006) History of aesthetics. Continuum, London

Tim Gabungan Pencari Fakta 1998_TGPF (1998) Laporan Ahir Peristiwa Kerusuhan Tanggal 13–15 Mei. Tim Gabungan, Jakarta

Tomkins S (2012) A short history of christianity. Lion Hudson, Oxford

Touwen-Bouwsma E (2013) The Japanese policy towards Chinese on Java 1942–1945. In: Kratoska P (ed) Southeast Asian Minorities in the Wartime Japanese Empire. Routledge, London, pp 55–96

Vic VM (2005) Kudeta 1 Oktober 1965: sebuah studi tentang konspirasi. Yayasan Obor, Jakarta

Wie TK (2006) Policies for private sector development in Indonesia. ADB Institute Discussion Paper 46. Asian Development Bank, Tokyo

Chapter 13
Patrimonial Figure and Historic Sites in Banda Aceh and Demak

Abstract In Indonesia, there are no other striking notions for the cult of pilgrimage to venerated sites than the ancient concept of *keramat*; this concept is considered a spiritually pivotal notion on the relationship between place and society in Indonesia. This religious tradition has been well known since prehistoric times that is deeply rooted in the animistic beliefs of most people in the archipelago. This sustenance is driven with respect to the patrimonial power of ancestors. Until today, this tradition has been surviving in various forms, ways, fashions, and manners that confirm the syncretism of ancient and modern values system within the framework of peacefully growing harmony. What is the significance of this concept of *keramat* for cultural sociology and urbanism? This paper investigates and unfolds the relationship between *keramat* and Islam in the context of place, culture, and society. Theoretically speaking, this study attempts to frame this relationship within the concept and practice of collective memory that constitutes Anderson's imagined community. Banda Aceh and Demak are chosen as case studies regarding their historical background as the sites where the teaching of Islam begun since the fifteenth century in the Southeast Asian archipelago.

Keywords Pilgrimage · *keramat* · Banda Aceh · Java · Demak · Mosque · Tomb

13.1 Place, Memory, and Pilgrimage

Theoretically speaking, place memory and pilgrimage are socially established by a community as a culture because they share something in common as a monumental history. Paul Ricoeur recognizes this concept as something that is not because of its excess but that of the usefulness of ideal that evokes and improves (Ricoeur 2004, 184); every past, as well as the present, deserves admiration or condemnation because both are perishable. Accordingly, conserving and venerating customs and traditions are worthwhile to live (Ricoeur 2004, 290). In this sense, Edward Casey puts the act of conserving and venerating of the past in terms of commemoration that is considered as intensified remembering with collective participation (Casey 2011, 184–7).

B. Wiryomartono, *Traditions and Transformations of Habitation in Indonesia*,
https://doi.org/10.1007/978-981-15-3405-8_13

For Casey, commemoration does not only look forward in looking back—thereby transmitting differed effects of the past—but it also affirms the past's selfsameness in the present by means of a consolidated reenactment with text or ritual; in doing so, commemoration assures the continuity of remembering into the future. The pilgrimage to venerated sites is nothing but the ritual of such reenactment of remembering based on the religious beliefs in the selfsameness of beings as a whole. The religious beliefs in selfsameness defy the separation of 'the self' and 'other'. This is because the nature of memory is memorializing that is an opportunity to redeem the perishable particulars and to consolidate and sustain the ideal.

The rationale behind the religious beliefs system is the necessity for evoking and preserving well-being and growth, which are impossible without unity and harmony. The religious beliefs in selfsameness are socially constructed as the impartial immortality and established throughout the historical participation in the ritual of commemoration that enables people to have a collective memory. The unity of place, memory, and pilgrimage are established and consolidated as the ritual of commemoration for the preserving and sustaining the ontological necessity for unity and harmony between the past, present, and future. The effectiveness of this necessity comes into being as an ecstatic happening such as the ceremony of commemoration day with communal participation.

The phenomenon of pilgrimage to venerated sites in Indonesia has been a focus of academic attention from various fields (Kruijt 1906; Pigeaud 1967; Chambert-Loir and Reid 2002; Guillot 2002; Kammerer and Tannenbaum 2003; Kartodirdjo 2007; Goh 2011; Mandal 2014, 128). The question leads this study to find out the relationship between people and place that matters for their identity in terms of cultural bonding and local affection. In order to unfold and divulge this relationship, this study is to investigate and examine the venerated sites in Banda Aceh and Demak. Both are considered as the historical centers of Islamic teaching in Southeast Asia during the fifteenth century with a unique legacy of mosque architecture (Nas 2006, 57; O'Neil 1993, 152).

Unlike Java, Aceh was well known as the region with a close relationship with the Ottoman since the fifteenth century. To what extent historical sites and mosques have something to do with the potent dead? And how do historical meanings of the places unfold? The materials of the study were carried out from the author's fieldworks (2013–2014) and literary studies from previous scholars on similar matters and sites. The study departs its analysis from historical data and fieldwork observation in both sites. From this point, the concept of *keramat* is deliberately confronted with Islamic customs, traditions, and religious beliefs system that is to unveil and divulge how syncretism has worked in the cultural works and human behaviors of Indonesian society today.

The focus of the study is to dismantle local concepts and understanding concerning historical artifacts, such as tomb and mosques, that matter for local cultural identity. Tomb and mosque are the main sites of this study regarding their significant relation to ancient religious beliefs and practices. Since animistic devoutness is by nature synthesizing, harmonizing, developing, and enriching, this study argues that this religious beliefs system has been underlying spiritual beliefs among most Indonesian populations despite they have embraced Islam as their religious affiliation and live in modernity (Forschee 2006, 154–5; Vickers 2013, 2).

There is no more arresting idea concerning historical sites in the Indonesian archipelago better than the concept of *keramat*. It would be narrow-minded to think of *keramat* as a concept growing up of and originating from unreasonableness. In the Southeast Asian archipelago, the concept of *keramat* has been always associated with historic sites and the cult of saints and local heroes. The cult of legendary ancestors and cultural heroes in the archipelago remains strong among Indonesian populations despite it is performed behind the scene of the daily lifeworld. The Indonesian and Malay word of keramat is derived from the Arabic word *karamat* meaning miracles of the God that work through saints for power and blessing of peace, well-being, and prosperity (Chambert-Loir 2002; Taylor 2012; Tarling 1999). Clifford Geertz points out the concept of *keramat* in similar meaning with the Sanskrit word *sakti* in Balinese culture and society, in terms of a divinely inspired gift of power and ability to perform miracles (Geertz 1980, p. 106).

The transformation of the subject matter of the cult happens, but the rituals and ceremonies remain the same. This has been constructed in alignment with the ancient religious beliefs that every being contains potent spiritual power. Accordingly, strong personalities do not come from anything but they are so because of their potent spiritual energy. Pre-Islamic religious beliefs in the archipelago paid highly regard to the sites with natural power and mysterious atmosphere; such places were considered sacred—*suci*—and spiritually daunting—*angker*—because they contain the potent energy of nature.

In the case of the graves of powerful personalities, the sacredness of places is conceived by most Indonesian pilgrims as the source of the spiritual power of the potent dead. As a whole, sacred places were believed that contain potent spiritual power so that they were magically powerful, *keramat*. The content that matters for sacredness is called *pusaka* or heirloom. The site that contains heirloom or *pusaka* is considered to be *keramat*. The concept of *keramat* is originally adopted from the Arabic term meaning marvel or saintly miracle and applying it to imaginatively every object or place associated with the supernatural powers of God as well as the host of sacred items (Renard 2008, 207). The concept of *keramat* is considered as the impartial immortality of the religious beliefs system of most people in the Indonesian archipelago.

The Pre-Islamic Javanese and the Chinese shared something in common concerning the animistic beliefs that there were spirits guarding and living in water wells, caves, pristine hills, deep riversides, lush big trees, forests, and strange landscape elements. In Chinese Mandarin, such spirits are called *Na Tuk Kong* or *Da Tok Kong* in Chinese Hokkien or *danyang* and *lelembut* in Javanese. In the Malay world, the Chinese populations pay homage to the guarding spirits by making effigy with a respectful name such as *Da Tok Samsuri* or *Datok Kongs* (Lee and Tan 2000, 289). All these confirm the idea that ancient religious beliefs system of spirituality has been working as the way of life. By practicing such a way of life, people in Southeast Asia feel the connection of all beings as a whole, regardless of their visible and tangible or invisible and intangible manifestation.

As a matter of fact, mysterious places are mostly associated with the unknown that are believed to have natural powers. Paying homage and regarding such places with rituals and offerings are nothing but wishful thinking and wise action in protecting and sustaining the ecological nature of the environment. Based on her experience in studying on ecological ethics in the Minangkabau land of West Sumatra, Davary comes to a conclusion that the Malay notion of *keramat* synthesizes culture, religion, and nature seamlessly (Davary 2002, 4). Among other ethnic groups in Indonesia, traditional communities still strongly hold and practice the concept of *keramat* in their customary laws—*adat*—and animistic beliefs. Among Sundanese populations, the concept of sacredness, *keramat*, is always bound with the concept of taboo, *pamali*. Sacred sites such as the water well with its forest and graveyards are considered as a *tempat larangan* meaning prohibited area for residential and agricultural activities. The traditional Javanese people commonly built special altars for sacred places. They used to put their offering flowers and foods on the altars with rituals of burning incense; some of them still maintain this ritual by reciting surah from the Al Qurán. Ritual visitation to spiritually venerated places by the Javanese and other ethnic groups in Indonesia has nothing to do with the polytheism but more about pantheism based on spiritual unity in the diversity of manifestations; the Javanese believe in the Supreme Being that is present in all-natural things (Zoetmoelder 1995; Geertz 1960, 16–29, 309–338).

13.2 *Keramat* in Pre-Islamic Aceh and Java

The Islamization in the Southeast Asian archipelago was a peaceful process through trade interactions between coastal people and foreign merchants from Persia, Arabian Peninsula, Gujarat, and Chinese mainland between thirteenth and fifteenth centuries. The reports of Italian traveler Marco Polo at the end of the thirteenth century and Moroccan Ibn Battuta in the mid of the fourteenth century confirm the practice of the ancestral homage in the Islamic state and principality in Samudra Pasai nearby Perlak, North Sumatra (Djajadiningrat 1958/1987, 376; Mohammad-Yatim 1988); tomb of heroes and ancestors was regarded as venerated sites.

The arrival and dissemination of Islam in the north part of the island was inveterate by the tomb and gravestone of Sultan Malikul Saleh dated 696 Hijiriah or 1275 A.D. The similar Arabic epitaph has been found in Gresik, a port town in East Java (Kayadibi 2011, 4). The gravestone was carved with Kufic script mentioning the name of Maulana Malik Ibrahim from Persian origin who died in 822 or 1419 A.D. However, the Chinese I Tsching reports in his book Nan-hai-chi-nai-fa-chuán that the Islamic Sultanate of Perlak or Peureulak has been established since the end of the twelfth century (Muljana 2005, 129–31). Accordingly, the merchants from Egypt, Gujarat, Persia, and Morocco had settled in the port later known as Peureulak since the early twelfth century and established an Islamic city-state of Perlak with Sultan Sayid Abdul Azis as their first ruler. Based on the Chinese reports and local narration: *Hikayat Raja-raja Pasai*, historian Slamet Muljana reconstructs the historical facts about Sultan Malikul Saleh that Sunni Muslim: He was the first Islamic statesman who was not the follower of Islamic Shia like his predecessors (Muljana 2005, 137).

Prior to the arrival of Islam, burying the deceased under the earth with gravestone and epitaph was uncommon and not familiar custom for most Indonesian populations. During Hindu–Buddhist Java, only dignitaries entitled to have mausoleum, *candi*, for their ashes. Most of deceased of common people were buried that was in order to wait for the ritual of cremation with their patrons or the landlords in the open area of the village center or town square. Islam and Christianity introduced the people in the archipelago to bury the corpses with epitaphs and gravestones; animistic indigenous religious beliefs, Hindu–Buddhist, and Chinese influence taught them to pay homage to the dead by visiting their tombs with offering flowers and burning incense.

In the pre-Islamic Java and other islands, graveyards were a special area on the hills or away from their settlement. In Java, such an ancient place is called *pralaya* or *pusara* meaning the site of demolishing and devouring. Such place was customarily constructed with terracotta or andesite stones in the form of *punden*, or ziggurat. In many cases, the prehistoric *punden* was constructed with steps from the earth. In Java, the graves of most dignitaries are customarily constructed with a shelter called *cungkup* and tombs or *kijing*.

In the context of the village, traditional Javanese villages are used to establish a *punden* for their ancestral shrine; this is the ritual center of the village for their annual festivals and other spiritual occasions that are intended to cleanse their community from bad lucks and pandemic diseases (Hefner 1985, 59). Most *punden* sites are associated with the supernatural power of nature and ancestors, while places of *keramat* have to do with the sacredness of saints. However, the relationship between the sacredness of *punden* and *keramat* never ends (Chambert-Loir and Reid 2002, 133) that is in a collective memory of people within the framework of spiritual power for well-being and growth.

The construction of the complete tomb, *ngijing*, is traditionally executed that is in commemorating of 1000 days after the person passed away. *Slametan* or the ritual for well-being takes place at every time of life transition, from birth, marriage to death (Hefner 1985, 5). The central activity of this ritual is a communal feast with a communal meal and solemn gathering (Geertz 1973, 142–5). The cult of ancestors has built the customary paying homage with the visits and offerings of flowers and foods to the graveyards on special days in Javanese calendars related to harvests and new monsoon. This pre-Islamic custom and tradition are still alive today. In the house, family members, relatives, and friends gather together to perform the ritual of *slametan*. Whatever religious affiliation the dead is, the ritual of *slametan* will take place with its central event of the communal feast with a special meal (Geertz 1960, 11–12). This special meal is called *tumpeng* made of yellow turmeric rice in the form of the sacred mount with a cone peak made of banana leaves. The mountain of rice, *tumpeng*, is decorated and embellished with various side dishes such as meats, tofu, *tempeh*, cooked vegetables, and fruits. All this symbolizes a wholeness of well-being and prosperity (Fig. 13.1).

Among Javanese populations, the cult of ancestors has been preserved as their homage and gratitude to the ancestors; some Islamic streams, such as Muhamadyah's

Fig. 13.1 Tombs and epitaphs of Sultan Malik Al Saleh (left) in Banda Aceh and Maulana Malik Ibrahim (Right) in Gresik, photographs by the author

and Wahabi's followers, forbade the rituals to pay homage to ancestors by visiting their tombs, *ziarahnyekar*. The rituals take place mostly in the days before the Ramadhan, on the Islamic New Year, on Prophet Muhammad's birthday, and other Islamic and Javanese special days. Today, the legacy of heroes and saints is commemorated and paid homage to them with the offering of flowers and reciting surah from Al Quran. Offering flowers on the tombs is common practice. Burning incense and offering food have been less done with the ritual of visitation to the tombs, *makam*.

13.3 Mosque and Tomb of Iskandar Muda in Banda Aceh

The main source of history of Acehnese Kingdom during the reign of Sultan Iskandar Muda is indebted to the work of Denys Lombard: *Le Sultanat d'Aceh au Temps d'Iskandar Muda 1607–1636* (Lombard 1991). There are several local sources such as *Sejarah Melayu* and *De Hikayat Atjeh* (Iskandar 1959) that provide this study with complementary data for the facts of the Aceh world in the seventeenth century. The mosque of Baiturrahman and the tomb of Sultan Iskandar Muda Meukuta Perkasa Alam (1583–1636) are the legacy of the first Islamic kingdom in the Indonesian archipelago. Iskandar Muda was in power between 1607 and 1636 that made his sultanate as the paramount authority over most trading posts in Northern Sumatra in the seventeenth century (Ricklefs 2001, 37); most of the historical facts on Aceh in the seventeenth century have been well documented by Lombard (1991).

One of the most important sources on the history of Acehnese Sultanate under Iskandar Muda reign was the reports from Turkish Ottoman scholar Saikh Davud Ismail Mustafa ar-Rumi or Baba Davud between 1650 and 1750 (Ozay 2011) and a French commodore, Augustin de Beaulieu in 1621 (Hirosue 2009, 178–9). Sultan

Iskandar Muda was well known in Acehnese history as a strong and charismatic leader who established territorially urbanized system—*mukim*—and a newly organized group of elites—*orang kaya*—in Acehnese Sultanate. Accordingly, the state territory was divided into districts of a settlement called *mukim*. This unit of territorial settlement was controlled and managed by a state chief or lord called *Uelebalang* who was responsible and reported directly to the Sultan of Aceh Darussalam. Each settlement district was organized as an economically self-contained settlement with an appropriate number of populations for having a mosque. A *mukim* consisted of several villages or *gampongs*. Iskandar Muda developed the city of Banda Aceh which was founded by his ancestor Sultan Ali Mughayat Syah (1496–1530).

As the head of state and government, the sultan was assisted by *qadhi* for Islamic law affairs, *qanun* for ethical manners and behaviors, and *muazin* for mosque maintenance matters. Such state officials were to find at the level of districts. The center of the community is the mosque which was established in each *mukim*. The mosque was led by an *ulama* called *imeum* who was with the lord of the district, *Ueleubalang*, sharing their duty in leading the people in his/her territorial control, *kaum*, toward civic society. The lord was in charge of defense and public security in his/her territory; he was the commander, *panglima kaum*, of armed forces, *bentara*.

Sultan Iskandar Muda was also the first state leader who assigned his first lady, Maharani, with the specific duty to be in charge of state rituals and ceremonies. Under his reign, Iskandar Muda drafted and documented his statecraft in several books on laws, taxes, defense, state officials, and administrations. One of them is *Qanun Mahkota Raja-Raja* comprising rules and principles of conducts for dignitaries. Based on this law, four women were able to lead the Acehnese Sultanate (Kahin 2015, 517): Safiyatuddin Taj-al-Alam (1641), Naqiyatuddin (1675), Inayat Zaqiyatuddin Syah (1678), and Kamalat Syah (1688–1699).

The original mosque of Baiturrahman was established under the patronage of Sultan Iskandar Muda. He assigned the state *imeum* to hire the best carpenters and stonemasons in the country in 1612. The architecture of the ancient mosque has not survived either in the graphical documentation or three-dimensional model after 1873 when the Dutch expedition attempt to take over the control over the Sultanate of Aceh. The original Baiturrahman mosque was burnt down together with other buildings of the sultanate palace of Iskandar Muda.

The mosque was built as the center of daily sociocultural activities with a public square. Local sources tell the story how the Sultan Iskandar Muda in the past celebrated communal festivity, *kenduri* for the Friday Prayer, Ied Adha, Ied Fitri and the birthday of the Prophet Muhammad S. A. W. with military parade of warriors with horses and elephants from the palace to the square in front of Baiturrahman mosque; the sultan was acting as the conductor of the whole ceremonial performance. This can be seen as a religiously instrumentalizing way to enhance the standing of the sultan in society (Wormer 2014, 118). In other words, Nanggroe Aceh Darussalam in the age of Iskandar Muda was not only a governmental body of the nation but also a theater state at which the ruler is in charge as the ecclesiastic conductor of the state ceremonies (Reid 2005a, b, 112–35; Geertz 1980).

Based on the accounts of European travelers in the seventeenth century, such as Thomas Best and James Lancaster, Friday prayer was the time for the Sultan, especially Iskandar Muda, to impress his subjects and visitors with splendid spectacle (Reid 2005a, b, 124–5). After the congregation in the mosque was done, people were gathered in the public square in front of the mosque, while the king instigated to the parade and procession of his warriors on elephants and horses. Even for this weekly event, the splendid spectacle of military display and procession was assembled with feverish and vibrant performance.

The mosque of Baiturrahman was rebuilt by the Dutch colonial authority in 1879; the Governor-General of East Indies: Johan Wilhelm van Lansberge (1875–1881) commissions the Dutch architect and Dutch military engineering team for the design and construction of the new mosque on the site of former Baiturrahman mosque. The demolition of the mosque during the Dutch colonial expeditions was likely collateral damage of the Dutch colonial control over Aceh territory. In his terms, J. W. Lansberge was well known for his expansionist military expeditions to establish territorial control over Sumatra, Borneo, and Celebes as well as in other islands in Eastern archipelago (Locher-Scholten 2004, 142); during his administration, geographical surveys and military expeditions had worked for hand in hand to gather and document information on land and people for administrative purpose of the Dutch colonialism.

For the Dutch, Aceh was a challenging local competitor for controlling spice supply in north Sumatra (Andaya and Andaya 2015, 198). Banda Aceh has been well known among the international merchants as one important trading post and port of North Sumatra region with peppers, rice, ivory, sandalwood, salt, candle, gutta-percha, and other spices (Poesponegoro and Notosusanto 2008, 31). After the opening of the Suez Canal in 1869, the Dutch and the British came into a new treaty in 1871. Prior to this treaty, both colonial sea powers were committed to their 1824 agreement mentioning the Dutch should not politically intervene in the trading ports along the Malacca Straits, including Banda Aceh (Kipp 1990, 40). After the invasion of Aceh in 1873 until 1903, the Dutch colonial rule had never come into a peaceful relation with Acehnese populations.

The Baiturrahman mosque was constructed by the Dutch colonial rule in 1886. The construction took place on the former site of the mosque, which was according to Beaulieu located inside the city wall (Lombard 1991, 41–4). The design of the mosque was made by Dutch architect Bruins, from a local source was believed that the Dutch *Corps der Genie*, *Burgerlijke Openbare Werlen*, BOW, and a native *ulama* from Garut were involved in this process (Kreemer 1920, 70). For the execution of the mosque, General Major Vander who was the military commander for Aceh was in charge (1886–1889).

The architecture of the mosque was developed by the Dutch with reference to the central building of Palladian villa Rotunda; according to Palladio's *Four Books on Architecture*, its plan and elevations used to apply the symmetrical and geometrical proportion of width-depth-height for 4:6:9, a harmonic proportion in music (Palladio 2012). The ground plan of the mosque main hall with 12 × 12 m is provided with four *loggias* with 6 m depth to four directions. The west *loggia* was dedicated to the *mihrab* of the mosque. The two other wing *loggias* were not provided with *piano terra* structure; only frontage *loggia* is constructed with such construction that enhances the transitional experience of visitors between indoor and outdoor. The stairs lead one to the ground floor so that the mosques loggias and its main hall are constructed as a stage 1.2 m off the street level. The main hall is architecturally developed with a rectangular ground plan and an octagonal plan for upper fenestration structure that supports the 10.5 m diameter dome. The use of dome, *qubah*, was architecturally and culturally strange in the sense of visually superseding to Aceh building tradition and landscape. However, the dome architecture has been making the Baiturrahman attractive and unique in establishing the landmark of the former ground of the Aceh Sultanate capital city. The response of the local populations to the new mosque was unknown. However, they were not indifferent because the mosque was in fully functional as the main urban mosque of Banda Aceh.

The Dutch version of Baiturrahman mosque was architecturally unchanged until 1935; in that year, the Dutch colonial rule extended the mosque due to the request of local growing populations for their urban mosque space. The extension was made on two wing directions; each wing was provided with the similar upper structure and dome as its main hall. In doing so, the final form of the 1935 mosque had three domes with a continuous hall for about 900 m^2 area. The growth of populations in the urban area of Banda Aceh demanded more space for their mosque in the early 1960s. The realization of the extension for the mosque came into reality in 1967 with a similar way of the 1965 mosque. Today, the mosque has five domes and 4000 m^2 area for about 12,000 prayers capacity inside the building.

The question concerning its architectural idea remains open whether the 1889 Baiturrahman mosque is an outcome of the Dutch orientalism (Kurniawan and Kusumawardani 2012; Kusno 2010, 211) or simply the best option of payback from the Dutch out of guilty feeling for flattening the mosque during their military expedition to Banda Aceh in 1873. As a matter of fact, the Dutch colonial rule under Governor-General Johan Wilhelm Lansberge (1875–1881) was under pressure for military expansionist control over the Outer Islands of Java. Prior to 1891, the Dutch scholar of Islam, Snouck Hurgronje (1857–1936), the Dutch East Indies colonial adviser for Eastern languages and Islam, had not been involved with the development of design and its construction between 1886 and 1889. His active role for Aceh's matter was effectively between 1898 and 1903 as a close adviser of Colonel Joannes Benedictus van Heutsz (1851–1924); Van Heutsz was the Dutch military commander for Aceh Besar between 1889 and 1891 (Reid 2005a, b, 293–9).

Even though Aceh Sultanate was a politically centralized authority, its regional territory, *mukim*, had established territorially independent governance. After the Dutch expeditions in 1873 and 1887 were able to destroy the sultanate capital city, the districts under the control of Aceh Sultanate were not automatically fallen and subsumed under the Dutch colonial rule. The lords of each *mukim*, *Uluebalang*, resisted to working under the Dutch colonial power until 1903. Regarding his golden age of legacy for Aceh urban tradition and culture, Sultan Iskandar Muda's royal tomb compound and his mosque are considered as sacred heirlooms—*pusaka keramat*—though both are located in 750 m away from each other.

Despite the devastation and flattening by the Dutch military expedition (1873) and Tsunami in 2004, pilgrimage to both sites has been a living tradition for people in the region and other parts of Sumatra and archipelago. The remnants of Acehnese Sultanate palace compound are still recognizable from its open public spaces; they are located in the front of Baiturrahman mosque and in the area between Jalan Balai Kota and Jalan Masjid Raya. The former encircling city wall was constructed along the main streets of Banda Aceh—Medan and Sultan Alidin Mahmud Syah that provided the $620 \times 820\ m^2$ city of Banda Aceh with a stronghold made of stones.

For most Indonesians, the royal tombs and the Baiturrahman mosques in Banda Aceh are not only about tourist destinations but also their targets for pilgrimage because of its association with the spiritual *keramat* of the person of Sultan Iskandar Muda. Accordingly, people in Aceh and other parts of the archipelago keep their collective memory about this person as a saint and a hero. The sense of *keramat* comes into reality in the ritual of pilgrimage that brings about the experience of spiritual resolution between the imagined and the traces of reality. Tombs and mosque are such traces. In Indonesian society, the saints and heroes are traditionally believed to be present as spiritual power, *semangat*, that unites people and heals the world in terms of the village, community, nation-state, and the universe with grace, bless, and joy.

Traditionally speaking, strong personalities and historical sites/artifacts are conceived as an heirloom, *pusaka* that contains a sacred power for reminding people of the necessity and alignment for unity and harmony toward well-being, prosperity, and growth. In this sense, the Acehnese and archipelago populations do not care about the historical facts that the mosque has been designed and constructed by the Dutch with the Western interpretation in terms of Orientalism or Palladianism or Colonial Imperialism on Aceh and Islam. Even after the Tsunami wiped out everything historically authentic for the mosque and the royal tombs in 2004, a pilgrimage to the sites of *keramat* takes place until today.

13.4 Mosque and Tomb of Demak

The Demak mosque is one of the historical legacies of the first Islamic state in Java from 1476. The mosque is an exemplary work of architecture that shows Islamic adaptation with the locality. Nancy Florida's work explores the historical context of the mosque from the Javanese poem: *Babad Jaka Tingkir* (Florida 1995). *Babad* is a Javanese word for the poetic narratives and accounts of the political affairs; literally, the word *babad* means clearing the forest for establishing a new settlement. *Babad Jaka Tingkir* is a Javanese of historical account on the political affairs and events in Demak prior to the establishment of *Jaka Tingkir* before he came to power as Sultan Hadiwijaya; he was the first ruler of Pajang Sultanate in Central Java between 1549 and 1582. According to Florida's study on *Babad Jaka Tingkir*, the mosque is the embodiment of historical realities of Java that have been shaped by the tensions and struggles between concentrating authorities and more disseminate power (Florida

1995, 331). In this sense, the deliberations of *ulama* and state ruling power were indispensable for the planning and design of the mosque. Ideological contestations and negotiations between Javanese mysticism and Islamic Sufism took place in the design process for what a Javanese mosque in Demak was.

The poem of *Babad Jaka Tingkir* confirms the Javanese interpretation on the architecture of the mosque, based on the philosophical position and consideration of the nine Islamic saints of Java—*walisongo*—that it is not a monumental edifice but a *pusaka*. Under the notion of *pusaka*, the Javanese and Indonesian people understand as a spiritually reminding site from which people are able to address, locate, focus, and draw their attention toward the necessity for wholeness—*utuh*—and harmony—*rukun*—both these concepts are conceived as the conditions for well-being and growth. The mosque of Demak is the embodiment of Javanese interpretation on Islam at philosophical level; the nine *wali songo* learned and taught Islam in Java from Tasawuf and Sufism perspective (Aceh 1994; Wormer 2014, 119).

The mosque is situated in the urban center and at the west part of the town square of Demak. Nearby, the mosque was the former site of sultanate palace; it was located on the hill at the south part of the town square. This area is known as *siti hinggil* meaning elevated site. The north side of the square is Chinese quartiers, *pecinan* while the east side was reserved area with the natural green area. Today, the east and south parts of the town square are occupied with various state buildings and commercial areas.

The south area of the city square was the site of the Demak royal palace. This site had been abandoned since the civil war after 1548. The site was redeveloped by the Dutch colonial rule for a post road across Java under the Napoleonian Dutch Governor-General Daendels in 1809 (Meilink-Roelofsz 1962, 83). The road runs across the site and turns around the city square. The main commodities of Demak Sultanate were rice, spices, and foodstuffs which were mainly supported by its hinterlands.

The traces of Demak as a state capital are indicated by domain names inside the city wall such as Setinggil, Krapyak, Taubayan, Betengan, Tembiring, and Bintara. The names for professional domains include Gendingan, Pedhalangan, Sampangan, Penjalan, Tirtoyudan, Kembangan, Domenggalan, Beguran, Pungkuran, Sorogenen, Tukangan, and Pandean. Other domains are associated with ethnic groups and the names of their landlords such as Sempal Wadak, Karangbayan, Kalisusukan, Kauman, Merbotan, Petengan, Bale Kambang, Wonosalam, Jogoloyo, Bogorame, Pecinan, and Mangunjiwan. The life in the city of Demak in the sixteenth century was reported by the Portuguese geographer Tom Pires in his Summa Oriental; he reported that the city of Demak was a vibrant capital city with an international port with about 14,000 houses, 8000 of them were in the center area (Pires and Cortesao 2005; Meilink-Roelofsz 1962).

The site of Demak mosque occupies 1.5-ha land; originally, it was enclosed by 2-m height masonry wall construction at the west, north, and south. The east wall delineates the boundary of the mosque from the public square. The main entrance of the mosque is from the square through a masonry gate with shelter at the east. From the entrance area, visitors will be received by a front open hall called *serambi*, of

31 × 17 m area. The main hall, *mushala*, of 31 × 31 m area, is constructed with a core area of 5.6 × 5.6 m area where four wooden pillars of 32 cm diameter stand to support its main upper structure.

In Javanese building tradition, these four columns are the cosmologically unified structure of the world within the congregation of prayers in the mosque. In Javanese house, the four *saka guru* are the main pillars for the reception hall, *pendapa*, living room, *pringittan*, and the inner house, *dhalem*. By having the four pillars, a Javanese house or hall is provided with the sense of settle, *mapan*; this structural system enables the pyramidal roof system, *tajug*, to point out the sky, *nyunduk langit*. At the pinnacle of the three-tiered pyramidal roof of the mosque is architecturally embellished with a crown called *mustaka*. Indeed, the four pillars and its ring beams are not simply supporting elements, but they work together as the core of the whole structural system of the hall. The constructional joints between them work in such a way to allow vibrations and torsions of lateral and vertical forces, such as earthquakes and monsoon winds.

The most crucial design process of the mosque was the determination for mosque orientation. Mosque in Javanese culture is perceived as the hall for spiritual unity. The traces of the ancient Javanese shrines and Hindu–Buddhist temples—Prambanan, Kalasan, Sewu—regard the East as the source of blessing and care for well-being. The main entrance of such buildings is located in the direction of the sunrise as a sign of life. On the west is the area of the dead.

In dealing with the necessity for the direction of the *mihrab*, the mosque of Demak is designed to align with the Javanese tradition for sacred buildings. The direction of Qibla for the Demak mosque has not been understood as a strictly oriented to Ka'bah of Mecca, in terms of geospatial orientation, but it is philosophically perceived as the direction of eternity: the west. The east–west direction is deeply rooted in Austronesian way of thinking that has been applied to the most Hindu–Buddhist temples in Java and other settlements in the archipelago (Degroot 2009, 135). On the west partition of the main hall where the niche for *imam* is, Sultan Bintara or Raden Patah and *walisanga* inscribed the white stone wall with signs and symbols of Majapahit sun and the white turtle figure *candrasengkala*; from this symbolic form, Javanese people can read: *sarira sunyi kiblaliting kawulaGusti*, meaning 1401 *saka* calendar or 1476 AD.

Regarding its design conception by the Sufi *walisanga*, the Demak mosque is a Javanese syncretic representation and interpretation on what Islam is. This syncretism is incorporated with the arrangement of the domain the mosque complex and the mosque attributes and decorations. The domain system of the mosque is divided into three categories: east, central, and west domains. The east domain is the gate area with its reception yard and hall, *serambi*. The main domain is for the *mushala utama* or main praying hall. Interestingly, in the south part of the main hall, there is an architecturally devoted domain for women called *pawestren*. The notion of *pawestren* is derived from the Javanese word *estri* meaning female counterpart or wife for wholeness and perfection. Since its very beginning, the mosque had been designed to accommodate women in the main hall. The spatial category of left for female and the right for a male is deeply rooted in Austronesian tradition (Fig. 13.2).

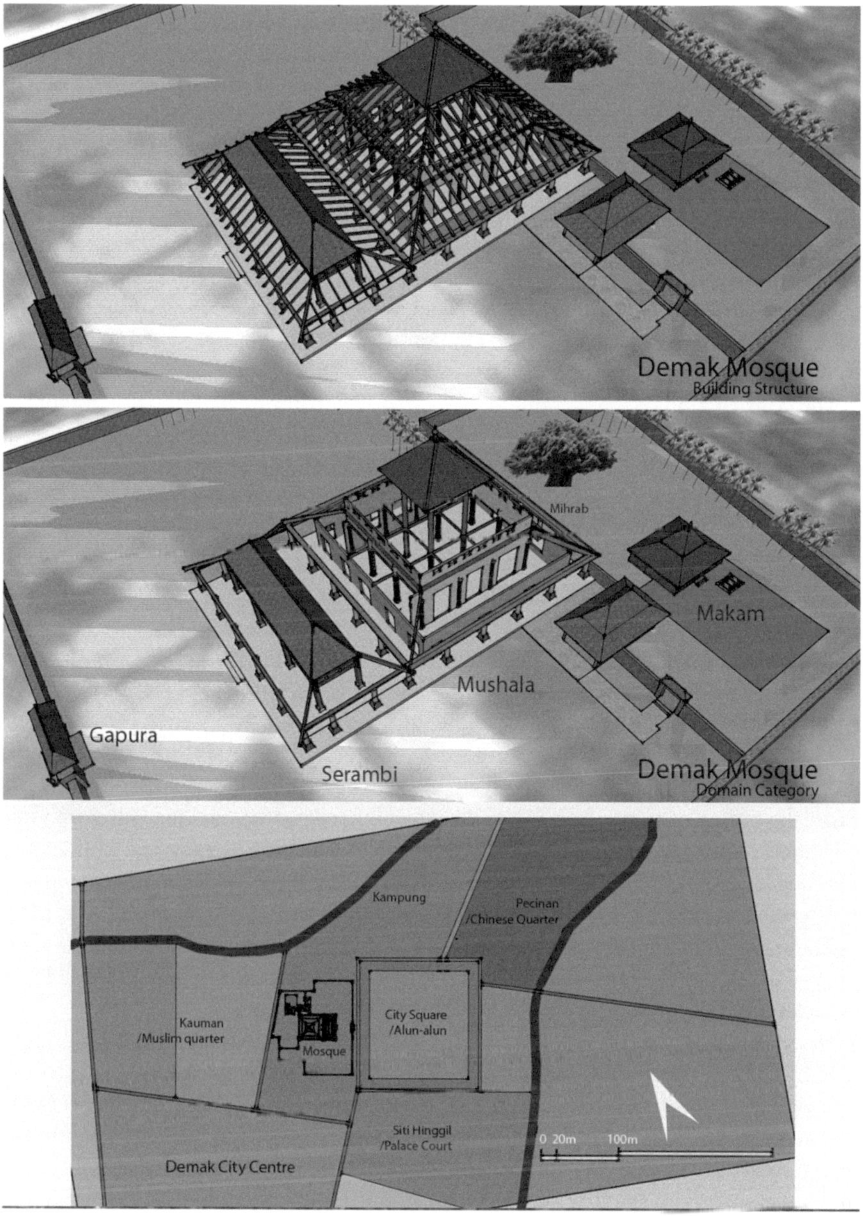

Fig. 13.2 Demak Mosque, building construction, domain layout, and site plan, drawings by the author

The other important element in the front hall of the mosque is the installation of the Chinese drum; such drum is commonly found in the Chinese Buddhist temples all over the Southeast Asian archipelago. Still, the three-tiered roof construction of the main hall is strongly indicative for architectural adaptation from the Chinese *pagoda* or temple which is also in alignment with the several tiered roofs of Hindu Balinese temples. The influence of Chinese culture under Ming Dynasty (1368–1644) is obvious in various fields of culture, from the mosque architecture of Java and other Indonesian islands to culinary (Lombard and Salmon 2001). The first sultan of Demak himself is the prince of Majapahit from a Chinese wife. He received the request of Chinese navy captain Gan Si Can to help the construction of the mosque; he mobilized his carpenters from the Chinese shipyard in Semarang (Muljana 2005, 197–8).

The construction of the mosque began in 1477 and completed in 1479, under the leadership of Raden Said or Sunan Kalijaga (Soekirno 1994, 8); it is still unclear whether the Chinese Gan Si Can is Sunan Kalijaga or someone else; because both persons are identified as the brother in law of Sunan Ampel or Bong Swi Hoo. In the sixteenth century, Demak was unlike the pre-colonial Banda Aceh or Kutaraja; most coastal ports and principalities of Java have been populated by Chinese migrants and traders since the fourteenth century. The Acehnese Sultanate (1496–1903) restricted their territories, *mukim*, and cities from non-Muslim foreigners.

The third domain is situated behind the main hall of the mosque on the west side. The west domain is designed for the royal cemetery. The sultans of Demak and their family members were buried with *punden* tombs in this area. Pilgrims pay homage to the historicity of Demak by visiting the mosque and the royal tombs. The most peak visitation for theses venerated sites takes place in Islamic holidays and the days before the month of Ramadhan. The relationship between the mosque and the people of Demak region comes into festive events and celebrations within the holidays of the Islamic calendar. The most important event is the *Gerebeg Besar*. The Javanese word *grebeg* means the sound of roaring devastating storm. This concept is applied to denote the grandeur of state ritual and ceremony for the well-being of nation-state, *wilujengan nagari*. This festival is also the Ied Adha that takes place on the 10th of Javanese month of *besar* or the Islamic month of Dzulhijjah.

The *grebegbesar* in Demak was conceptualized by the *walisanga* and accommodated by Sultan Bintara of Demak as their cultural strategy for integrating differences and diverse groups into a single grandeur event. This festival has been conducted in Demak since 1506 (Pursidi 2005). The purpose of the incorporation of Islamic Ied Adha into the state ritual and ceremony of Demak Sultanate was to establish and consolidate the power of the first Islamic state and its ruler in Java in the predominantly Hindu–Buddhist populations of Java.

The syncretic incorporation of all religious elements—animism, Hindu–Buddhism, Islam, and Confucianism—was considered necessary for the newly established state of Demak Bintara. In doing so, the festival was able to gather all people from diverse backgrounds and religious affiliations. Few days before the grand event, all royal heirlooms and regalia should have been ritually purified with clean water containing seven variety of flowers. Prior to the day of celebration, the royal palace of Demak prepared grand *tumpeng sanga*—nine mountains of rice—in front of the reception hall; the *grebegbesar* is a grand ritual of well-being *slametan*.

While the royal kitchen was busy with the food preparation for the grand feast, in the central hall of the palace, the sultan, *walisanga*, royal dignitaries and family members gathered; they were together here to watch the performance of solemn dance of *bedhaya*, the solemn and spiritual dance by nine girls of Sultan's family members. The celebration of *grebegbesar* was traditionally commenced with the ritual of paying homage to the sultan who sat on the throne in his royal reception hall, *bangsal witana*, of the sultanate palace. From this ritual, the festival was continued with the state procession, *kirab*, of the sultan, the nine saints, *wali sanga*, state dignitaries, religious leaders, and members of the royal family, royal guards and warriors to the front yard of the mosque.

Today, the similar procession mentioned above takes place from the reception hall, *pendapa*, of the Demak regent, *bupati*. The procession is to perform the Ied Adha Pray in the mosque. The culmination of celebration is the feast of communal meal from the nine *tumpeng* that takes place after the Pray. The feast is followed with the Sacrifice Feast with the ritual of slaughtering goats and cows in the town square; the distribution of meats has been an interesting part of the feast. In the sultanate era, the king sat on the throne in the square, *alun-alun*, spreading out small changes for the people; the act was known as *undik-udik*. For the whole rituals and ceremonies, the sultan was in charge as the inspector and the main conductor of the rituals and ceremonies of the celebration; today, the regent of Demak becomes the central figure of the festival and the conductor of the whole rituals.

Several decades later, Javanese Sultanate palaces have been performing their *grebegbesar* with similar rituals and ceremonies like in Demak until today, the best examples are in Yogyakarta and Surakarta; in these both cities, the grand feast of *grebeg* is celebrated three times on the Islamic holidays: the Ied Adha, the Islamic New Year, and the Birthday of the Prophet of Muhammad S. A. W. In course of centuries, the grand feast of *grebeg besar* has been becoming the folk festival events of the cities of former Javanese sultanates. Prior to the grand feast, the local government organizes temporary day and night market celebrated with folk performing arts.

13.5 Concluding Remarks

Keramat in Indonesia is a concept that connects people to historical artifacts and sites. This concept is considered significant for historical study in the context of the relationship between people and place for their sense of home, well-being, and growth that is sustainable because they share something in common in the form of religious beliefs system concerning the selfsameness of beings as a whole. In doing so, the past, present, and the future are ontologically experienced as an ecstatic happening of the commemoration of community. This is the event when the glory of past, which is associated with the achievement of their heroes and ancestors for the sense of unity and harmony, is evoked, recollected, and celebrated with the communal ritual in public realm as *kenduri* in Banda Aceh and *grebegbesar* in Demak. The sense of pilgrimage to the historical sites lies in the content of the venerated locations that are under the notion of *pusaka*. The concept of *pusaka* is considered as the objective immortality of their religious beliefs system concerning the symbolic selfsameness of beings as a whole. In this sense, the whole participants of the commemoration of *kenduri* and *grebeg besar* experience the sense of unity and harmony in the festivity.

Moreover, the event brings forth the relationship between personalities and their traces in the works of architecture in unity that is reenacted and demonstrated with the rituals and ceremonies of commemoration. These rituals and ceremonies are nothing but the cultural strategies of syncretism, by which differences are dissolved and synergized toward usefulness and worthiness for the community. The goal of syncretic strategies is to dissolve perishable particulars toward ideologically united immortality for communal well-being and growth.

The sense of history in the Indonesian context is inscribed as the ritual of pilgrimage to the venerated sites. In doing so, the collective memory of people is preserved, consolidated, and sustained with communal participation. The purpose of this ritual of commemoration is to keep the collective memory alive. This is because the collective memory of the relationship between personalities and the places is considered more important than the authenticity of the historical artifacts and sites which could be demolished by natural disasters or other force majors.

Historic monuments in most Indonesian archipelago are considered negligent, but the pilgrimage to the venerated sites with specific reference to the past, heroes, and ancestors remains alive until today. All this is because of the religious beliefs system that historical places are the traces that lead them to connect with the spiritual power of ancestors. In order to achieve the collective memory, there is a necessity for a cultural strategy toward unity and harmony. Syncretism in the Indonesian context is an ancient cultural strategy for inclusivity of sociocultural diversity toward the wholeness of beings and their growth. This strategy has been working since ancient times in Indonesia that enables people in the archipelago to establish and sustain a united nation-state of diverse ethnic groups.

References

Aceh A (1994) Pengantar Sejarah Sufi & Tasawwuf. Ramadhani, Jakarta

Andaya BW, Andaya LY (2015) A history of early modern Southeast Asia, 1400–1830. Cambridge University Press, Cambridge

Casey E (2011) From remembering: a phenomenological study. In: Olick JK, Vinitzky-Seroussi V, Levy D (eds) The collective memory reader. Oxford University Press, Oxford, UK, pp 184–187

Chambert-Loir H (2002) Saints and ancestors: the cult of Muslim saints in Java. In: Chambert-Loir H, Reid A (eds) Potent dead. Allen and Unwin, Crows, p 139

Chambert-Loir H, Reid A (2002) The potent dead: ancestors, saints and heroes in contemporary Indonesia. University of Hawaii Press, Honolulu, HI

Davary B (2002) Islam and ecology: Southeast Asia, adat, and the essence of Keramat. Asia Netw Exch 20(1):12–22

Degroot V (2009) Candi, space and landscape: a study on the distribution, orientation and spatial organization of Central Javanese temple remains. Sidestone, Leiden

Djajadiningrat PAH (1958/1987) Islam in Indonesia. In: Morgan KW (ed) Islam, the straight path: Islam interpreted by Muslims. Motilal Banarsidass, Delhi, India, pp 375–402

Florida N (1995) Writing the past, inscribing future. Duke University Press, Durham, NC

Forschee J (2006) Culture customs of Indonesia. Greenwood, Westport

Geertz C (1960) The religion of Java. University of Chicago Press, Chicago

Geertz C (1973) Interpretation of cultures: selected essays. Basic Books, New York

Geertz C (1980) Negara, the theatre state in the nineteenth century Bali. Princeton University Press, Princeton

Goh BL (2011) Spirits cult and construction sites: trans ethnic popular religion and Keramat symbolism in contemporary Malaysia. In: Lauser A, Endres KW (eds) Engaging the spirit world, popular beliefs and practices in Southeast Asia. Berghahn, New York, Oxford, pp 144–162

Guillot C (2002) The Tembayat Hill: clergy and royal power in Java between 15th and 17th century. In: Chambert-Loir H, Reid A (eds) The potent dead: ancestors, saints and heroes in contemporary Indonesia (trans: Couteau J). University of Hawaii, Honolulu, HI, pp 141–159

Hefner R (1985) Hindu Javanese Tengger; Tradition and Islam. Princeton University Press, Princeton NJ

Hirosue M (2009) The role of local informants in the making of the image of cannibalism in North Sumatra. In: Bonatz D, Miksic J, David Neidel J, Tjoa-Bonatz ML (eds) From distant tales: archaeology and ethnohistory in the highlands of Sumatra. Cambridge Scholar, Newcastle upon Tyne, pp 169–195

Iskandar T (1959) De Hikajat Atjeh. Nederlandsche Boek-en Steendrukkerij V. H. H. L. Smits, sGravenhage

Kahin A (2015) Historical dictionary of Indonesia, 3rd edn. Rowman & Littlefield, Lanham

Kammerer CA, Tannenbaum N (2003) Founders cult in Southeast Asia, ancestors, polity and identity. Southeast Asian Program, Yale University Monograph 52, New Haven

Kartodirdjo S (2007) Agrarian radicalism in Java, its setting and development. In: Holt C (ed) Culture and politics in Indonesia. Equinox, London, pp 71–125

Kayadibi S (2011) Ottoman connections to the Malay world: Islam, law and society. The Other Press, Kuala Lumpur

Kipp RS (1990) The early years of a Dutch colonial mission: the Karo field. University of Michigan Press, Michigan

Kreemer J (1920) De Groote Moskee te Koeta Radja. Niderlandsch Indie Oud & Nieuw 69–101

Kruijt AC (1906) Het Animisme in den Indischen Archipel. Nijhoff, sGravenhage

Kurniawan KR, Kusumawardani RA (2012) The influence of 19th century Dutch colonial orientalism in spreading Kubah (Islamic Dome) and Middle Eastern architectural styles for mosques in Sumatra. Des Built Environ 11

Kusno A (2010) The appearances of memory: mnemonic practices of architecture and urban form in Indonesia. Duke University Press, Durham

Lee KH, Tan CB (2000) The Chinese in Malaysia. Oxford University Press, Singapore

Locher-Scholten E (2004) Sumatran Sultanate and colonial state: Jambi and the rise of Dutch imperialism, 1830–1907. Cornel University, SEAP Publication, Ithaca, NY

Lombard D (1991) Kerajaan Aceh Jaman Sultan Iskandar Muda (1607–1636) (trans: Arifin W). Balai Pustaka, Jakarta

Lombard D, Salmon C (2001) Islam and Chineseness. In: Gordon A (ed) The propagation of Islam in Indonesian and Malay Archipelago. The Malaysian Sociological Research Institute, Kuala Lumpur, pp 181–208

Mandal S (2014) Popular sites of prayer, transoceanic migration and cultural diversity: exploring the significance of Keramat in Southeast Asia. In: Harper T, Amrith S (eds) Sites of Asian interaction: ideas, networks and mobility. Cambridge University Press, Cambridge, pp 127–143

Meilink-Roelofsz MAP (1962) Asian trade and European influence: in the Indonesian archipelago between 1500 and about 1630. Martinus Nijhof, The Hague

Mohammad-Yatim O (1988) Batu Aceh: Early Islamic gravestones in Peninsular Malaysia. Museum Malaysia, Kuala Lumpur
Muljana S (2005) Runtuhnya kerajaan Hindu-Jawa dan timbulnya negara-negara Islam di Nusantara. LKiS Pelangi Aksara, Yogyakarta
Nas P (2006) The past in the present: architecture in Indonesia. Nai Uitgevers Pub, Rotterdam
O'Neil H (1993) Islamic architecture under the new order. In: Hooker VM (ed) Culture and society in new order Indonesia. Oxford University Press, Oxford, pp 151–164
Ozay M (2011) Baba Davud: a Turkish scholar in Aceh. In: Kayadibi S (ed) Ottoman connections to the Malay world: Islam, law and society. The Other Press, Kuala Lumpur, pp 32–55
Palladio A (2012) The four books on architecture. Second (trans: Tavernor R, Schofield R). The MIT Press, Cambridge, MA, London
Pigeaud TG (1967) Literature of Java: synopsis of Javanese literature, 900–1900 A.D. bd. fl 35. Martinus Nijhoff, sGravenhage
Pires T, Cortesao A (2005) In: Cortesao A (ed) Suma oriental of Tome Pires and the book of Francisco Rodrigues, vol 1 & 2, 2 vols. Asian Educational Service, New Delhi
Poesponegoro MD, Notosusanto N (2008) Sejarah nasional Indonesia: Zaman pertumbuhan dan perkembangan kerajaan-kerajaan Islam di Indonesia. III. Balai Pustaka, Jakarta
Pursidi A (2005) Grebeg Besar Demak sudah ada sejak 1506. Accessed 01/30/2016. https://suaramerdeka.com/harian/0501/20/kot18.htm
Reid A (2005a) An Indonesian frontier: Acehnese and other histories of Sumatra. NUS Press, Singapore
Reid A (2005b) Asal mula konflik Aceh: dari perebutan pantai Timur Sumatra hingga akhir kerajaan Aceh abad ke-19 (trans: Maris M). Yayasan Obor Indonesia, Jakarta
Renard J (2008) Friends of god: Islamic images of piety, commitment, and servanthood. University of California Press, London, Berkeley, Los Angeles
Ricklefs MC (2001) A history of modern Indonesia since c. 1200. Stanford University Press, Stanford
Ricoeur P (2004) Memory history forgetting (trans: Blamey K, Pellauer D). University of Chicago Press, Chicago, IL
Soekirno A (1994) Sunan Kalijaga: asal usul mesjid agung demak: cerita rakyat Jawa Tengah. Grasindo, Jakarta
Tarling N (1999) The Cambridge history of Southeast Asia. CUP, Cambridge, p 216
Taylor JG (2012) Global Indonesia. Routledge, London, New York, p 96
Vickers A (2013) A history of modern Indonesia. Cambridge University Press, Cambridge
Wormer P (2014) The spread of Islam in Asia through trade and sufism (nine-nineteenth century). In: Turner BS, Salemink O (eds) Routledge handbook of religions in Asia. Routledge, London, New York, pp 109–121
Zoetmoelder PJ (1995) Monism and pantheism in Javanese Suluk literature, Islamic and Indian mysticism in an Indonesian setting (trans: Ricklefs MC). KITLV translation series 24, Leiden

Chapter 14
Politeness and Patrimonial Communities in Indonesia

Abstract Despite patrimonial traditional, culture is a dynamic process for refinement of what is to be human? This study argues that politeness is one of the cultural efforts that matter for the evolution of the patrimonial practice. This chapter attempts to deal with the question by studying the concept and practice of politeness with reference to ethnic groups in Indonesia with a strong patrimonial practice. This study investigates and explores the sense and practice of politeness based on the relationship between the spatial layout of the house and sociocultural values in various Indonesian communities. This relationship is alluded to by the concept of engagement with proper manners, *tata krama,* and sitting, *duduk*. The materials of the study were carried out from the author's observation in the archipelago from 1993 to 2014. The purpose of the study is to rectify the questions concerning what is a human being in a sociocultural context.

Keywords Politeness · Living · South Nias · Batak Toba · Toraja · Sumba

14.1 Politeness and the Art of Living

Pragmatically speaking, politeness is the cultural phenomenon of the linguistic and behavioral act in terms of the form of social practice (Kádár and Haugh 2013, 251). So that, the necessity for politeness is understood as the need for socially appropriate behavior in the context of a social contract that is ethically binding between the participants with rights and obligations (Frasser 1990). Pragmatically speaking in domestic habitation, politeness manifests as good manners in terms of highly respected behaviors while linguistically, it is expressed and articulated with face-to-face soft-spoken form and indirect contents of utterance; accordingly, politeness is considered as a cultural strategy to preserve congruence or stability in face-to face social interactions and as a preventive measure from potential tensions and aggressive communication. Grice observes empirically that the concept of politeness is founded on a cooperative principle for conversational behavior based on maxims of informative, truthful, relevant, and clear (Grice 1975).

B. Wiryomartono, *Traditions and Transformations of Habitation in Indonesia*,
https://doi.org/10.1007/978-981-15-3405-8_14

Robin Lakoff develops this principle within the rules of pragmatic competence that is to exemplify respect to others in terms of: Do not impose, give options, and make others feel good (Lakoff 1973). From this point, Brown and Levinson work on Goffman's theory of face-work within the framework of Lakoff's pragmatic principles for Negative and Positive Politeness; it is everyone's necessity to deal with each other's face either through the desire to impose one's ideas without resistance or the desire to one's ideas to be appreciated and approved of (Brown and Levinson 1978).

Terkourafi moves forward the theory of politeness from two premises: face-constituting and rationality that emphasize on the frequency of usage rather than lexical meaning (Terkourafi, Beyond the micro-level in politeness research 2005, Terkourafi, Toward a unified theory of politeness, impoliteness, and rudeness 2008); what is frequent is polite, which dynamically constructed through recurring social behavior in the context of normative nature. Studies on politeness in sociocultural context by Ide (1989) and Gu (1990) confirm certain norms and conventions in a culture apply in order to conform to the expectations of appropriate behavior based on respect.

To a certain extent, such polite behavior is potentially used as a strategic tool or obligatory courtesy for achieving communication goals. In the context of political communication, appropriate behavior is not about the polite or impolite statement but concerning the marked and unmarked of politic behavior (Watts 1992, 50); politeness in this sense is understood as positively maintaining the equilibrium of interpersonal relationship.

Within the framework of social behavior of habitation, politeness and impoliteness are not impartial from the daily world experience and socialization, so that politeness and impoliteness are the pragmatic outcomes of common sense, and necessarily regarded as the socially shared norms that guide language and social behavior within a specific sociocultural system. Despite its various approaches and preferences, Eelen's studies of the literature reveal a coherent and consistent line of thought underlying the whole field but also demonstrate how this thought has caused the scientific theory to miss out on many important aspects of the reality of everyday life (Eelen 2014). His analysis suggests the connectivity between politeness and impoliteness that common sense plays an important role in the socially constructed norms in a contextual locality in terms of Bourdieu's habitus and field. Habitus and social field in this study are derived and drawn from the works of Pierre Bourdieu (1930–2002). Both concepts are intertwined and interrelated to each other in the context of agents and their world.

The essential condition of the existence of both concepts is the socially constructed reality of the contextually local world. Habitus is defined by Bourdieu as an organizing mental/cognitive structure that manages concepts and practices of social behavior (Bourdieu 1984, p. 170); habitus is learned, developed, maintained, and reproduced with socialization. This process takes place in certain sociocultural context and locality that determines the actions and responses of the agent in an intuitive way. Habitus is considered as social behavior which is collectively memorized by people that construct a pattern of practices in the field concerning how various resources, such as a credential, reputation, network, capital, labor, and material, are organized, intertwined, synergized, transformed, and utilized for well-being and growth.

The relationship between agents and their world is demonstrated and recognized by their habitus while the field is the phenomena of the world where the social interactions, power struggles, negotiations, productions, and reproductions take place in dealing with the resources. Accordingly, politeness and impoliteness within the framework of sociocultural habitus in a certain field are considered as durably installed generative principle of customarily regulated improvisations that engender practices (Bourdieu 1977, 78); such practices tend to socially reproduce regularities in terms of customs and traditions that are inherent in the impartial conditions of the social production of rules, values, and dispositions, which are gained from cultural history. The question is what is the purpose of all these?

In light of respect and dignity, sitting posture during the social interaction is considered as an important sign and act of politeness that matters for the functionality of communication and interaction. Nevertheless, there is the power struggle for the pride that drives people to mobilize economic, social, and cultural capital for their symbolic capital of pride in their society. What is a human being if he or she does not have pride in his/her society? Is politeness a sociocultural strategy and mechanism to manage the Bourdieu's social field of power struggle so that cruelty, malice, brutality, and violence are avoided and diminished?

Even though politeness and impoliteness are contextual norms and conventions based on a sociocultural system of values, all these represent the social need for the preservation of respect and dignity that concerns the question of what human is in a sociocultural context. The origin of respect is the awareness of beings as a whole that everything in the world is working together for the good and growth because there are law and principle above us. Immanuel Kant understands the concept of respect as the categorical imperative of human conducts that leads to human dignity (Kant and Korsgaard 1785/1997). Socioculturally, the question of what human is leading this study to unfold and divulge how respect and dignity are perceived and practiced in the daily lifeworld and habitation in Indonesia. It is hard to find the most relevant social behavior in Indonesia for studying politeness than the concept and practice of sitting as social behavior. The concept of politeness in this study will be unfolded and divulged from local notion and practice of appropriateness in their everyday world in the context of sitting; this is about the politeness as the practice in everyday interaction (Watts 2003). Traditionally speaking, politeness in the Indonesian culture has been regarded either as obedience or negotiation of power relations within the framework patrimonial culture.

This study concerns the relationship between politeness and daily life of peasant communities in Indonesia from the appropriateness of seat posture and arrangement. In the archipelago, sitting, *duduk* is the other word of dwelling and living that signifies the state of being fulfilled and ready to grow. In view of that, sitting enables people to have the state of mind at peace and stillness as well as the state of being aware of their whereabouts. Development in this context is necessarily conceived as the growth of what has been achieved and settled, that is not simply economic activities without cultural outlook. In Buddhist tradition in China, Japan, and India, a proper sitting position is a condition for meditation; the significance of sitting lies probably in its capacity to reach the state of mind of tranquillity.

This study attempts to unfold and divulge the relationship between manners and sitting in the context of living houses in Indonesia. This study argues that the manners of sitting in the house reflect and signify the art and way of living within the framework of nonverbal communication (Frie 2003, pp. 137–58; Birdwhistell 1970; Sharma 1971, pp. 72–85; Poyatos 2002, pp. 199, 233). Regarding its body posture, sitting enables man to have a spatial position, a point of orientation, and a relation to others. Moreover, the sitting arrangement in a formal meeting shows a symbolic order that demonstrates the behavioral act of stay in the relational system, by which power relation is exercised and displayed as cultural production. Seat position and arrangement are cultural instruments and signs that enable people to know their social position and authority in the context of deliberation and gathering.

From seat position, cultural content, social position, and agency come into play to establish a relationship in the linkage between social structure and agencies' action as mediated by their habitus, Bourdieu locates this relationship within a system of durable and transposable dispositions that work as the generative basis of socially structured and objectively unified practices in terms of habitus. Politeness in a sociocultural context is regarded as the symbolic production and skillful social activity of habitus that enable its people to embody, preserve, develop, sustain, and reproduce the cultural refinement of respect in their social world.

What is the relationship between sitting and politeness in the Indonesian context? At first glance, the historical and cultural root of the idiom of 'sitting' is probably one important source in understanding Indonesian tradition. The source of sitting tradition here is deeply rooted in the history of the Austronesian culture that 'sitting' has something to do with the standing of household or family in a village assembly. Having a seat in the assembly confirms one's respectful social status and position in the community and society. Accordingly, to sit is not merely to be aware of one's social standing, but probably also about his/her power relation to others.

In daily Indonesia, the hierarchical social structure is commonly applied to seat position (Forschee 2006, 89). Accordingly, the seat position designates the power relation in the local realm of community. Knowing one's seat position in society is knowing who he/she is as a sociopolitical person, by which seniority by age, gender, and social status plays an important role in the whole group. In Malay Indonesian culture, asking the name at first encounter is uncommon, but asking the place one comes from is more acceptable. In this culture, the seat position and formation demonstrate the hierarchical order when a formal gathering takes place. In the house, the formal arrangement of seat position is demonstrated as well when the family receives guests. An old custom in the archipelago taught the children to treat their visitors or strangers as their respectful guests. Strangers are not seen as enemy or danger. This is not surprising if one understands the nature of culture in the archipelago. Hospitality is necessary regarding the geography of their open homeland, which is situated between two oceans and two monsoons.

The people of archipelago enjoy their presence and have nothing to protect but their smile and hospitality; they have to make friends to foreigners; politeness in this sense is exercised as the ritual of Brown and Levinson's Negative Face strategy that alludes personal boundary and freedom (Brown and Levinson 1978, 61). In a matter of fact, their land is geographically surrounded by open shores which are defenseless and permeable from any direction. They have likely nothing to defend but the ultimate presence and the tenseless language. Why should they not be nice to foreigners?

Like other East and Southeast Asian cultures, politeness, respect to the elders and higher status persons is evident in almost Indonesian traditions (Low 2010, 54; Bacon and Collins 2010, 91; Farhadian 2005, 130). Inviting people to have a seat in the house is nothing but making a friend with respect and dignity. Elsewhere in Indonesia, guests are respected and will be entertained with the best drink and food of the house. Of course, guests will not be asked about what they would like to drink or eat. Instead, the host just wants to entertain you what they think the best for you. They just want to treat their guests with beverages and snacks as offering or gift; it does not mean that the host does not respect or care for their guests' interest. The tradition tells them to do so and the guests usually accept it with honor and gratitude. In old Malay tradition, it is a common practice when encountering foreigner they will be asked with the question: where do you sit or have you already eaten. The question signifies the importance of your social status and health. Asking the question: who are you by name and origin, is traditionally uncommon. Your name is not that important but your seat by occupation, family background, and achievement does matter.

In the house, however, visitors are to be seated in the front domain, known as *ruang tamu*, guest room; this is considered as the sociological Goffman's notion of the face of the house (Goffman 1967). Despite the shortage of space, this room will be kept available as long as possible. This is the room where the dignity of sitting is upheld, and it is also the place of a show-off for the pride of family heritage and wealth. As a matter of fact, the room is mostly useless and empty daily. For Indonesian people, this room is the altar of the family. The best and memorable things about the house are on display in this room (Forschee 2006: 91–92). The images of the family's ancestors from father and mother side are usually displayed in this room. The guest room for most Indonesian families is a highly regarded room and conceived as a sacred place.

Being a guest in an Indonesian family house is having a seat in the family with respect and dignity. Having a guest is a sign that its host is a well-established family. Therefore, to invite a guest is not simply because of courtesy, *sopan santun*; hospitality, *ramah tamah*; and friendship, *persahabatan*. Rather, to have guests is to gain the honor, pride, and reverence of others. In short, a house with many visitors is a powerful home. Therefore, the more powerful a person is, the larger is his/her guest room is. Most powerful persons usually have a reception hall for their guests in their home. Providing a guest room or hall is not without interest, the room is a sign of the generosity and prosperity of the host.

Most middle-class families have their own house with a guest room and a spare bedroom. The reception room and guest bedroom are located in the front domain; it is considered the 'face' of the family. The back part of the house called *dalam rumah*—inner house—is the space where the family lives day-by-day. The division of rooms inside the back domain varies depending on the family wealth. However, living rooms, bedrooms, and dining rooms are seen as the places, which are associated with the idea of center (Geertz, The Religion of Java 1960, 46; Rutherford 2003, 42).

The actual back domain of the house is the kitchen, bathroom/washroom, and maid quarter. In daily life, going to the back means going to the washroom; it is taboo for Indonesians to address going to the restroom directly. However, the actual back domain is something that the family does not want to show to their guests. It is conceived as the area of weakness and shameful for the family.

As mentioned earlier, spatially, the house is divided into two zones: the front and back. It is common in modern Indonesia to associate the front with the public appearance and the back with domesticity. Again, the front is for pride and glory while the back zone is for retreat and secrecy. The origin of this division is probably ancient; this dated back to the Austronesian tradition (Watson-Andaya 2000, 320). What is important from this common phenomenon of the dualistic spatial division is the fact that this imaginary category still works and is in function until today. Indeed, modernity introduces the functionality of activities in the house, but Indonesian families adjust all these rooms to fit within the ancient system of the spatial category. They hardly accept the kitchen in the front part of the house, or a house is without a guest room. Cooking is traditionally sacred because of its relation to the nourishment for vitality; it is a sign of weakness. Moreover, cooking and eating are supposed not to be seen from the outside if they are not able to offer or share with others.

Traditionally, the front area of the house is a male category, which is associated with public encounter, audacity, strength, and pride. On the other hand, the back is a female domain which denotes retreat, reproduction, recovery, and consolidation. Even though in contemporary Indonesia such gender category has been gradually dissolved, the dualistic division of home zone is still in existence and subtly present that connotes the front for the sense of pride and the back for that of resource and support. Accordingly, the house as the 'self' represents the seat of the family in society. Moreover, concerning the sitting position, the group of the layout is not simply about the front/back category, but also the above/below and right/left category. The above/below category in Indonesia, as well as Malay and Indonesian culture, is in association with superiority and inferiority (Carsten and Hugh-Jones 1995, 123).

Still, the dualistic 'front/back,' 'right/left,' 'male/female,' above/below,' and inside/outside categories of spatial system could have been already in existence from the ancient Austronesian family and kinship system which was based on social role of gender as described by Cunningham (Cunningham et al. 1999). In many ways and throughout history, the Javanese culture has tried to overcome this dualistic category from gender to symbolic category. In doing so, the idea of unitary harmony is subtly achieved with the category of left—*gandok kiwa*—and right pavilion—*gandok tengen*—for their house, *dalem*. Architecturally speaking, the traditional house plans of Java, Bugis and Palembang are more centrist than the rest ethnic groups in the archipelago.

Errington (1989: 54) puts the Javanese culture falling into the category of Centrist Archipelago because its societies are preoccupied with oneness and view break and disintegration as either the product or source or both, of personal disease, community bad luck or political malfunction. Politically speaking, the majority of ethnic groups in Indonesia fall into the Errington's centrist category that values the importance of integrity, unity, and harmony. Indeed, the Javanese society is neither egalitarian nor individualistic. In this point, the house is a sign of a seat with dignity because it contains the power of the self or family. Especially for the Javanese people, the house is the self, *dalem*, which is inclusive in terms of family and kinship that values the unity and the harmonious wholeness of all antagonistic elements. All this condition is made possible because of its capacity to keep knowledge, nourishment, health, and prosperity in the secrecy of as signs of esoteric power (Pamberton 1994, 102).

Even though it is not literally articulated, the house for Indonesian family is an extended family house. It is seldom to find an Indonesian house purely inhabited only by a nuclear family. Other family members from both side parents are common to see in the house. Sleeping for them is never a problem of privacy. Being together in the crowd is more important than being lonely. Moreover, the relation between parents and children is traditionally mutual. Children are due to caring for their parents when they are getting old. Sending them to the senior house—*panti jompo*—is shameful, ungrateful, and disrespectful. Extended family is real-life insurance for most Indonesian families. Members of the extended family are expected to be helpful and supportive of each other as a whole either in good or bad times. Across ethnic groups in the archipelago, such a mutual help is known as '*gotong-royong*' (Anheier and Toepler 2010, 1222).

Nevertheless, in order to have a profound and insightful understanding of the diversity of dwelling traditions in the archipelago, it is necessary to survey and study case by case of building and dwelling in various regions. The importance of the in-depth study is to find the unique and particular aspects of sitting culture with respect to specific circumstances and history of each ethnic group. This investigation will be focused on the dignity of sitting in communities without or with less influence of Hinduism and Buddhism to that of Islam and modernity.

14.2 Politeness and Living in Banua South Nias

Studies on South Nias art, culture, building, and dwelling tradition have been made by several scholars (Schroeder 1917; Feldman 1977; Marschall 1978; Domenig 1980; Viaro 1980; Beatty 1992). However, the relationship between sitting and dwelling as the art of living has not yet been explored. The South Niasan context of sitting is unique, which is because of not only geographical remoteness but also that of its living megalithic tradition. South Nias traditional community is living in a village called *banua*.

The district of South Nias is the area where megalithic tradition leaves its traces in stone memorial monuments. The villages of South Nias are mostly erected on the hills as indicated by their first name, *hili* or *bawoe*. Houses in the village stand in a row on the left and right sides facing a linear open space, *ewali*, paved with stones. Coconut trees stand behind the settlement guarding the village against the storm and strong wind blow from the Indian Ocean. From ancient time, most villages are a self-sufficient settlement.

Traditionally, the dignity of sitting in South Nias is a prestigious achievement, which is associated with feasts of merit, *owasa*. *Owasa* means power. It is, however, a privilege only the aristocrats. Only those who are able to perform the series of feasts are traditionally powerful. Gold and pigs play an important role in the offering of feasts with a communal meal and village celebration. In short, *owasa* is the qualifications to achieve the most prestigious seat in the village. However, the glory and pride of sitting do not stop at the highest rank of the series of feasts of merit. There is another splendid event for the dead. Still, it is the demonstration of power to immortalize the seat in the community of South Nias. The seat stone, *daro-daro*, is traditionally installed to commemorate this achievement. The feast of *owasa* is considered as the highest aim of mobilization of economic, cultural, and social wealth for the demonstration of power in terms of Bourdieu's symbolic capital.

Erecting the memorial stone for powerful aristocrats was a grand and stunning celebration of the past. As we see when arriving in their village, the glory of pre-Christianity South Nias was immortalized by the memorial stones which are erected at the edges of the village street, *ewali banua*. A South Nias man has never achieved a highly respected seat in their community without having offered the series of feasts of merit. All this is for pride and glory which are in South Nias known with one word '*soekhi*.' One important thing about this achievement is to build a traditional Chief house called *Omo Sebua*. How big and high the house is subject to his social class and wealth. There are two social strata in South Nias community: *si ulu*, the aristocrats, and *sato*, commoners. The elites of commoners by skills and wealth become to be the elders of the village. Building a traditional house, *omo hada*, in the village is in South Nias to gain a seat in the village assembly, *orahua*.

The village meeting in Bawomataluo takes place in the village square. All members of aristocrats and elders sit together on the stone monuments facing the same direction. The chief sits at the highest position, while others sit in a row according to their relational social rank and status to the village head. Deliberations in the village meeting are chaired, regulated, and ordered by the village chief with his assistant called *sanema gorahua*; the role and function of this person are to shout between speakers. His rhythmic shouting is to maintain the mood of the meeting at a conducive atmosphere. In doing so, the village meeting as Bourdieu's social field is managed under control as the ritual of deliberation. Senior members are ritually asked to contribute to the meeting according to their rank and status after the village chief; only this person has the right to talk beyond the sequence of speakers. The village meeting is actually the social field for the exposition of the chief's patrimonial power.

House or *omo* in South Nias is a family house. However, other members of the family are usually around in the house. House in South Nias is for the extended family. Father and mother in the house have strict different role and function in the house as well as in the village. The public world and the village political life are considered a male domain, whereas domestic and family care belongs to the female sphere. Spatially, the house is architecturally divided into front—*tawolo*—and back—*foeroema*—areas. A floor-to-roof wall stands between the two domains. A single door—*golu*—connects both domains. The function of both domains is sleep areas for male and female family members. The parents usually sleep in a cubicle in the back area. A traditional house is distinguished from a common one not because of its design location, but also by the rituals of its construction and the pride of its feasts. Being a socially respected person in the village is being an owner of the traditional house. The architecture of each house is traditionally necessary to adapt similar form but its dimension, material quality, and perfection need to comply with the social rank and status of the owner in relation to the village chief.

Modernization in most regions of Indonesia has been transforming the politics in the villages since 1970s where most traditional villages have been restructured and integrated into the modern bureaucratic governance of Indonesia. Since then, the village chief has been elected by villagers and the power struggle among the aristocrats often has been taking place for hegemonic leadership every other 5 year period. The liberation of the village social field from the hegemony of patrimonial domination is demonstrated by the architecture of the house from traditional form to pragmatic contemporary building with modern material and construction system. The logic of economy for less expensive construction challenges the symbolic conformity and obedience of village courtesy throughout architectural form without traditional ordering system. Even though modernization in South Nias has transformed most of the structures, elements, functions, and processes of social field in the village, the concept and practice of politeness in the context of social behavior and language remain less changing. Respect to seniority and expertise is still in their habitus.

Having built a traditional house validates and confirms the status and the seat in the village assembly as respectable aristocrat, *si ulu*, or as highly regarded elder, *si ila*. The assembly represents the dualistic cosmic power of *lawalangi*—god of the sky—and *latura danoe*—god of the earth. The dynamic interplay of both powers is believed by South Nias people as the life of the world *banua* and in the house, respectively. Harmony and consensus—*fahasara doedoe*—are highly regarded as the purpose of life as well as meeting in any level of case and form.

Architecturally speaking, the house form is a raised floor on piles or stilts system. Row grids of round vertical and oblique columns enact as the substructure of the boat-form house erected in a row and standing side by side with neighboring houses. The symbolic image of the boat form is architecturally expressed with the extension of the side beam—*sikhoeli*. The beam runs from back to front simulating the blend image of snake and bird as one, flying ahead. The beam acts as a structural element that brings about the front and the back areas as a single form.

The guarding the spirit of the house is represented by the altar and the image of ancestors from both parents. The images called *adu zatua* and *adu nuwu* are installed on the wall of the front domain facing the entrance door. Entering the house, one needs to go up by stairs and reach the foyer—*edhuo*. The front room is constructed uniquely to articulate a platform system. The design signifies the way and manner of sitting at different level off the floor. Since the South Nias people were converted from indigenous beliefs to Christianity, the images of ancestor were replaced by the cross and image of Jesus Christ. Only the ancestors' photographs are still on the wall signifying the house altar in modern time.

The front form of the house is constructed with two levels of seat platforms: *batoe* and *farakhina*. The first platform is actually a raised floor 40 cm off the ground and used for sleeping in the night. Beams of this platform can be opened, and storage is provided for sleep accouterments. Daily, the platform is used for the seating area of the family. The dignity of sitting is performed when the house has visitors. The male guests will be seated on the second platform—*farakhina*—at the front door side. The female guests are pleased to sit on the same side with their male fellows but on the first platform. The centerline dividing guest and host domain comes into the light during the visit. Ritually, they open their conversation with a betel chewing ceremony—*masirih*. Meeting between the guests and the hosts is the sign that a house is a healthy and alive home. This implies that the house is blessed by their ancestors.

The highest rank of the villagers is the *si ulu baroe banua*. As a traditional ruler, he was the successor of his father. However, to gain respect and pride of the whole villagers under his rule, he had to offer a series of feasts of merit, including to build or restore the great house—*omo sebua*. His seat in the village assembly was never fully recognized without his achievement to build his own house in a traditional way and form. Until today, the village ruler's house is the most elaborate house in the village. It is not only par excellence by grand scale and style but also extraordinary by artistic refinement and high quality of building materials. Today, the village chief is not hereditary. Village assembly elects the chief from the aristocrats of the village who is considerably capable as a leader.

Home for South Nias people is *banua* meaning living in and with their village community as the world. The village is for them the world, community, and home as well. Accordingly, the village, as well as community, is perceived as home because they are able to sit and work together based on their customary laws. Even though emigration to other parts of Indonesia has been familiar among younger generations since the 1960s, most populations still maintain well their culture and tradition. To a certain extent, the pride and glory of village citizenship are an important aspect of community life in South Nias village. However, village citizenship is never enough without having a seat with dignity in their village assembly. They love likely to pay the pride and glory of the village seat with everything in their power. Indeed, the dignity of sitting has its price that is necessary to be demonstrated with a series of feasts delivered to the village community. Prosperity and success are meaningless without being able to sit and work together with their village fellows.

14.3 Politeness and Living in Huta of Batak Toba

Batak Toba art, culture, building, and dwelling tradition have been studied from various perspectives by numerous scholars (Domenig 1980; Waterson 1990: 242, Forschee 2006: 96; Rodenburg 2008: 114–128). Nevertheless, there are few studies that deal with a specific question concerning politeness and sitting and their living house and daily life. *Horas* is the first word that opens communication among Batak people; the word means how are you and May God bless you? Batak is an ethnic group that has been well known for their tradition with a strong kinship bind and clan tie as well as with a patrilineal custom called *adat*. Politeness in Batak Toba is perceived as the social behavior based on respect to this patrimonial tradition known as *hahormaton*. To sit in this community is to exercise this courtesy, which is indivisible from socially accepted and having a seat in the clan and kinship forum. In doing so, one becomes a person with a voice in deliberations and making decisions for the clan's and kinship's interests and rituals.

Sitting in the Batak land is living in the area of North Sumatra surrounding the lake Toba and on the isle of Samosir. Culturally, Batak is agriculture rural-based community with a trace of Hindu influence. Contacts with Indian and Chinese traders were recognizable from some artifacts, Batak alphabet, and wet rice cultivation. The Batak Toba ethnic group lives in groups of clan compound called *huta* consisting 5–12 *rumah bolon*, clan house; traditional house, *jabu*; and *sopo*, rice barn.

The traditional subsistence of Batak people is based on wet rice cultivation—*sawah*. Their dwelling organization is not determined solely by their settlements which are built in different locations and scattered out in several areas. The living house is possibly erected anywhere, but the location of the ancestor house is the center of rituals and ceremonies of their patrilineal clan members. Sitting together in the ancestral houses and tombs is more than just participating in rituals and ceremonies. It is for most Batak people validation and confirmation of their membership in the clan—*marga*—and kinship.

Sitting with dignity for Batak Toba people is being able to contribute to the pride and glory of their clan—*marga*. The principle of sitting in Batak culture is to be in function according to the kinship of *dalihan nan tolu* as a whole. Accordingly, the whole of kinship is conceived as a system of communion consisting of three components: the wife's giver, *hula-hula*; the wife's receiver, *boru*; and the member of the clan, *dongan sabutuha*. Relations with other persons and how to address them —*partuturan*—is to identify his/her seat in the kinship—*dalihan nan tolu*. As a member of the kinship, one is necessarily a person by married status and active in any kinship's occasion. The absence from the clan's activities without reasonable cause is subject to isolation and alienation. Being polite is being obedient and conform to play their role and function in the ceremonial meeting by which the patrimonial power relation is demonstrated with the seat position and arrangement based on seniority and relational appropriateness to wife's giver.

The ancestral house, *jabu*, in the Batak context is the center of the cosmological home. Members of the family sleep and occupy the main floor of the house according to the cardinal position. The parents sit and sleep in the back right area (*jabu bona*) while girls and children under age occupy the back-left area of the house (*jabu soding*). The front right part of the house, *jabu suhat*, is dedicated to the guests, while the front-left domain is for the sons of the family, *jabu tampar piring*. At the center area of this cardinal, domains are meeting place of the family called *jabu tongatonga*; the family does their breakfast and dinner at this center domain. The center area is considered as the place of meeting, *rai*, in the broadest meaning of the word for making decisions, family deliberations, and reception of family guests. The ground floor of Batak house is not divided by any wall, but an open area in terms of Bourdieu's social field of power, in which the patrimonial hegemony of father is demonstrated and exercised in various conversations and deliberations while all family members sit together.

The Batak people can be anywhere on the globe, but their home is where their ancestral clan house is. All this is directed toward connecting lineage ancestors and ensuring their blessings. From this center, the kinship and family tree is drawn to identify who they are. Any occasion concerning the rites of passage of clan members is morally compulsory for being participated by any clan or kinship member, either as a sponsor or a volunteer.

Wedding and funeral occasion plays a very important role in the reunion of all clan members. Such a ritual event is the time when young members meet their elders and listen to traditional wisdom. Generation to generation from rural to an urban center, the idea of sitting in Batak tradition is to be fit in their clan and kinship customs and patrilineal structure. Obviously, seniority is not the only the determinant factors for the leading position of kinship or clan. The hegemonic lineage of their parents plays a decisive role in holding the lead in the community. During the traditional rituals and ceremonies, the sitting arrangement of people is laid out according to their elderly status of lineage and seniority. Rituals and ceremonies in the house are the acts to produce and reproduce Bourdieu's habitus that construct a collective memory and identity of the Batak patrimonial family and its extended kin.

The Batak traditional house is recognizable from its expressive pitch gable roof and stands on piles. The roof form has curvilinear shape standing on the raised floor structure; the form emulates an image of sea vessel sailing on the ocean. The main structure is made of timber construction. However, the roof structure is made of bamboo. In doing so, the roof has an expressive curve line form with flexibility and curvilinear jutting peaks at the back and front edges. Palm fibers or leaves cover the roof with natural color. The house and the rice barn of Batak have a similar roof form. At least three couple of houses and barns stand face to face in a cluster constituting the idea of the home of clan's origin called *huta*. Sitting in a compound, *huta* is to integrate and align with the patrimonial order of architectural scale, dimension, form, and style with the existing structures in the relational seniority of house owners. Today, most Batak people go abroad and leave their compound empty. The descendants of the founding fathers live in the house to take care of its function as the clan's shrine. In some cases, the house is left mostly empty without inhabitants. However, waiters who usually live in the nearby area take care and maintain the house on a regular basis so that clan and kinship members pay homage the house as clan's place of pilgrimage on the occasional and regular basis. In other words, *rumah bolon* and its *huta* become a whole living shrine of the clan and kinship.

Since mostly Batak Toba people traditionally emigrate from their ancestral land to other places of Indonesia, their *huta* and traditional buildings are left in negligent. Going home is not coming to the ancestral land but to their seat in the congregation, *huria*. The gathering of the members of clan and kinship is the homeworld. One important reason why Christianity has been appropriately adopted by Batak Toba culture and tradition was that it provides them with the sustainable seat of the congregation, *huria* that consolidates and validates their membership of clan and kinship on regular basis. Anywhere the Batak people live what they build is their seat for a congregation that collects and gathers families, relatives, and people from their ethnicity and homeland.

The allusion to sitting as a home is signified by this sustainable seat of the congregation, *huria*. However, to get the highest power in this congregation, one necessarily needs to demonstrate his active social engagement, sponsorship, and contribution of economic capital to the events that gather all members of clans and kinships associated with him. In other words, Batak Toba societies put the highest value of habitus for social capital in their Bourdieu's social field of *huria*. Those who are very active in social engagement, financial sponsorship, facility contribution, and event organizing activity for the happening of *huria* deserve the highest respect and admiration.

14.4 Politeness and Living in Banua Toraja

Unlike other ethnic groups in Indonesia, the allusion of sitting for Toraja people is explicitly articulated with their clan and kinship house, *tongkonan*. Studies on Toraja art, culture, building, and dwelling tradition have been done from various approaches (Tangdilintin 1974; Nooy-Palm 1979; Domenig 1980; Volkman 1985; Kobong 1988; Nooy-Palm et al. 1988). However, the question of sitting is still open and leaves us a room for this investigation.

The native land of Toraja people is situated in the interior highlands of Sulawesi. Most of them live along the river Sa'dan. Besides animal husbandry of pigs, chickens, and water buffaloes, wet rice cultivation is their main subsistence. Toraja people build their settlement on the hill or on the plain surrounded by bamboos and rice fields, *sawah*. Bamboo is an important building material for Toraja custom and tradition. Traditionally, Toraja lives in scattered areas according to their clan. The settlement unit is known as *tondok* meaning literally home or place of origin.

Socially speaking, Toraja community is originally divided into two hereditary classes: the aristocrats or landowners, *puang*, and the commoners, *to makaka*. The other group beyond the two classes is slaves, *kaunan*. In the course of time and interaction with Hinduism as well as Islam and Christianity, social classes become more complex and sophisticated. However, grandeur rituals and ceremonies mostly belong to aristocrats. The significance of the social status comes into play when the meeting of lineage elders, *kombongan*, takes place.

Traditional *tondok* consists of several traditional houses, *tongkonan*, and rice barns, *alang*. Access to the *tondok* is commonly from the east direction. On the West, we usually find a pathway leading to the graveyard, *liang*, or memorial field, *rante*. Settlements in Tana Toraja are not territorially organized but genealogically coordinated through rituals and ceremonies of merit and funeral. The living house could be anywhere, but the place to seat is certain. It is the place where their *tongkonan* from both parents are.

Tongkonan means the place of the seat. The *tongkonan* house is actually the ancestral place of the seat (Parinding and Achjadi 1988, 73). From this seat, Toraja clan and kinship is developed. From this ancestral house, the Toraja lineage is drawn to somewhere else. Every Toraja person has to know his/her place of origin by identifying the name of his/her *tongkonan* house. Sitting, *matongkon*, in Toraja culture is to know their way back to the place of origin. This implies that one is due to do regular visits and concrete contributions for the feasts.

The confirmation of their seat is made possible by participating and contributing to rituals and ceremonies in the place of origin, especially for the feast of ascending and that of descending smoke or the funeral, *rambu solok*. Since the adoption of Christianity, the feasts of ascending smoke, *rambu tuka*,' or the feasts of harvest, have been gradually dissolved because they are mostly related to gambling, alcoholic drinking, and betting.

For most traditional feasts, there is usually a permanent committee consisting of the elders from the clans and kinships of the sponsors. The committee of the feasts is the core of the feast association called *saroan*. The association is established as an organizing team that prepares and conducts the feasts for the collection of funds and the distribution of meats during the occasions. The distribution of means during the funeral is the show for the relational power among the elites of the kinship world in terms of Bourdieu's social field. The distribution of the meats is considered as the climax of the funeral rituals that single out its essential purpose: the show the rank and status of sponsors for the vent with the symbolic currency of meat. Struggle for pride in their kinship world is considered the fight for their social existence and hegemonic position in the ritual meeting of *matongkon kombongan*; pride in the Toraja's kinship world is regarded as Bourdieu's symbolic capital that is the highest than other capitals (cultural, economic, and social wealth.

Pride, shame, and competition play a very important role in the role of the contributors or sponsors of the feasts (See also Volkman 1985: 77). The consequence of such a play for personal and communal pride could drag people in deep debt and serious financial trouble. All this is for one's own seat in his clan and kinship as a socially respected person. However, the show must go on with a sense of glory and abundance. Success for being somebody abroad is nothing if one has nothing to offer to his/her place of origin and clan.

The architecture of Toraja is characterized by their saddleback roof and expressively jutting front façade with a roof supporting column, *tulak somba*, full of a chain of buffalo's horns. The four or six side piles, *bantuli*, stand beside the housing body to support the edge of the roof on the right and left aside. The common plan of the *tongkonan* house is 4.6 × 6.8 m with 12–16 m in height from the ground. The scale, size, number, decorative elements, and craft of the building tell us about the pride and wealth of the clan and kinship who build it.

Most powerful settlements, such as Nanggala, Kete' Kesu, Pallawa, are decorated with architecturally elaborated ornaments and decoration on their structures. Obviously, decorating the facade is not only for an artistic purpose. All decorative motifs on the building tell us about something associated with values, history, beliefs, cosmogony, and customs they uphold. Black, yellow, white, and red play an important role in their ornaments and decorations. Most importantly, rich ornaments are a sign of more power, pride, and prestige for their lineage's seat.

Symmetrical composition is mostly applied to house and rice barn decoration. Effigies and images of buffalo head belong to decorative elements of the house and rice barn. However, the right and left categories indicate a slightly different value and meaning. It is not about good or bad, but more about complementary pairs like dark and light. The ancient tradition holds the journey of the human soul in terms of smoke in ascending for life and in descending for dead. For Toraja people, life and death are in a circle to find the seat in the ancestral land, *puya*. The feasts with the communal meal are demanded by the power of smoke, *rambu*, for moving forward. All this is to drive their seat on earth to heaven with pride and glory.

The *tongkonan* house is accessed by a ladder under the roof or beside the building. The interior of the house is dominated by a central room, *Sali*, with front, *sumbung* or *paluang*, and back platform, *tando*, simulating boat stern and boat bow. In many cases, small windows are provided at the left and right sides. A hearth, *dapo*, is installed at the center of the house. Sleeping is around the fireplace and at the back part. Usually, only elderly people sleep in the *tongkonan* house. According to ancient beliefs, *aluk to dolo*, the balsamized deceased was seated on the floor of *sali* waiting the best time for the funeral ceremony. The funeral is actually time for the dead for the accumulation of power for having a seat in heaven.

The living house in Toraja culture is called *banua*. Most people do not live in *tongkonan*. Their daily living house is a simple building with a gable roof structure known as *banua barung-barung*. In daily life, *tongkonan* and *alang* remain empty. The difference between both is that the common living house does not have a raised floor system, but simply standing on the ground. The space under the *tongkonan* floor is used for their livestock or buffaloes. Pigpen is usually located away back from the *tongkonan* house. During the days of harvest time, men and women are busy taking care of their paddy. They dry up the paddy by the sunlight on the ground on *tondok*, and in the late afternoon, they store their paddy inside the rice barn. The storeroom for paddy is provided at the upper structure of the barn, *alang*, while the main part of the building is used for an open and a commonplace for relaxation and siesta during the day. The raised floor of the rice barn is a favorable area for children and elderly people.

The function of the *tongkonan* house is the seat place of Toraja clan and kinship. This function comes into the light during the rituals and celebrations such as the funeral feasts. The main traditional house called *tongkonan layuk* becomes the magnet of a homecoming for the kinship members. They make a pilgrimage to their homeland with the *tongkonan layuk* as their destination. During the feasts, the hosts provide the temporary guest halls, *lantang* for their guests.

During the funeral events, elders of the kinship sit together in one of the rice barns; they form and conduct an assembly of elders. Prior to their deliberations, one of the elders performs liturgical songs and poems delivering ancient messages on the voyage to heaven. In the past, the traditional priest, *tomina*, recited those matters. Since the adoption of Christianity, Protestant priests take over *tomina* leading and arranging the funeral ceremony. The event will be commenced with the ritual of respect to the dead for several days and taking it to the grave, *liang*. The Christian missionaries such as Adriani, Krujt, and van der Veen had done well their job on the adaptation and adjustment of animistic rituals to the Christian contents.

During the days before the funeral event, slaughtering buffaloes is a special attraction for all guests and the hosts. Another dramatic scene is watching the fight of buffaloes in special filed called *rante*. This place is actually dedicated to the commemorative monuments, *batu simbuang*, of the deads. The erection of the stone is the sign that the deads have already been seated peacefully in heaven, *puya*, with pride and glory. However, the megalith stones represent only the seats of the most powerful persons in the kinship.

Home for Toraja people is never without the place where their ancestors and their descendants sit together. It is not simply a homeland and the house of origin. Rather, it is the occasion where clan and kinship members come together for funeral feasts. Tana Toraja as a homeland for Sa'dan Toraja people is more than just their native country, but it is the port for their symbolic sailing toward the ancestral seat in the sky, *puya*. Nevertheless, pride, fame, and glory are the most important part of the feasts. Gathering and homecoming are also the occasions to show the clan's and family's pride with a magnificent spectacle of colorful costumes and banners. After the show, there are many families in deep debt and bankruptcy for paying their vanities and splendid spectacles of funeral feasts.

14.5 Politeness and Living in Kabisu West Sumba

Compared to other parts of Indonesia, Sumba as an area of cultural studies has been unfortunately neglected in many ways. However, there are some significant works that cover structural aspects of village and house in West Sumba (Needham 1987, 1988). Prior to the Pacific War 1942–1945, the island and its society were relatively isolated from the outside world. The reason is explainable in part because there are fewer resources found on the island as economic commodities beyond the sandalwood. Besides, Sumba had been in suffering for the absence of centralized government and perpetual internal conflicts of tribes until the end of the Pacific War in 1945 (Needham 1980: 21).

The Sumbanese people call their island *tana humba*. It is the homeland where life and death, as well as the children of mankind and their ancestors, are present eternally as visible and invisible and never drifting apart; everywhere, social behavior is circumscribed by the presence of the invisible. Sumba is one of Indonesia site where megalithic tradition is still alive. According to this tradition, the presence of ancestors and heroes are represented by their stone monuments in their village. The island is believed by the native as the meeting place of light and dark, sky, and earth, male and female, right and left as well as antagonistic cosmic powers in one and nameless (Toichi 1974, 129). Linguistically, Sumba is divided into regions. Each clan has its own language and dialect such as Anakalang and Kodi. Similar cultural artifacts and ideas could be differently articulated and expressed. All this has something to do with local pride and cultural isolationism of the Sumbanese clan tradition.

Sandalwood is probably the most important commodity of this island. Cultivating rice on the dry and wet field, *sawah*, is a basic subsistence besides planting corn and sweet potatoes. For their protein, chicken, pig, and water buffalo are their livestock. Breeding horse and cattle is a long-standing tradition in Sumba. Horse and cattle play a very significant role in their ritual and ceremony based on indigenous beliefs, *marapu*. Nonetheless, fishing and hunting are their ancient tradition. In short, all these resources build the Bourdieu's economic capital for Sumbanese populations.

The landscape and natural environment of Sumba are arid under two monsoons: dry and wet seasons. Albeit there are few areas above 1000 m from the sea level, the terrain of the island is mountainous and featured by monsoon forests, savannah, and grassland. Geologically, the terrain of Sumba is composed primarily of sandstone and mudstone. Traces of mixed fauna and flora of Sumba are apt to Australian continental crust. Physically, the geology of the island is rocky, comprising deeply dissected plateaus. Shortage of water and geological terrain prescribes Sumbanese settlement in a linear form with less opportunity for large concentration.

Even though the Sumba community is divided into two social strata which are the aristocrats—*marimba*—and the commoners—*kabihu*—the social life of the people is egalitarian. Wars among clans and kinships are an ancient tradition in which the bravery, courage, leadership, and fairness become to be their values. The consequence of traditional battle is the emergence of prisoners of war. The prisoners are traditionally treated as slaves—*ata*—and work for aristocrats. In their social organization, Sumba community is structured in a confederation of compounds—*paraingu*. Aristocrats who are commonly also landlords—*manu tamangu*—are in the role of rulers and political leaders. Sitting with them as advisers are the elders, *kabihu*; this is the Bourdieu's social field of Sumbanese communities, where the elites of the village struggle for their social standing.

The ruling class and the elites of the community build an assembly known as *paraingu*. The elders are qualified by their speech skill, martial skill, and knowledge of custom and tradition. In the daily activity, the elders of *kabihu* are assigned by their tradition to run specific role and function in their community. The first elder is responsible as the parent—*ina-ama*—of the community. The second has to undertake the job for social and customary laws affairs—*luku uru*. The third is in charge of defense and warfare—*makaborangu*. The fourth is the priest—*ratu*—of the community. All these social roles and functions are demonstrated by the seat arrangement during the village assembly.

The houses of the aristocrats and four main elders have their strategic positions in the compound. In front of their house, people erect, carve, and decorate the stone monuments with specific signs and symbols representing their seat position in the village. Coupling the opposite role and function as the ruler-elder system is an essential principle in the building layout. In order to seat properly as a settlement, the number of the house is believed to be even numbers. However, it does apply too for the number of compounds surrounding the parent compound. A parent compound has to be surrounded by at least two other derivative compounds that establish a *paraingu*.

Regarding local terrain and climate, traditional settlements are built on the hilly sites with a central linear open space surrounded by houses. Architecturally, the site layout of the traditional settlement is adaptive to local topography and compact to accommodate their less dense populations. In Anakalang, the building placement on the site is erected around the central plaza, *talora*. The settlement is developed gradually according to population growth. The expansion of settlement is traditionally constructed through the establishment of new satellite settlements surrounding the first settlement with a well-connected path and space to the parent compound. Architecturally speaking, new settlements are constructed with a similar structure to the parent settlement.

The spatial organization of settlement is a compound, *kabisu*, consisting of 4–10 houses. Spatial orientation in the compound is adjusted in accordance with the upstream—*bondo*—and downstream—*lola*—directions. The parent compound, *ina ama*, is situated at the higher site with an ellipse open space, *natara*, used for the graveyard. The next generation of compounds, *uhu ana*, are usually developed adjacent to the compound origin with a continuous pathway. The center plaza, *natar*, is believed to be the center of the cosmos where their ancestors seat with their pride and glory.

The *natar* as the central plaza of the compound is the place where commemorative stones are erected. The more powerful the dead is the bigger and more elaborated is his/her stone. A group of stones is established as the place of village guarding spirit, *marapu wano*, at the center area. The plaza is built with different levels. The highest level, *natara kaba*, is for the ancestral stones and the lower one, *natara podu*, for the graveyard. Right at the direction, they erect a stone monument of *bina bondo* indicating the upstream direction while at the other side is the stone of *bina lola* signifying the downstream. The compound is believed as a safe and clean area. To keep this in perfect shape, the compound area is provided with two gates: entry, *bina tama*, and exit, *bina louzo*.

The ancestral house, *uma*, is home of family and ancestral clan. The smoke of hearth, *rabuka*, is the sign of living and nurturing as well as caring. The house is not the machine for living, but the seat and home of the family, and the shrine for having a blessing from the ancestors. The seat is about social achievement and wealth measured with the number of cattle they own. The seat place for the ancestors, *uma dana*, is the roof room, *toko uma*, which is signified by the architecturally peaked rooftop. The seat for the family is the living area, *doku-gawi*, with hearth, *rabuka*, as its center. The substructure, *kaki kambuga*, of the house is the place for livestock.

Vertically, the three spatial domains reflect the value of the category from low to high. Accordingly, the living area is in between the animal and ancestral realm.

The plan of the house is spatially divided into two categories: left from the front house for male, *doku*, and right for female, *gawi*. Symmetrical opposition or dualistic category is applied to signify the balance of gender though they have different role and function in Sumba community (Wouden et al. 1968, 25). To enter the house, male and female people are provided with separate entrances on the raised floor terrace, *baga*, under the roof. The pride and prestige of the owner are demonstrated with the display of buffalo horns on the wall and house entry. All horns show how many sacrifices animals the house owner has offered for rituals and ceremonies for merits of him and his wife. Inside the house is dark, cool and smoky. To be eligible as the guest of the house, one will be invited by the host with betel chewing ceremony; this is considered as the courtesy of friendly engagement.

The center of the house is the hearth, *rabuka*, constructed within the area established by four pillars: *parii mbali tonga, parii kere pandalu, parii koro kalada, and parii mata marapu*. Living in the house is accomodated in two areas: family room, -*halai katongu*-, and sleeping room, -heda kabali mata and keri penuang-. The two areas are differentiated from each other at a different level. The sleeping area is set 30–40 cm above the living floor.

Home for West Sumba people is unbelievable without the existence of their compounds, *kabisus*, with their ancestral clan houses, *uma*. The village is conceived as the spiritual unity of the livings and the dead. Fracture and disintegration of the community and village structure are signs of social illness, political weakness, and leadership failure. All this in West Sumba is focused on the ancestral house, *uma*, from which lineage is drawn from generation to generation. Funeral feasts take place around the *uma* that signifies and designate the rebirth and restoration of the family home as the world.

Socially speaking, the structure of the world is shown as a pyramidal form with the chief of the priest, *rato*, at the apex and the commoners at the base. The execution of the feast demonstrates how this hierarchical system of spiritual authority works. The main purpose of the religious feast is to confirm the dignity of the seat of the dead in the community of *kabisu*. The peak of a feast for the establishment of the dignity of the seat is demonstrated with the erection of the stone monument, *woleku*. Despite its religious meanings, the other purpose of the feast is likely to demonstrate the clan's and family's pride in a competitive manner with various currencies, from gold, pig, horse, cattle to money. Unfortunately, in this case, the sense of home must be paid with sacrifices in the name of pride and self-glory. The feast is the Bourdieu's social field for the struggle of power relation in terms of symbolic capital for the highest pride among other elites in their community. For this pride, all other capitals are mobilized and transformed that display how many resources have been offered to hold the event for the erection of a stone monument for the dead.

14.6 Concluding Remarks

Politeness among peasant communities in Indonesia has been conceived as the necessity for humility and obedience under the authority of patrimonial tradition; pragmatically speaking, this politeness is practice within the principle of Brown and Levinson's principle of Positive Face politeness. Accordingly, the conversational politeness is exercised to manage disagreements and differences to preserve the face of authority and status quo. The power relations of persons in social life in the house and village have been traditionally arranged and configured. Politeness, in this case, is not about face-work in the framework of individual freedom, but within the hierarchical system of wealth, seniority, social class, and expertise that build and establish the symbolic capital of pride.

The most vivid representation of such power relations is demonstrated in the seat arrangement of ritual gathering or meeting in the house or in the village hall. The source of politeness is respect to the tradition from which the sense of unity and harmony is believed to be put in jeopardy. Tradition and ancestors are believed and considered as the presence in absence that is always there whenever they meet together because they believe in the wholeness of beings that include the visible and invisible, as well as the mortals and the immortals. In this sense, politeness is considered as the ritual way and art of keeping this wholeness sustainable that is experienced as the well-being of their world in unity and harmony. Accordingly, the unity and harmony of their family, neighborhood, and village are considered as the structures of their well-being. Cross-ethnic boundaries, working together voluntarily, gotong-royong, for the community is considered as the ritual of politeness for the revitalization, replenishment, and rebirth of cosmological order and their community as the world.

Regardless of the diversity of architectural expressions and styles, building and dwelling tradition in Indonesian archipelago demonstrates the similar principle of spatial organization. The division of inner home into front and back categories is conspicuous. The subtlety of this division in Indonesian modern home is manifested in various ways but the existence of the guest room as the family altar is evident. House in Indonesia is not the machine for living, but the container of benevolent spirits with blessing and gratitude. There is mostly a room which is spiritually dedicated to allowing the spirits of ancestor present in the house. Besides, rituals surrounding the building and dwelling process are part of the way to preserve the traditional values of the home as the seat of a family with respect and dignity. The other important aspect of dwelling culture is hospitality by offering guests with a ritual of sitting together.

Above all, pride, prestige, vanity, and self-importance are present in various allusion of sitting that demonstrate power, gender, authority, seniority, and superiority in terms of Bourdieu's symbolic capital while at the same time they reveal themselves as the signs of personal weakness, lack of self-confidence, and immaturity. The story of the allusion of sitting for pride continues on and on in dealing with modernity. To sit in this case is not only to adopt modern products for local pride and lavish lifestyle, but also to deal with the ideology of individual liberty, equal opportunity, and human right. All these are challenging for the traditional concept of sitting in Indonesian society and its ethnic communities.

References

Anheier HK, Toepler S (2010) International encyclopedia of civil society. Springer, Berlin & New York

Beatty A (1992) Society and exchange in Nias. Clerendon Press, Oxford

Birdwhistell R (1970) Kinesics in context. University of Pennsylvania Press, Philadelphia

Bacon D, Collins T (2010) CultureShock! Jakarta: a survival guide to customs and Etiquette. Marshall Cavendish International Asia, Singapore

Bourdieu (1977) Outline of a theory of practice (trans: Nice R). University of Cambridge, Cambridge, UK

Bourdieu P (1984) Distinction: a social critique of the judgement of taste (trans: Nice R). Harvard University Press, Cambridge Massachussetts

Brown P, Levinson SC (1978) Universal language usage: politeness phenomena. In: Goody EN (ed) Questions and politeness: strategies in social interaction. Cambridge University Press, Cambridge, UK, pp 56–289

Carsten J, Hugh-Jones S (1995) About the house Levy-Strauss and beyond. Cambridge University Press, Cambridge, UK

Cunningham CE, Aragon LV, Russel SD (1999) Structuralism's transformations: order and revision in Indonesian and Malaysian societies. Papers written in honor of Clark E. Cunningham. Vols. Program for Southeast Asian Studies, Monograph Series. Arizona State University, Tuscon, AZ

Domenig G (1980) Tektonik im primitiven Dachbau. ETH, Zurich

Eelen G (2014) A critique of politeness theory, vol 1. Routledge, London & New York

Errington S (1989) Meaning and power in Southeast Asia. Princeton University Press, Princeton, NJ

Farhadian CE (2005) Christianity, Islam, and nationalism in Indonesia. Routledge, London & New York

Feldman J (1977) The architecture of Nias, Indonesia, Ph.D. thesis,. Microfilm. Columbia University, Ann Arbor Michigan

Forschee J (2006) Culture and customs of Indonesia. Greenwood, Westport CT

Frasser B (1990) Perspectives of politeness. J Pragmatics 14:219–236

Frie R (2003) Understanding experience: psychotheraphy and postmodern. Routledge, London

Geertz C (1960) The religion of Java. University of Chicago Press, Chicago

Goffman E (1967) Interaction ritual: essays on face to face behaviors. Aldine, Chicago IL

Grice HP (1975) Logic and conversation. In: Cole P, Morgan J (eds) Syntax and semantics 3: speech acts. Academic Press, New York, pp 41–58

Gu Y (1990) Politeness phenomena in modern Chinese. J Pragmatics 14:237–257

Ide S (1989) Formal forms and discernment: two neglected aspects of universal linguisitic politeness. Multilingua 8(2/3):223–248

Kádár DZ, Haugh M (2013) Understanding politeness. Cambridge University Press, Cambridge UK

Kant I, Korsgaard CM (1785/1997) Groundwork of the metaphysics of morals (Gregor M, ed). Cambridge University Press, Cambridge, UK

Kobong T (1988) Evangelium und Tongkonan. Peter Jensen, Ammerbeck, Hamburg

Lakoff R (1973) The logic of politeness. In: Corum C, Cedric A, Weiser A (eds) Papers from the ninth regional meeting, Chicago Linguistic Society. Chicago Linguistics Society, Chicago, IL, pp 291–305

Low PKC (2010) Successfully negotiating in Asia. Berlin & New York: Springer Business Media.
Marschall W (1978) Der Berg des Herrn der Erde. DTV, Munich
Needham R (1987) Momburu, history and structure of a domain of Western Sumba. Clarendon Press, Oxford
Needham R (1988) Principles and variations in the structure of Sumbanese Society. In: Fox J, Adams M (eds) The flow of life: essays on Eastern Indonesia. Harvard University Press, Cambridge MA, pp 21–47
Nooy-Palm H (1979) The Sa'dan Toraja. Martinus Nijhoff, The Hague
Nooy-Palm H, Reimar S, Ursula SD (1988) Banua Toraja: changing patterns in architecture and symbolism among the Sadan Toraja, Sulawesi Indonesia. Royal Tropical Institute, Amsterdam
Pamberton J (1994) On the subject of Java. Cornel University Press, Ithaca NY
Parinding SC, Achjadi J (1988) Toraja, Indonesia's Mountain Eden. Times, Singapore
Poyatos F (2002) Nonverbal communication across disciplines; culture, sensory interaction, speech, conversation. John Benjamins, New York
Rodenburg J (2008) Toba Batak women as custodian of the house. Nordic Institute of Asian Studies, Denmark
Rutherford D (2003) Raiding the land of the foreigners. Princeton University Press, Princeton
Schroeder EEWGs (1917) Nias Volume I & II. Brill, Leiden
Tangdilintin (1974) Toraja dan Kebuidayaanya. Lembaga Sejarah Antropologi, Ujug Pandang/Makasar
Terkourafi M (2005) Beyond the micro-level in politeness research. J Politeness I:237–262
Terkourafi M (2008) Toward a unified theory of politeness, impoliteness, and rudeness. In: Bousfield D, Locher M (eds) Impliteness in language: studies on its interplay with power in theory and practice. Mouten de Gruyter, Berlin, pp 45–74
Toichi M (1974) Ethnology of the Southwestern Pacific: the Ryukuyus Taiwan Insular Southeast Asia. Chinese Association for Folklore, Taipei
Viaro A (1980) Urbanisme et Architecture Traditionelle du Sud Nias. Unesco, Paris
Volkman A (1985) Feast of honor: ritual and change in Toraja highlands. University of Illinois Press, Urbana
Waterson R (1990) The living house: an athropology of architecture in Southeast Asia. Oxford University Press, Singapore
Watson-Andaya B (2000) Other past, women, gender, and history in early modern Southeast Asia. Center for Southeast Asian Studies, UH Manoa, Honolulu HI
Watts R (2003) Politeness. Cambridge University Press, Cambridge, UK
Watts R (1992) Linguistic politeness and politic verbal behaviour: reconsidering claims for universality. In: Ide S, Ehlich (eds) Politeness in language: studies on its history, theory and practice. Mouton de Gruyter, Berlin, pp 43–69
Wouden FAEV, Needham R, Locher GW (1968) Types of social structure in Eastern Indonesia. Nijhoff, Dordrecht

Postscript

This study understands patrimonialism as an ideological system that is practiced by a person or regime within the traditional paternal authority. Instead of a pluralistic society and a multiparty political system, the goal of patrimonialism is the establishment of a monolith cultural identity under a highly concentrated authority appointed to one person or regime. In order to achieve this, the practice of patrimonial authority needs to follow the logic of instruction, direction, and tuition through a top-down hierarchical structure of authority. In many ways, patrimonial authority works within a highly centralized, formal, and bureacratic system; all this delimits options of democracy, civil liberty, and individual right forTop of FormBottom of Form people in the category of a subordinate.

The excess of patrimonial practice is inevitably the abuse and misuse of power for its superiority of control of the ruling group over the ordinary populations at large, mostly at any cost. In the daily lifeworld, the patrimonial culture and tradition are eager to show off the privilege and specialness of social status and pride. Accordingly, income inequality is perceived more like the logical consequence of historic legacy than as the challenging outcome of socioeconomic discrepancy in the patrimonial capitalist system of production. Therefore, poverty elevation through the trickle-down economy is more preferable than a radical restructuring of the production system based on the social-democratic economy.

Notwithstanding modernism and democratization, political culture has been playing a momentous role in Indonesian politics since power is not impartial from populations' insight, comportments, and conducts. The democratic structure provides people with the freedom to elect their bests. Yet, since Indonesia's politics is predominantly predisposed by the Javanese patrimonial tradition, the sociocultural values system cannot afford to be overlooked. The relationship between urbanism and society is a complex interwoven system that works for a collectively imagined concept of better living in the socioeconomically cohabitating realm. Studies on urbanism in Indonesia cannot afford to ignore the sociocultural reality of patrimonial culture and tradition. The patrimonial tradition provides Indonesian society and urbanism

B. Wiryomartono, *Traditions and Transformations of Habitation in Indonesia*,
https://doi.org/10.1007/978-981-15-3405-8

Fig. A.1 Thamrin city shopping center, Jakarta 2012. Photograph by author

with the subconsciously dualistic division of socioeconomic strata: the ruling group or *priyayi* and the ruled or *rakyat*. The reality of dualistic strata is experienced on a daily basis. The urban scenes of the dichotomy of *priyayi*'s lifestyle and ordinary people's daily routine are demonstrated with the contrasting world of the luxury malls, cars, and *gedongan* and the traditional markets, public transit, and *rumah kumuh* (Fig. A.1).

Income inequality in urban areas as a development issue has been perceived by the Indonesian government from 1970 until today with a poverty reduction approach. Such a method is considered fractional that is not to review and rethink the patrimonial system of corporate economic capitalism. Rather, poverty mitigation does not actually transform the current economic system from patrimonial capitalist production to democratic manufacturing. In other words, the poverty reduction is to elevate the group of urban poor populations from the poverty line with policy and program interventions, but it does not change the working people's position in the economic production system; the system remains patrimonial that divides humanity into two categories: the economically ruling capitalist, who sit in the boards of directors, and the economically ruled working people, who operate and run the business of productions and services with labor. In real political life, government intervention into the economic system of production is constrained by the labor law in terms of human rights, worker safety, and health conditions. However, the intervention does not go further to the rights of the workers in the decision-making process.

Economically speaking, Indonesia is not independent of the global network of the capitalist economic system. The global capitalist economy prescripts indirectly Indonesian politics on urbanism from regional to the national level in terms of foreign investments and other businesses such as real estates, global retails, tourism, energy, and infrastructure. This is inevitable because of the domestic demands for luxurious goods and fancy things are high; the affluent ruling group consumes such expensive imported commodities as the symbols of their pride and privilege. For most *priyayi*, consuming exclusive and expensive things is not only a matter of lifestyle but also demonstrative proof of privileged social status. In order to uphold an exclusive and high social status of *priyayi*, the practices of graft, bribery, and corruption among public servants and politicians become common activities.

Unsurprisingly, running for public offices is not free from mone politics. Political campaign contributions by large corporations to certain figures and parties are not considered charities but their investments for long-run projects. At the end of the day, all these need to be paid off. Everybody knows that the economic power system is behind all political agenda on economic policies and regulations. In this sense, the outcome of political leadership is clear that they have to serve and accommodate the interest of their sponsors: the large corporations. The logical consequence of the political process under the patronage of large private corporations is crystal clear: the patrimonial economy in favor of the rich.

The real economy of urbanism cannot afford to ignore the interest of the capitalist machinery of productions and services for more and more profit. The main issue is not about good or bad concerning profit but the social justice of the wealth distribution. The patrimonial culture and tradition are considered not conducive to urbanism based on social justice. The riches are the owners of such enterprises who are only less than 20% and enjoy half of the total income of the whole population. Income inequality in this country will not be solved overnight by any newly elected government from provincial to a national level even though they focus on three policies: eradication corruption, creating new jobs, and better social protection. Al these options do not touch the root of all evils in the global capitalist economy. The root lies in the greed and selfishness of the capitalist corporation; they do not realize that there is an inherent contradiction of the capitalist economy. The growth of wealth depends on the increase of consumption by the larger populations. If the large populations do not have enough purchasing power for the products and services of capitalist corporations, how can the wealth grow? As a matter of fact, the concentration of wealth only in a few of populations and the trickle-down economy does not significantly increase economic activities at large.

Index

B. Wiryomartono, *Traditions and Transformations of Habitation in Indonesia*, https://doi.org/10.1007/978-981-15-3405-8

MIX
Papier aus verantwortungsvollen Quellen
Paper from responsible sources
FSC® C105338

If you have any concerns about our products,
you can contact us on
ProductSafety@springernature.com

In case Publisher is established outside the EU,
the EU authorized representative is:
Springer Nature Customer Service Center GmbH
Europaplatz 3, 69115 Heidelberg, Germany

Printed by Libri Plureos GmbH
in Hamburg, Germany